Kant's Modal Metaphysics

Kant's Modal Metaphysics

Nicholas F. Stang

OXFORD
UNIVERSITY PRESS

OXFORD
UNIVERSITY PRESS

Great Clarendon Street, Oxford, OX2 6DP,
United Kingdom

Oxford University Press is a department of the University of Oxford. It furthers the University's objective of excellence in research, scholarship, and education by publishing worldwide. Oxford is a registered trade mark of Oxford University Press in the UK and in certain other countries

First Edition published in 2016

British Library Cataloguing in Publication Data
Data available

Library of Congress Control Number: 2015942949

ISBN 978–0–19–871262–6

Preface

The Unicorn and the Narwhal

The cover of this book reproduces a plate from *Museum Museorum* (1704), a kind of 'cabinet of curiosities' in book form by the eighteenth-century naturalist Michael Bernhard Valentini.[1] It depicts, on the upper right-hand side, a unicorn. Below that, it depicts a narwhal, a species of whale (*Monodon monoceros*) that lives mainly in Arctic waters near Canada, Greenland, and Russia, recognizable by the helical 'horn' that protrudes from the foreheads of adult males.[2] There is a natural visual association of unicorns and narwhals, reinforced by their German names: *Einhorn* and *Seeeinhorn*.

The left-hand side of the drawing depicts a skeleton unearthed in 1663 by lime-miners in Quedlinburg, near the city of Mecklenburg, now in the state of Mecklenburg-Vorpommern. It became known as the 'Quedlinburg unicorn', and over the next hundred years came to the attention of many of the great minds of Europe. Its discovery is described by Otto von Guericke, the mayor of Mecklenburg and a noted seventeenth-century naturalist, in a work on vacuum tubes.[3] From von Guericke's account Leibniz learned of the Quedlinburg unicorn, while working in the Harz Mountains in the 1690s; a drawing of it (not by his hand) is included in *Protogaea, or A Dissertation on the Original Aspect of the Earth and the Vestiges of Its Very Ancient History in the Monuments of Nature*, a work of natural history Leibniz began during this period, but never finished. Interestingly, Leibniz himself seems to have taken it to be the remains of an actual unicorn. The Quedlinburg unicorn is even mentioned in an article by Baron d'Holbach in Diderot's *Encyclopedia*.[4]

While Leibniz followed Thomas Bartholin's influential debunking of unicorn horns in *De unicornu observationes novae* (1645) and accepted that most 'unicorn horns' are narwhal tusks, he seems to have believed that the Quedlinburg skeleton was that of an actual unicorn: "However, we should not hide the fact that a quadruped unicorn of the size of a horse can be found in Abyssinia, if we have to believe the Portuguese

[1] Its complete title is *Museum Museorum, oder vollständige Schau-Bühne aller Materialen und Specereÿen, nebst deren natürlichen Beschreibung, Election, Nutzen und Gebrauch.*

[2] In fact, the 'horns' of narwhals are not technically horns but *tusks* (essentially, very long teeth), but I will ignore that complication.

[3] *Experimenta nova (ut vocantur) Magdeburgica de vacuo spatio* (1672). Amsterdam.

[4] A more complete history of the Quedlinberg unicorn can be found in the editor's comments in Leibniz (2008), 103; see also Cohen (2002) and Ariew (1998). I first learned of 'Leibniz's unicorn' from the blog of John McKay; see McKay, J. (2013). "Leibniz's Unicorn." Retrieved February 20, 2015, from http://mammothtales.blogspot.com/2013/06/leibnizs-unicorn.html.

Hieronimus Lupo and Balthasar Tellesio; and similarly, the skeleton extracted from limestone in 1663 on Mount Zeunikenberg, next to Quedlinburg, looked more like a terrestrial animal than anything else."[5] This interest in the Quedlinburg unicorn is just one part of a larger early modern project of finding a place for 'unicorn horns' and other buried animals remains (e.g., mammoths) within a changing picture of the natural world. If these are unicorn remains, then a species of animals has gone extinct, which means that the natural world is not a static creation but something that can change and thus have a *history*. If they are not unicorn remains, then some other explanation must be given of their provenance (e.g., that they are horns of narwhals, or the tusks of elephants, etc.).

Kant would have been well aware of this literature. He possessed another book by Valentini, *Armentarium artis et naturae* (1709), to which is appended a summary of recent scholarly literature, including a capsule summary of *De unicornu fossili* (1666) by Johann Lorenz Bausch. Bausch (who draws on Bartholin) argued that buried 'unicorn' horns, like that found in Quedlinburg three years earlier, are the fossizilized tusks of mammoths and narwhals, not the remains of actual unicorns or the *sui generis* products of geological processes, as some had maintained.[6] It is thus noteworthy that in his 1763 work *The Only Possible Ground of Proof* [Beweisgrund] *for a Demonstration of the Existence of God*, Kant uses the narwhal and the unicorn as, respectively, his examples of what does and does not exist:

> [...] when existence occurs as a predicate in common speech, it is a predicate not so much of the thing itself, but of the thought which one has of it. For example: existence belongs to the narwhal [*Seeeinhorn*] but not to the land-unicorn [*Landeinhorn*]. This simply means: the representation of a narwhal is an empirical concept; in other words, it is the representation of an existing thing. (*OPG*, Ak. 2:72f)

In writing this passage, Kant likely had in mind cases like that of the Quedlinburg unicorn and what, at the time of his writing, was a scientific 'discovery' of fairly recent provenance: buried ivory is the tusks of narwhals or other animals, *never* the fossilized remains of unicorns.[7] In addition to the many criticisms Kant levels against Leibniz's metaphysics in that work, then, we can add another, more zoological objection: unicorns, unlike narwhals, have never existed.

The Only Possible Ground of Proof, which I will refer to as '*Beweisgrund*' for short, will be a main focus of this book. But before we descend to the details of Kant's arguments, it is worth pausing to reflect on the many differences between narwhals and unicorns: narwhals are aquatic, unicorns are not; narwhals have little hair, while unicorns have flowing manes; narwhals eat fish, while unicorns are herbivores.

[5] Leibniz (2008), 101.

[6] Valentini (1709), on p. 15 of the supplementary volume, *Historia literaria*; see Warda (1922), 30.

[7] Within a few years Kant would have known of the Quedlinburg unicorn itself, if he did not already in 1763, for he owned Hans Blumenbach's 1769 work *Handbuch der Naturgeschichte* (Warda (1922), 27), which contains an account of it, as well as a reference to Leibniz's *Protogaea*.

However, one difference between them stands out: narwhals exist, while unicorns do not. Intuitively, though, unicorns are *possible*; from a twenty-first-century perspective, if the evolutionary history of horses had been different, or if the ancestors of narwhals had evolved into land mammals, then there would have been unicorns, and some fossilized remains, like the Quedlinberg unicorn, might have been skeletons of actual unicorns.[8] How should we understand the difference between what actually exists (e.g., narwhals) and what does not actually exist but is possible (e.g., unicorns)? Is this a difference between objects that have, and objects that lack, a property, the property of 'existence'? Or is it wrong to say that *there are* non-existent objects in the first place? What does it mean to say that unicorns do not actually exist but possibly could have? And if unicorns are possible, how do we know this? Merely by imagining them, or do we need to know something about the laws of nature to know whether unicorns are possible? This book is the story of the development of Kant's thinking about these, and other, issues about existence and possibility. In other words, this is a book about the difference between narwhals and unicorns.

[8] However, the possibility of unicorns is not universally granted. In fact, it is the subject of a famous debate in the semantics and metaphysics of modality. Saul Kripke denied that unicorns are possible (Kripke (1980), 24, 156–8), while Michael Dummett upheld their possibility (Dummett (1983)).

Contents

Acknowledgments

I would like to thank the publishers of the following journals for permission to use some material from three previously published papers:

- Stang, N. (2010). Kant's Possibility Proof. *History of Philosophy Quarterly*, 27(3): 275–99.
- Stang, N. (2011). Did Kant Conflate the Necessary and the *A Priori*? *Noûs*, 45(3): 443–71.
- Stang, N. (2015). Kant's Argument that Existence is not a Determination. *Philosophy and Phenomenological Research* 91:3, 583–626.

I have been working on this project, in one form or another, for over ten years; consequently, I have incurred a number of debts, great and small, to people who have read and commented on various versions, numerous audiences to whom I have presented various parts of various chapters, and all of the people who have helped and encouraged me along the way.

It began its existence ten years ago as my Ph.D dissertation at Princeton University, under the supervision of Béatrice Longuenesse and Desmond Hogan. I am especially grateful to Béatrice, who has supported this project ever since I proposed, in my second semester at Princeton, to write my thesis on Kant on modality. I continued to work on this book during my six years at the University of Miami, where I developed most of the arguments and interpretations in these pages. Two consecutive Chairs of the Miami Department, Harvey Siegel and then Otávio Bueno, spurred me along in this process, constant sources of encouragement and advice. Since coming to the University of Toronto, Arthur Ripstein has been generous enough to read the whole manuscript and give me invaluable advice on how to organize and structure the final product to give readers a 'way in', though I am certain I have not done nearly enough on that score. Thanks also to Ian Drummond for some excellent proofreading that caught many of my numerous typos. Those that remain are my fault, for which I beg the patience of the reader.

Along the way many people have been generous enough to read and comment on drafts of various chapters. Thanks to Rae Langton, Eric Watkins, Lanier Anderson, Beau Madison Mount, and Catharine Diehl for their comments on various parts of the manuscript. Karl Schafer has read and commented on various chapters over the years (including, thanks to Karl's excellent comments, a rejected version of Chapter 6). Lucy Allais gave me much-needed feedback on most of the chapters of the book; by allowing me to read the page proofs of *Manifest Reality* before it was published, Lucy also helped me massively improve the discussion of concepts and intuitions in Chapter 6. Three anonymous readers from Oxford University Press gave me extensive notes on the whole manuscript, which formed the basis of an extensive revision beginning in the spring of 2014. Finally, Colin Marshall and Mike Raven were generous enough to read, and extensively comment upon, almost the whole manuscript of the book; thanks to Colin and Mike for helping me make the whole work much better. I would also like to thank various audiences to whom I have presented this material over the years, both for the opportunity and for their insightful questions: grad seminars at UMass and Toronto; Departmental colloquia at UMass, Irvine, Santa Barbara, Miami, Florida-International, Oxford, Stanford, Michigan, Toronto, McMaster, and Western Ontario; as well as conferences in São Paulo, Berlin, Atlanta, and Vancouver.

I would especially like to thank Tobias Rosefeldt, who generously invited me to present the ideas from this book at the Berliner Kant-Kurs in July 2015. I learned an enormous amount from the insightful comments and questions of the course participants, but the book was already in production by that point. Where possible I have tried to add a note or two responding to an objection, or adding a note of clarification. I would like to thank everyone who came, but especially Andrew Stephenson, Max Edwards, Noam Hoffer, Catharine Diehl, Karl Schafer, Lucy Allais, Stefanie Grüne, Jim Kreines, Damian Melamedoff, Manish Oza, and Tobias himself for their excellent comments.

The last ten years have been a particularly exciting time to be writing on Kant's theoretical philosophy, due to a generation of (I hesitate to call us 'young' anymore) scholars whose work and whose feedback on my ideas has been a constant source of inspiration and stimulation. I have been benefited in particular, over the years, from reading the work of, and conversing with, Andrew Chignell, Des Hogan, Anja Jauernig, Dina Emundts, Tobias Rosefeldt, Stefanie Grüne, Colin Marshall, Colin McLear, Karl Schafer, Lucy Allais, James Messina, Dai Heide, Jessica Leech, Uygar Abaci, Tim Rosenkoetter, and Clinton Tolley.

In addition to my intellectual debts, I would also like to acknowledge the many friends who helped and encouraged me along the way. Eli Chudnoff, Brad Cokelet, and Whitney Bauman offered friendship, good cheer, and someone to complain to, over the years. Thanks to everyone who read and commented on the more extensive version of the 'Narwhals and Unicorns' preface: Jennifer, Catharine, Tim, Dino, Omer, Sanmay, Radu, and others. The first draft of this book was written during a sabbatical year, when I lived at 75 Stewart Avenue in Bushwick, Brooklyn, where I was fortunate enough to live with the best roommates of my life: Nancy, Jack, Meg, and Barry. Thank you for all the great times, for letting me get away with never cleaning the dishes, and for throwing me a surprise party when I 'finished' this book in early 2013 (ha ha). By far the best thing that happened to me during my year in Brooklyn, though, was meeting Michael Moss. I have been working on this book for all the time I have known Michael and I offer my deepest thanks for all the love and patience he has shown at every stage in its writing, rewriting, revision of the rewriting, complaining about the revision of the rewriting, etc.

I would also like to acknowledge several scholars whose work either appeared, or became known to me, too late for me to engage with it in this book. Lanier Anderson's *The Poverty of Conceptual Truth: Kant's Analytic/Synthetic Distinction and the Limits of Metaphysics* (Oxford University Press, 2015) appeared while this book was already in production. Omri Boehm's *Kant's Critique of Spinoza* (Oxford University Press, 2014) covers several topics (esp. *Beweisgrund*) of this book from a different direction; I became aware of the relevance of Boehm's work too late to include discussion of it here. Giuseppe Motta is the author of two highly significant works on Kant's theory of modality—*Die Postulate des empirischen Denkens überhaupt. Kritik der reinen Vernunft, A 218–235/B 265–287. Ein kritischer Kommentar* (De Gruyter, 2012) and *Kants Philosophie der Notwendigkeit* (Peter Lang, 2007)—that deserve to be better known to English-language scholars. However, I found I could not discuss Motta's work in the depth it deserves without making this book, already rather long, much longer. I have engaged critically with Anderson's and Boehm's books in the pages, respectively, of the *Australasian Journal of Philosophy* and the online journal *Critique*. I look forward to engaging with Motta's work in detail in the near future.

Finally, this book is dedicated to Charles Parsons, who taught my first class on the *Critique of Pure Reason* in the Fall of 1999. I did not know at the time how much of my intellectual life would be devoted to that book, but I am enormously grateful for the start he gave me. Prof. Parsons has always been, and remains, my model of a scholar and a philosopher.

List of Figures

List of Tables

Notes on the Text

Sources. At this end of this book you will find a Note on Sources, which explains the abbreviations I use to refer to the works of Kant and other important primary sources, as well as a complete Bibliography.

Translations. Quotations from Kant are mainly from published translations (where available), with some modifications by me; quotations from other thinkers are a mixture of published translations and my own. The Note on Sources also contains information on which translations were used.

Emphasis. The *Critique* was printed originally in *Fraktur*, using roman letters for Latin phrases (like '*noumenon*') and *Fettdruck* for emphasis. When quoting from Kant, I have followed the editors of the Cambridge Edition in rendering the former in italics and the latter in bold. Within quoted Kantian texts, underlining indicates my emphasis.

Concepts and propositions. I use angle brackets to denote the concepts that would normally be expressed by the italicized expression within, e.g., <*substance*> is the concept of substance. I use italicized sentences to denote propositions; the sentence 'Gold is a yellow metal' expresses the proposition *Gold is a yellow metal.* I refer to the relation between an object and a concept whose extension includes that object as 'instantiation', for example, particular substances instantiate <*substance*>. I realize this will strike some readers as odd, since instantiation is usually taken to be a relation between objects and properties (universals). However, the expression usually used by Kant scholars for the relation between an object and a concept (an object is said to 'fall under' a concept), leads to many awkward constructions, and talk of 'subsuming' an object under a concept brings in a misleading reference to the agent who does the subsuming.

Optional sections. Certain sections are not necessary to understand the larger argument of this study; they are usually further developments of a line of thought, or consideration of further texts. I have indicated such sections with an asterisk (*) in the Table of Contents.

Supplementary. In order to keep this book at a reasonable length, I have left out certain additional material, largely engagements with other scholars and more involved technical developments of my own ideas. These can be found on my website: https://sites.google.com/site/nickstang/.

Introduction

1. From Ontology to Transcendental Philosophy

This book is a study of Kant's theory of possibility, starting from the so-called 'pre-Critical' period of the 1750s and 1760s, through to the Critical system of philosophy Kant published in the 1780s and the early 1790s. It discusses Kant's confrontation with the ontological theory of all possible beings shared by his predecessors (mainly Leibniz, Wolff, and Baumgarten), his own positive theory of possibility in the 1760s, and his distinctively Critical theory of possibility in the 1780s. Its guiding orientation will be the idea that the way to understand both the change and the continuity between Kant's pre-Critical and his Critical theory of possibility is that he comes to believe that the ontological question about possibility must be transformed into a question in 'transcendental philosophy': how do we represent possible objects *a priori*? My goal will be to explain Kant's theory of possibility before he reoriented his theory around that question, why he believed that reorientation was necessary, what this question means, and his own answer to it. Possibility cannot be studied in isolation, for it becomes one of the three 'categories of modality' in the *Critique*, along with actuality/existence and necessity. Consequently, I will also investigate Kant's views about what distinguishes actuality from mere possibility, his theory of the closely related notion of existence, and his theory of necessity, including, crucially, his theory of necessary existence. Throughout, however, possibility, and the transition from ontology to transcendental philosophy, will provide the unifying thread that runs through this whole study. However, since that transition occurs at a particular point in the book (Chapter 6), I want to begin by explaining what 'ontology' and 'transcendental philosophy' are and preparing the reader for the transition from the former to the latter.

As every reader of the *Critique of Pure Reason* knows, that work is an attempt to answer the question, how is metaphysics possible? Kant undertakes a 'critique of pure reason,' an investigation into the nature and scope of our cognitive capacities, in order to determine the scope of what we can cognize independently of sensory experience, and whether it includes the concepts and claims of metaphysics. It is also well known that Kant makes a traditional distinction between *general metaphysics* and *special metaphysics*. General metaphysics, sometimes called *ontology*, is the science of all possible beings *qua* possible; ontology studies the properties shared by all possible beings as such. Special metaphysics, by contrast, is the science of particular kinds of beings, and divides into three specific sub-disciplines: psychology (metaphysics of souls), cosmology (metaphysics of the universe), and theology (metaphysics of God). Clearly, *possibility* is an important topic in ontology, for it is something all possible beings share, and an adequate ontology will include some account of what possibility is (what all possible beings have in common in virtue of being possible) and what principles govern possibility. The delineation of the proper subject matter of ontology requires us to distinguish possible beings (e.g., thinking substances) from impossible ones (e.g., round squares), and the major metaphysics textbooks of the eighteenth century make precisely such a distinction.[1]

Kant regularly lectured from such textbooks of metaphysics, mainly Baumgarten's *Metaphysica*, and some of his students transcribed these lectures.[2] In the metaphysics lecture transcripts available to us Kant makes this traditional separation between *metaphysica generalis* (ontology) and *metaphysica specialis*. But he often begins his discussion of ontology by claiming that the distinction of possible from impossible beings is a distinction within a more general concept, the concept of an 'object' [*Gegenstand*, sometimes the Latinate *Object*]. One such instance comes from the Mrongovius transcripts from the winter semester 1782–3, between the publication of the two editions of the *Critique*:

> [...] if two concepts are opposed—here [in Baumgarten's *Metaphysica*], for example, possible and impossible—then they always stand under a higher concept, for opposition always presupposes a disjunctive proposition. Now there must be a divided concept that has the opposing concepts as members of the division, and this is a higher concept. *Object* [...] is surely the highest concept in ontology. The possible we call 'thing' [*Ding*], 'something' [*Etwas*], and opposed to it is the impossible, nothing [*Nichts*]. (MM, Ak. 29:811[3])

This remark, and others like it, is mirrored by a short passage in the *Critique* that is easy to overlook:

> The highest concept with which one is accustomed to begin a transcendental philosophy is generally the division into the possible and the impossible. Since all division, however, presupposes a concept that is divided, a yet higher concept must be given, and this is the concept of an object in general [*Gegenstand überhaupt*] (taken problematically, and unspecified whether it is something or nothing). (A290/B346)

I take the concept <*object*> here to be the concept <*object of representation*>. I will argue for this in the course of this study; for now it will constitute an interpretive hypothesis.

In the *Critique* Kant describes <*object*> as the highest concept of "transcendental philosophy." That the readers of the *Critique* can be "accustomed" to something in

[1] See Ch. 1.2 for specific references. [2] Ak. 28–9.

[3] Cf. MV (Ak. 28: 410 f.), MvS (Ak. 28: 477–9 f.), ML_2 (Ak. 28: 542 f.), and esp. MK_3 (Ak. 29: 960).

"transcendental philosophy" means that the latter phrase does not refer to the distinctive philosophical project inaugurated by that work (as it does in other contexts), but to some more general type of philosophy, of which both pre-Kantian ontology and the *Critique* are instances. Kant defines this generic notion of transcendental philosophy in the Introduction to the A edition: "I call all cognition **transcendental** that is occupied not so much with objects but with our *a priori* concepts of objects in general" (A11).[4] This applies to pre-Kantian ontology, since the distinction between the possible and the impossible is not a distinction between two kinds of *objects* (i.e. round squares are not a *kind* of object) but between two concepts: the concept of the possible and the concept of the impossible. From a Kantian perspective, the metaphysics textbooks of Wolff and Baumgarten are catalogues of *a priori* concepts, principal among them <*possible*> and <*impossible*>.

In addition to this general idea of transcendental philosophy (the study of *a priori* concepts of objects, not directly of objects themselves), Kant also uses this term more specifically to refer to the philosophical project inaugurated by the *Critique* itself: "not every *a priori* cognition must be called transcendental, but only that by means of which we cognize that and how certain representations (intuitions or concepts) are applied entirely *a priori*, or are possible" (A56/B80). The guiding question of transcendental philosophy properly so called is: how is it possible (for us) to represent an object *a priori* (independently of sensory experience)? This is what Kant means when he says that the concept <*object of representation*> comes before the concepts <*possible*> and <*impossible*> in transcendental philosophy: we must first inquire into our capacity to represent objects, specifically, our capacity to represent them *a priori*. The ontology of Wolff, Baumgarten, and others erred, according to Kant, in directly considering the concepts of possibility and impossibility, without first inquiring into the cognitive capacities by which we represent objects in the first place.

But the difference between ontology and its successor, Kantian transcendental philosophy, is not merely that the second considers an antecedent question about a more general concept (<*object of representation*>). By prefacing its inquiry into possibility with that question, transcendental philosophy transforms both the original question of ontology (the nature of all possible beings as such) and its basic distinction (between possibility and impossibility). By subordinating <*possible*> and <*impossible*> to the more general concept <*object of representation*>, Kant transforms these into the concepts <*possible object of representation*> and <*impossible object of representation*>. Consequently, the fundamental question of ontology becomes the following question in Kantian transcendental philosophy: how is it possible for us to represent possible objects *a priori*? Ontological questions about the nature of possible

[4] This sentence is emended in the B edition to: "I call all cognition transcendental, which is not concerned so much with objects but with our mode of cognition of objects, insofar as this is supposed to be possible *a priori*" (B25–6). This is the specific concept of transcendental philosophy that does not apply to e.g., Wolff and Baumgarten.

beings as such are transformed, in Kant's hands, into cognitive-semantic questions about our *representation* of possibility. It will take the rest of this book for me to explain why Kant thinks this transformation is necessary, and its larger implications for his theory of possibility.

2. Breakdown of Chapters

I begin, in Part I, with Kant's confrontation in the 1750s and 1760s with the ontology of Wolff and Baumgarten. Wolff and Baumgarten share with Leibniz, their great predecessor in the German rationalist tradition, a view about possibility which I call *logicism*: anything that is not logically contradictory is possible. Consequently, the principle of non-contradiction is the highest principle of ontology. The logicists also hold a corresponding principle about necessity: a necessary truth is one whose negation entails a contradiction. The space of possible being, for the logicists, is a logical space, consisting of logically consistent concepts of possible beings and logical relations among those concepts.

The logicists hold the traditional view that God exists necessarily. Combined with the logicist view of necessity, this entails: the proposition *God does not exist* logically entails a contradiction. The reason this proposition logically entails a contradiction is that existence is contained in God's essence, his mere possibility. God is the unique being whose mere possibility entails his existence. I refer to the doctrine that God exists in virtue of his essence as *ontotheism*. A term coined by Kant himself, *ontological argument*, refers to arguments that purport to derive God's existence from his essence.

Chapters 1 and 2 focus on Kant's critique of ontotheism and ontological arguments. Although Kant's objections to the ontological argument in the *Critique of Pure Reason* are well known, they are a continuation of a line of argument he had been pressing, in one form or another, since 1755. After some preliminary background on the metaphysics and epistemology of logicism, I explain why logicism, paired with the assumption that God exists necessarily, commits these thinkers to the doctrine that God exists in virtue of his essence. I then show that Kant rejects ontotheism: there can be no ontological arguments because there cannot be a being whose essence grounds its existence. His objection to ontological arguments—that existence is not a 'real predicate' or 'determination'—is one of his most famous, but also one of his least well understood, doctrines. I argue that this doctrine means that existence is not a predicate that applies to a subset of objects: objects do not divide into the *existent* and the *non-existent*. Existence can be used as a predicate of *objects* themselves, in which case it necessarily applies to all objects there are, or it can be applied to *concepts*, in which case it applies to concepts that are instantiated by at least one object. I argue that ontotheism as such is committed to denying this; the ontotheistic explanation of why God necessarily exists—his essence contains existence—entails that existence is a 'real predicate' that potentially applies only to a subset of objects. If Kant is right that existence is not a real predicate, then ontotheism is false. If ontotheism is false, then logicisim is false, for it entails ontotheism (given the assumption of divine necessity). If Kant can show that God does not exist with logical necessity, then he can show that there is a distinction between what is necessary (e.g., that God exists) and what is logically necessary, and, correlatively, between what is possible and what is *merely* logically possible (e.g., that God does not exist).

In the first half of Chapter 2 I reconstruct in detail the ontological arguments of Descartes, Leibniz, and Baumgarten to show they are implicitly committed to the doctrine that existence is a real predicate. Although Descartes is not a logicist, he is an ontotheist, and, since his ontological argument was deeply influential on Leibniz, and hence on Wolff and Baumgarten, I include him in my discussion of (otherwise logicist) ontological arguments. In the second half of Chapter 2 I reconstruct in detail Kant's arguments that existence is not a real predicate, in both his 1763 work *The Only Possible Ground of Proof* [Beweisgrund] *for a Demonstration of the Existence of God* and the better-known arguments in the *Critique of Pure Reason*. I argue that none of Kant's arguments should trouble a sufficiently sophisticated ontotheist. No matter how intuitive many of us post-Kantians may find this doctrine, Kant fails to prove that existence is not a real predicate. However, once Kant adopts this view, it requires a new metaphysics of existence, one in which possibility-facts and existence-facts are insuperably different: no being exists in virtue of its essence (its possibility). This divorce between possibility and existence, between concept and object, will have far-reaching consequences for Kant's philosophy.

Chapter 3 concerns Kant's criticism of another key logicist doctrine. The logicists hold the natural view that certain pairs of predicates (properties)[5] are metaphysically incompatible: it is impossible, for instance, that one and the same being *thinks* and *is extended*. Logicism entails that if a pair of predicates is metaphysically incompatible, they are logically inconsistent (their co-instantiation entails a contradiction). I reconstruct Kant's sophisticated argument in *Attempt to Introduce the Concept of Negative Magnitudes into Philosophy* (1763) that there is non-logical metaphysical incompatibility between predicates. That argument depends upon a prior Kantian argument, that the relation between a cause and its effect cannot be modeled as a logical relation. I critically reconstruct both arguments and show how Kant's argument that there are metaphysically incompatible but logically consistent predicates offers further support to his anti-logicist doctrine that there are logical possibilities that are not possible *simpliciter*, or as we might now say, not *metaphysically* possible (e.g., that an extended thing thinks).

Kant's term for 'metaphysical' possibility as distinct from logical possibility is *real* possibility. Chapters 4 and 5 concern Kant's own positive theory of real possibility in

[5] On the relation between predicates, properties, and concepts see Ch. 3.1.

the 1750s and 1760s, focusing on *Beweisgrund*. The core idea of Kant's pre-Critical theory of real possibility is that real possibility requires not only the logical consistency of its elements, but that the atomic predicates of which real possibilities are composed be possible in their own right. This in turn requires that there be an existing being that *grounds* or makes possible these atomic predicates. Kant argues at length in *Beweisgrund* that God is the ultimate such ground of possibility, without which nothing would be really possible. Before reconstructing that argument, though, I examine in Chapter 4 what it means to say that God is the *ground* of real possibility. I canvass various interpretations that have been offered in the secondary literature and argue that none of them is compatible with Kant's texts. I do not, however, go on to offer my own *positive* interpretation, for Kant explicitly states that we cannot understand the grounding relation between God and real possibility. In fact, Kant expresses puzzlement about our ability to so much as *think about* (mentally represent) this relation. In Chapter 5 I reconstruct in detail Kant's argument that there is a being that grounds all real possibility. I argue that, although he explicitly claims we cannot understand this grounding relation, his argument implicitly requires us to think of God as grounding real possibilities through his *powers*. We are simultaneously required, and forbidden, to represent God as grounding real possibility through his powers.

Part II concerns Kant's modal theory in the Critical period, and it begins, in Chapter 6, with an account of the role of modality in the 'Critical turn.' I point out that in *Beweisgrund*, and the nearly simultaneous 'Prize Essay,' Kant presupposes that real possibilities are 'given' to us (they are cognitively available to be thought about by us) but fails to develop an adequate philosophical account of how and why they are given. This further compounds a problem already diagnosed in *Beweisgrund*: Kant lacks an account of how we can mentally represent the grounding relation between God and real possibility. Kant's pre-Critical modal theory lacks any adequate account either of the epistemology or cognitive semantics of real possibility (how we can mentally represent it). But the lack of such an epistemology (or semantics) is not a minor flaw, for it entails that Kant lacks an adequate account of how we can mentally represent his own modal metaphysics, much less know it to be true.

A key step on the road to the Critical philosophy comes in the Inaugural Dissertation (1770), when Kant distinguishes between two fundamentally different kinds of representation: intuitions and concepts. An intuition is an immediate presentation of a singular object, while a concept is a general representation that may or may not have any instances. This requires a reformulation of the questions in modal semantics and epistemology from above. Since intuitions immediately 'give' us objects (make them available for singular thought), the questions about possibility become: how do we know *a priori* which of our concepts are really possibly instantiated (by an object)? And what is the content of the thought that one of our concepts is really possibly instantiated (by an object)?

In a famous 1772 letter to Marcus Herz, Kant poses a question ("what is the ground of the relation of that in us which we call 'representation' to the object?") that, according to one popular narrative, motivates the project of the *Critique*. I argue that this is a question in modal semantics and epistemology: what explains our possession of *a priori* concepts that are really possibly instantiated by objects? And, how do we know *a priori* that our concepts are really possibly instantiated? The epistemic question becomes especially pressing when applied specifically to the categories (the fundamental concepts of metaphysics), because Kant shares with his ontologist predecessors the view that metaphysics concerns *possible beings* and that we must be able to prove that its fundamental concepts are possibly instantiated. If we cannot know *a priori* that the categories are concepts of possible objects, then metaphysics is impossible. Kant answers this question by distinguishing between concepts and intuitions, and dividing all objects *überhaupt* into those that can be objects of sensible intuition (phenomena) and those that cannot be such objects (noumena). I reconstruct Kant's positive explanation of how we can have *a priori* knowledge that the categories are possibly instantiated by phenomena and his negative argument that we cannot have such knowledge of noumena. Consequently, there can be *a priori* metaphysics but its generality is limited: it applies only to phenomena. Because there can be no *a priori* metaphysics of noumena, metaphysics of all possible beings (ontology, *metaphysica generalis*) is impossible for discursive intellects like ours (intellects that sensibly, or passively, intuit objects).

Chapter 7 treats Kant's Critical theory of possibility, necessity, and contingency within the empirical world. Through a close reading of the 'Postulates of Empirical Thinking in General,' I argue that Kant implicitly distinguishes the *formal* possibility of an empirical object from its *causal* possibility. An empirical object is formally possible just in case it is compatible with the forms of our experience. An empirical object is *causally* possible just in case there is an actual object that can causally bring it about. However, Kant's transcendental idealist doctrine that empirical objects are appearances of noumena entails that they have two kinds of grounds: other empirical objects, and the noumena that affect us in experience. Thus, an empirical object is empirically-causally possible if and only if another empirical object can causally produce it, and is noumenally-causally possible if and only if noumena can cause that object to appear. The clearest case where the modalities diverge, I argue, is Kant's theory of freedom. If I perform an action, empirical determinism entails that the omission of that action is empirically-causally impossible. If empirical objects had only empirical grounds, we would lack freedom. I am free to omit an act I actually perform because it is noumenally-causally possible for me to cause that act not to appear: as a noumenon, I have the power to cause that act to be omitted.

Chapter 8 examines in greater detail Kant's theory of modality as applied to the empirical world. Kant claims *both* that empirical causal laws (e.g., the inverse-square law of attraction) are not formally necessary *and* that all laws as such are necessary. In what sense, then, are laws necessary? I call this kind of necessity 'nomic necessity' and examine, and reject, various alternatives proposed by earlier commentators. Through an analysis of Kant's theory of matter in *Metaphysical Foundations of Natural Science*, I argue that Kant accepts the existence of natural kinds (constituted by objects that share an essence) and that nomically necessary truths are truths grounded in the essences of natural kinds of empirical objects. In particular, the laws of motion are grounded in the essence of matter, the natural kind picked out by the concept <*matter*>, the most general empirical concept, which applies to all outer objects. I defend this interpretation in light of Kant's apparent skepticism about our knowledge of the essences of empirical objects and their kinds.

In Chapter 9 I return to Kant's pre-Critical modal theory and show that it is largely retained in the Critical system, bringing an unexpected degree of historical and systematic unity to Kant's modal metaphysics. His modal theory is historically unified because he retains within the Critical system the pre-Critical view that God is the absolutely necessary ground of all real possibility, but demotes its epistemic status: what he considered an object of *a priori* rational knowledge in *Beweisgrund* is now an article of 'necessary theoretical belief.' We are rationally required to postulate such a first ground of real possibility in order to explain why anything is really possible, but we do not thereby come to know that such a being exists. This historical continuity in Kant's modal metaphysics also shows the systematic unity of that metaphysics in the Critical system. The various different kinds of real possibility he accepts are unified by a common scheme: real possibilities are grounded in an absolutely necessary first ground. For each kind of real possibility, there is a corresponding conception of an absolutely necessary ground: space (formal possibility), the laws of nature (empirical possibility), real essences (nomic possibility), and God (noumenal possibility). Kant retains his pre-Critical conception of real possibility, generalizes from it to a set of general conditions on any concept of real possibility whatsoever, and then instantiates it at the level of various particular kinds of possibility. Far from being a mere aggregate of different, or even incompatible, conceptions of modality, Kant's Critical modal metaphysics is unified by his conception of possibility as grounded in an absolutely necessary being.

In Chapter 10 I diagnose and resolve what I refer to as the 'antinomy' of Kant's Critical modal metaphysics. His theory of freedom rests on attributing modal properties to the wills of finite rational agents (considered as noumena), specifically the possibility that they act otherwise than they actually do, but in his influential discussion of the 'intuitive intellect' in §76 and §77 of the *Critique of Judgment*, he claims that an intuitive intellect would not cognize its objects as *possibly* being otherwise than they actually are. If there is a God, he has an intuitive intellect, and what is more, an *omniscient* intuitive intellect: he cognizes all of the properties of noumena as such; it follows that noumenal wills lack modal properties. In this chapter I resolve this modal antinomy by arguing that, although God does not cognize modal properties in noumena (hence they do not have such properties), he does cognize a 'noumenal correlate' of modal properties: dependence on his essence and dependence on his will. These noumenal correlates are not 'in themselves' modal, but we represent them modally as possibility and actuality, respectively, when we apply the concepts of noumenal possibility and actuality to them. I develop a theory of these correlates and the content of the concepts of noumenal real possibility and actuality that explains how those concepts represent those correlates. I argue that we can retain the freedom of noumenal wills while holding that possibility is an artifact of our modalized representation of them by developing a non-modal account of noumenal freedom as the power to act otherwise. I explain how this non-modal power to act otherwise is represented modally by us as the ground of the possibility that we act otherwise.

I then use this account of noumenal correlates and our concepts of noumenal modality, retrospectively, to clarify a tension that arose in Kant's pre-Critical modal metaphysics: Kant denies that God grounds possibility through his powers (because his powers are among the possibilities that need grounds), but implicitly thinks of God as grounding possibility causally through his powers. From a Critical perspective, there is no modality at the purely noumenal level, so the possibility of God's powers does not need to be accounted for; we modalize these powers by representing them as the ground of the possibility of the properties they are the powers to cause to be instantiated. What is more, we resolve another outstanding issue from *Beweisgrund*: our ability to so much as represent God as the ground of all real possibility. We can represent God as the ground of possibility because possibility is an artifact of our modalized representations, not something in noumena 'in themselves.' To return full circle, this completes Kant's transition away from modal ontology (*metaphysica generalis*) to the study of modal representation: possibility does not characterize noumenal beings in themselves but is an artifact of how we represent them. The first question in modal metaphysics is not: *What is a possible being?* but: *What is it to represent an object as possible?*

PART I

Kant's Pre-Critical Modal Metaphysics

1

Logicism and Ontotheism

1.1. Introduction

As Kant developed his own positive views in modal metaphysics in the pre-Critical period, he did so primarily in reaction to the 'school metaphysicians' Christian Wolff and Alexander Gottlieb Baumgarten[1] and the available writings, such as they were, of Gottfried Wilhelm Leibniz.[2] Kant's modal theory was also influenced by Christian August Crusius, but since Crusius was himself critical of what he saw as the excessive rationalism of Leibniz and Wolff, I discuss him later, in Chapters 3 through 5. While Leibniz, Wolff, and Baumgarten differed significantly over central metaphysical issues like the possibility of inter-substantial causation, the nature of the 'pre-established harmony,' and the relation between simple substances and extended bodies, they shared a set of core doctrines about modality that I will call *logicism*.[3]

Since logicism will form the background of the next three chapters, I devote the first two sections of this chapter to exploring the logicists' metaphysics (section 2) and epistemology (section 3) of modality. Section 4 introduces a key pair of concepts: ontotheism and ontological arguments. *Ontotheism* (my term) refers to the view, held by many early modern thinkers, that God exists in virtue of his essence. An *ontological argument* is one that purports to derive the existence of God from his possibility, from his essence. Ontotheism entails therefore that there is a sound ontological argument, an ontological *proof*.[4] The logicists accept the traditional view

[1] For instance, in his critical discussion of philosophical theories of existence in *Beweisgrund* Kant mentions only Wolff, Baumgarten, and Crusius (Ak. 2: 76–7).

[2] When discussing Leibniz in this historical context, it is important to bear in mind that only a fraction of Leibniz's writings were available in the eighteenth century. Kant and his contemporaries engaged with the Leibnizian doctrines presented in those texts. Texts available in this period include the *Theodicy*, *New Essays on Human Understanding*, the Clarke correspondence, *Monadology*, "Meditations on Knowledge Truth and Ideas," "A New System of Nature," Part I of the *Specimum Dynamicum*, "On an Emendation of First Philosophy," and "On Nature Itself." We must therefore keep separate Leibniz *as it would have been reasonable for Kant or his near contemporaries to read him* and Leibniz *as we now know him*.

[3] For an overview of this period in Kant's development, see Schönfeld (2000) and Watkins (2005); see esp. Watkins's discussion of the differences between Wolff and Leibniz in his (2005), 46–50, as well as his (2006). For a general overview of German philosophy in the centuries and decades before Kant, see Beck (1969) and Wundt (1939) and (1945).

[4] Kant typically speaks of the 'ontological proof [*Beweis*]', but since *proof* entails soundness I want to separate ontological proofs (which, according to Kant, are impossible) from ontological arguments, which are putative ontological proofs.

that God exists necessarily, and this commits them, given their logicist account of necessity, to the claim that God necessarily exists in virtue of the logical containment of existence in his essence. In section 5 I argue that ontotheism entails that there are, or could be, merely possible but non-existent objects. I call this view *possibilism*. In section 6 I argue that Kant's famous objection to the ontological argument—that existence is not a real predicate—means that 'exists' is a predicate that necessarily applies to every object there is: it is not possible for there to be non-existent objects. If Kant is right that existence is not a real predicate, then ontotheism is false and there can be no sound ontological arguments, for nothing, not even God, can exist in virtue of its essence. If ontotheism is false, then logicism is false, for there is a necessary truth (that God exists) that is not logically necessary.

1.2. Logicist Metaphysics

Wolff and Baumgarten conceive of metaphysics in modal terms. They divide this philosophical science into general metaphysics (ontology), and special metaphysics, consisting of the rational sciences of theology, cosmology, and psychology. Following Aristotle's definition of 'first philosophy,' they define ontology as the science of beings *qua* beings.[5] Wolff writes, in the preface to the *Ontologia*: "ontology is the science of a being insofar as it is a being [*scientia entis in genere, seu quatenus ens est*]."[6] Baumgarten begins the 'Ontology' section of *Metaphysica* with this definition: "Ontology (*metaphysica, metaphysica universalis, philosophia prima*) is the science of the most general predicates of a being [*entis*]."[7] 'Being' (*Ding, ens*) refers to any possible entity.[8] Thus, a science of "all possible beings [*Dinge*], insofar as they are possible," as Wolff defines philosophy [*Welt-Weisheit*] in the so-called 'German logic' is also a science of beings, insofar as they are beings, that is, "the science of the most general predicates of beings."[9] 'Being' is the highest genus in ontology. It

[5] See *Metaphysics* 1026a16. Note, though, that while Aristotle defines first philosophy as the science of being *qua* being, Wolff and Baumgarten define ontology as the science of beings (*entia*) *qua* beings, a difference that would later be emphasized by Heidegger; see Heidegger (1975), §3.

[6] *Ont.* §1.

[7] *Meta.* §4.

[8] See *Dt.Met.* §16 and *Meta.* §8. As Dirk Effertz notes in his translator's introduction to Wolff (2005), Wolff treats the German *Ding* as equivalent to the Latin *ens* (Eberhard follows him in this in his German translation of Baumgarten, Baumgarten (1783)); compare the definition of *Ding* in *Dt.Met.* §16 ("*alles was seyn kann, es mag wirklich oder nicht*") to the definition of *ens* in *Ont.* §134 f. ("*quod existere potest consequenter cui existere non repugnat*"). This is somewhat unfortunate, because in later German philosophy (including, in some but not all passages, Kant himself) *Ding* refers specifically to an *ens* insofar as it has concrete reality and (in Kant's case) forces. It is also unfortunate, because, to the English ear, *thing* (the cognate of *Ding*) has a more concrete connotation than *being* (*ens*). Cf. Honnefelder (1990), 315. Thanks to Abe Stone for drawing my attention to these issues, and to Corey Dyck for further discussion.

[9] *Dt.Log.* §1. Compare Wolff's definition of philosophy in *Discursus Praeliminaris*: the "science of the possibles, insofar as they can be" (§1). The equivalence of these definitions is confirmed by Johann August Eberhard, a contemporary of Kant, who gives the following gloss on Baumgarten's definition of ontology: "because the predicates which appear in ontology apply to beings [*Dinge*], not because they belong to a particular kind, but because they are beings: so ontology can be explained by Aristotle, the Scholastics, and

includes everything there is: God, finite substances, their modes, space, time, etc. To be a being is to be possible; whatever is not possible is a non-being, and does not fall within the purview of ontology, or any science for that matter. Ontology is the science of possibles *qua* possible, and therefore the most general of all sciences. Kant himself quotes this definition of ontology and of 'being' approvingly; in the Mrongovius metaphysics lectures, he says "we now begin the science of the properties of all beings [*Dinge*] in general, which is called ontology" (Ak. 29: 784).[10]

Wolff and Baumgarten share with Leibniz a core set of views about modality I will call *logicism*. First, they distinguish between what is possible in itself (*in se*) and what is possible in connection with what actually exists (*hypothetice*). I will focus almost exclusively on the former notion, which corresponds with Leibniz's idea of *strict* or *metaphysical* possibility; unless otherwise noted, *possibility* (*necessity*) refers to *in se* possibility (necessity), not possibility (necessity) *hypothetice*.[11]

Wolff and Baumgarten define (*in se*) possibility, impossibility and necessity as follows:

Wolff:
[...] one can further see that whatever contains [*enthält*] nothing contradictory in itself is possible, that is, whatever not only can itself exist next to other things that are or can be, but also contains in itself only those things that can exist next to each other is possible.[12]

When what is opposed to a thing contains in itself something contradictory, the thing is necessary. Now since what contains something contradictory in itself is impossible (§12), that which is opposed to something necessary is impossible. And if what is opposed to a thing is impossible that same thing is necessary.[13]

Baumgarten:
Every thing that is representable, that does not contain a contradiction, which is not A and not-A, is POSSIBLE.[14]

NECESSARY is that whose opposite is impossible; the non-necessary is CONTINGENT.[15]

However, these definitions of possibility and necessity are hard to interpret. While they purport to be definitions of what it is for any 'being' (*Ding*, *ens*) in general to be possible or necessary, they are given in terms of the (non-)containment of a contradiction in it,

Wolff as the science of beings in general, insofar as they are beings" (see Eberhard's editorial comments to Baumgarten (1783)).

[10] Cf. Kant's announcement for his lectures for the winter semester 1765–6: "I shall then proceed to ontology, namely, the science which is concerned with the most general properties of all beings [*Dinge*]" (Ak. 2: 309).

[11] For their theories of possibility *hypothetice* and its role in their defense of contingency, see *Dt.Met.* §§ 574–5 and *AzDM* §195, 197; and *Meta.* §§15–18, 102–5. Baumgarten and Wolff inherit from Leibniz the distinction between what is 'in itself' or 'strictly' possible, and what is possible 'in connection'; see "On Freedom and Possibility" (AG 19–23).

[12] *Dt.Met.* §12. Translation from Watkins (2009), 10. Cf. *AzDM* §6.

[13] *Dt.Met.* §36. Translation from Watkins (2009), 13.

[14] *Meta.* §8.

[15] *Meta.* §101.

but only concepts, or other logically structured items, can contain contradictions.[16] A concept contains a contradiction just in case it contains both a concept C and its negation ~C.[17] Wolff and Baumgarten appear to be claiming that:

(*Possibility*$_C$) A concept is possibly instantiated if and only if it contains no contradictions.

However, their definitions of necessity are given in terms of 'opposites' and only propositions have 'opposites'—their negations. These are most naturally read as definitions of the necessity of a proposition:

(*Necessity*$_P$) For any proposition p, it is necessary that p if and only if $\neg p$ entails a contradiction.[18]

Entailing a contradiction is the natural propositional equivalent of a concept's containing a contradiction.[19] How then do we unify this definition of possibility and this definition of necessity? There is a natural way of converting a claim about a concept's possible instantiation into a claim about the possibility of a proposition being true: C is possibly instantiated if and only if the proposition *C is instantiated* is possibly true. So we might define possibility for propositions as follows:

(*Possibility*$_P$) For any proposition p, it is possible that p if and only if p does not entail a contradiction.[20]

One might wonder how well this agrees with the original definition of possibility in terms of the instantiation of concepts, (*Possibility*$_C$). I think they are co-extensive. If the proposition *C is instantiated* entails a contradiction, this must be because the

[16] It is somewhat unclear what these thinkers took to be the exact ontological status of concepts, propositions, and other logical entities. For ease of exposition, I will be assuming they all share the following views, which are derived from (a certain reading of) Leibniz. First, they are nominalists: everything that exists is an individual (including individual particulars, and individual properties, or modes). The thoughts of an individual mind are modes of that mind, not shareable by any other mind. So my token concept of water is numerically distinct from God's token concept of water. However, we can talk about 'the' concept of water by talking about what all token concepts of water have in common in virtue of being token concepts of water. Since all minds express the complete truth about the world (universal harmony), this means that all token concepts of an object will have the same content (but vary in their clarity and distinctness). Talk of concepts is a *compendium loquendi* for talk of individual token concepts in individual minds. Likewise, talk about 'the proposition' that p is a *compendium loquendi* for talk about individual propositionally structured thoughts inhering in individual minds: what all such token thoughts have in common in virtue of having the same content. See Mates (1986), 170–88, for discussion.

[17] I use ~ for the negation of a concept and ¬ for the negation of a proposition or judgment. I explain why this difference might be significant in Ch. 2.4. For more on this distinction, see Stang (2012).

[18] Baumgarten defines the impossible as "that which contains [*involvens*] a contradiction, or which implies [*implicans*] a contradiction" (*Meta.* §7); given that the necessary is that whose 'opposite' is impossible (§101), the definition given in the body of the text follows.

[19] After all, the German word '*enthalten*' can mean both 'contain' and 'entail.'

[20] Cf. Wolff's definition of the possible as that from which nothing impossible follows [*fliesst*] (*Dt.Log.* §1.35).

concept C itself contains a contradiction.[21] Conversely, if concept C contains a contradiction—it contains, say, concept C_1 and $\sim C_1$—then the proposition *C is instantiated* entails a contradiction, for it entails that there is something that is both C_1 and $\sim C_1$. If this is correct, then a concept C is possibly$_C$ instantiated if and only if the proposition *C is instantiated* is possible$_P$. I propose therefore that we interpret Wolff and Baumgarten as accepting the definitions of possibility and necessity in terms of propositions, (*Possibility*$_P$) and (*Necessity*$_P$).

These are definitions of what Baumgarten and Wolff call '*in se*' or '*an sich*' possibility, and they correspond to Leibniz's category of 'strict' or 'metaphysical' possibility, of beings possible in 'themselves' or 'in their own nature':

> Provided also that it is understood that necessity and possibility, taken metaphysically and strictly, depend solely upon this question, whether the object in itself or that which is opposed to it implies contradiction or not.[22]

Leibniz's definition is more straightforward than Wolff's or Baumgarten's because he defines possibility and necessity in terms of 'implying' a contradiction. Since only propositions (judgments, etc.) can imply anything, this suggests that possibility and necessity are to be defined in terms of the possibility and necessity of *propositions*, and that the possibility or necessity of the instantiation of a concept should be assimilated to that model: concept C is possibly instantiated if and only if the proposition *C is instantiated* is possible (i.e. it does not entail a contradiction). Leibniz too accepts (*Possibility*$_P$) and (*Necessity*$_P$).

Some readers might wonder what precise notion of 'entailment' is involved in these definitions of possibility and necessity. While it is reasonably clear that for these thinkers the relevant notion of entailment is syntactic rather than semantic, this distinction is not precisely drawn in this period. Wolff analyzes 'inferences' [*Schlüsse*] in terms of their agreement with the traditional Aristotelian syllogistic forms, which are syntactically defined.[23] Leibniz worked extensively on what he called a 'universal characteristic,' a universal symbolic language for all reasoning. The symbolism of the universal characteristic is so constructed that whether an argument is valid can be mechanically checked, without attending to the meanings of the individual symbols

[21] The other potential source for contradictory consequences of the proposition *C is instantiated*, aside from contradictory content in C, would be if the instantiation of a C violated some other necessary truth. In the context of logicist metaphysics, the main such example would be that the existence of a less than maximally perfect possible world violates the necessary truth that God creates the most perfect possible world. However, Wolff and Baumgarten both maintain that the proposition *a less than maximally perfect world exists* is *in se* possible (possible in its own nature), but impossible *in nexu* with God's perfect will. For their theories of possibility *in nexu* and its role in their defense of contingency, see *Dt.Met.* §§574–5; *AzDM* §§195, 197; and *Meta.* §§15–18, 102–5.

[22] *Th.* §367. See also *Th.* §§44, 45, 228, 230–2, 234, 235; *NE* 497; AG 19, 44, 95, and 98. In this book I am primarily concerned with what is possible and necessary 'in itself' or in its own nature, not with possibility *in nexu*.

[23] See *Dt.Log.* §§4.165–72; esp. his emphasis on "*förmliche Schlüsse.*"

(which concepts they stand for). In terms of the contemporary distinction, this is the idea of a purely syntactic definition of provability.[24]

The definitions of possibility and necessity given so far apply to whole propositions (*dicta*); in contemporary terms, they give an account of *de dicto* modality. What is *de re* possible and necessary for a being (*ens*) is determined by its essence. An essence is a logically complex concept that determines what it is to be that being, and which explains, at least partly, all of the other predicates of the being (which concepts it falls under).[25] Leibniz, Wolff, and Baumgarten identify the possibility of a being with its essence.[26] A being has a logically consistent essence. The predicates contained in its essence, its essential predicates, are necessarily had by that being. For instance, a triangle is necessarily three-sided and necessarily a plane figure. The predicates that follow from a being's essence, but are not contained in it, are its attributes. The predicates of a being that do not follow from its essence are its accidents. Attributes and essential predicates pertain necessarily to a being, and accidents pertain only contingently. The essence constrains the range of accidental predicates a being can have, because the accidents must be consistent with the essence. Consequently, any predicate possessed by a being is at least partially explained by its essence: wholly explained in the case of essential predicates and attributes, and partly explained in the case of accidents.[27] Essences of finite beings are incompletely determinate; they underdetermine the accidents of beings, which are partly grounded in their essences (which constrains the range of accidents beings can have) and partly grounded in the (causal) relations beings bear to one another. Those very beings could (in the '*in se*' sense of possibility) exist with different accidents.[28]

[24] Cf. "Samples of the Numerical Characteristic" (AG 10–18, esp. 18), "On the General Characteristic" (L 221–7, esp. 224–5), and "Two Studies in the Logical Calculus" (L 235–47); cf. *P* 42–3.

[25] As I am using them, 'predicate' and 'concept' are co-extensive terms with different senses: every concept is a predicate because it can be the predicate of some judgment, and every predicate is a concept. One refers to a concept as a predicate in a context where that concept is being predicated of an object or is one of the marks (constituents) of a larger concept. Thus, predicates are not linguistic entities (because concepts are not). For more on Kant's use of the term '*Prädicat*' see Ch. 3.1.

[26] *Dt.Met.* §35, *Meta.* §40, and *NE* 293. For more on Leibniz's identification of the possibility of a thing with its essence, see AG 21, L 146 f., and A II.i.390; for discussion, see Adams (1994, 136–8).

[27] *Dt.Met.* §§33–8, 44, 175, 176; *Meta.* §§41, 42, 50, 52, 65, 195; and A VI.iii.574. See also *Ont.* §§143, 144, 153, and 168.

[28] The question of the determinacy of possibilities is one on which Wolff and Baumgarten depart from Leibniz. Leibniz famously held that all of the predicates that apply to an individual substance follow from its complete individual concept, and no individual substance can exist with a different complete concept than it actually has (AG 11, 30, 31, 40–1, 95, 98; L 226, 231–2; and *NE* 486). He distinguishes the essential predicates of an individual substance from its contingent ones—both of which are contained in its complete concept—by claiming that the former follow from a *finite* analysis of the concept, while the latter require an *infinite* analysis. See "On Contingency" (AG 28–30) and "Primary Truths" (AG 30–4). Cf. Mates (1986), 108–17; Sleigh (1990); and Adams (1994), 25–30. In my discussion of Leibniz, therefore, whenever I talk about 'entailment' or 'demonstration' I mean entailment *in a finite number of steps* and demonstration *by finitely many steps*. Without this qualification, Leibniz's view is that *every* truth (even contingent ones) about an individual substance can be demonstrated from its complete concept (though it may require an infinite demonstration), and the negation of every truth (even contingent ones) about an individual substance entails a contradiction (although it may require an infinite demonstration to uncover

The contemporary reader might be puzzled by talk of the logical consequences of *predicates*; after all, one might think, only propositionally structured entities stand in logical relations, and predicates are not propositionally structured (i.e. they cannot have truth values). However, this is not how thinkers in this period see matters. Leibniz, Wolff, Baumgarten, *and* Kant frequently talk about non-propositional entities—especially predicates (concepts)—entailing one another. This is not an insuperable barrier to dialogue. We can translate the earlier idiom into the contemporary one by understanding talk of logical relations between concepts (predicates) as shorthand for talk of logical relations between propositions that predicate those concepts of objects. For instance, we can say that the predicate <*triangular*> entails the predicate <*trilateral*> because, for any object *x*, the proposition *x is triangular* entails *x is trilateral.*

In some texts Leibniz defines necessity slightly differently:

> ($\mathit{Necessity}_{\mathrm{Dem}}$) For any proposition *p* it is necessary that *p* if and only if there is a (finite) demonstration of *p* from identities and definitions.[29]

One might wonder whether these definitions of necessity are equivalent. In particular, one might wonder whether, if there is a demonstration of *p* from identities and definitions, $\neg p$ entails a contradiction. I will assume that these definitions of necessity are equivalent. If this is correct, then we can also define possibility in terms of demonstration, using the interdefinability of possibility and necessity:[30]

> ($\mathit{Possibility}_{\mathrm{Dem}}$) For any proposition *p*, it is possible that *p* if and only if there is no (finite) demonstration of $\neg p$ from identities and definitions.

If I am correct, this definition of possibility is equivalent to the earlier definition of possibility in terms of entailment. Sometimes it will be convenient to adopt the definitions of these modal notions in terms of demonstration.

It is crucial to note, though, that the identities and definitions that can be appealed to in a demonstration of a putatively necessary proposition (or the negation of a putatively possible one) are limited to the identities and definitions of the concepts involved in that very proposition. This is because we are concerned with what is necessary (and what is possible) 'in its own nature.' In other words, we need to restate these principles slightly:

> ($\mathit{Necessity}_{\mathrm{Dem}}$) For any proposition *p* it is necessary that *p* if and only if there is a (finite) demonstration of *p* from identities and definitions of the constituents of *p*.

this contradiction). For critical discussion of Leibniz's views on the determinacy of possibilities, see Mondadori (1973), (1975), (1985); Mates (1986), 137–51; Sleigh (1990); Adams (1994), 53–110; Nachtomy (1997), 51–103; and Cover and O'Leary-Hawthorne (1999), 87–142.

[29] "On Contingency" (AG 28–30), "Primary Truths" (AG 30–1), and "The Source of Contingent Truths" (AG 98–101). Cf. L 226, AG 21, and 95. Cf. M. Wilson (1990).

[30] $\Box p \leftrightarrow \neg\Diamond\neg p$. Cf. *Meta.* §101.

($Possibility_{Dem}$) For any proposition p, it is possible that p if and only if there is no (finite) demonstration of $\neg p$ from identities and definitions of the constituents of p.

If there is some being or beings whose concepts are not involved in p, but from whose definition there is a demonstration that p, p can still count as '*in se*' contingent (not necessary) as long as there is no demonstration of p from the definitions of the very concepts contained in p.[31] Translating this point back to the earlier principle ($Possibility_P$) we can impose the requirement that a proposition p entails q only if this relation is grounded in the syntactic structure (see previously) of both propositions and the concepts contained in them (not in any 'outside' concepts).[32]

The principle of non-contradiction (PNC)—that no being is both A and not-A—is the principle that governs all of possibility, all beings, and hence is the highest principle of ontology. Since ontology is the highest and most general science, the PNC is the highest principle of all sciences and all thinking whatsoever.[33] Possibility, for Leibniz, Wolff, and Baumgarten, ultimately reduces to logical self-consistency. The space of possibilities is a logical space, governed by logical relations of compatibility and entailment.[34] This also means that the space of possibilities is governed by a form of the principle of sufficient reason. If it is possible that p, there is an explanation of why p is possible: it does not entail a contradiction. Likewise for necessities: if it is necessary that p, this is so in virtue of the fact that $\neg p$ entails a contradiction. There are no brute, ungrounded facts about possibility or necessity.

Before continuing, I want to pause and note that logicism is a deeply plausible view about the metaphysics of modality. One core idea behind logicism is that whenever two predicates are incompatible, there is a reason for this incompatibility, and this reason lies in the logical structure of the predicates in question. For example, many thinkers in the early modern period held that *thinking* and *extension* are incompatible; it is impossible for an extended being to think. This much is familiar from Descartes' Sixth Meditation. But whereas Descartes appeals to our intuitive knowledge of a brute incompatibility between the natures of thought and extension, logicism holds that wherever there is such an incompatibility, there is a logically sufficient explanation. On the logicist view, the natures of thought and extension

[31] This is crucial to secure the '*in se*' possibility of non-actual worlds. For if it is necessary that God creates the best possible world (which, in some contexts, all three of these thinkers are willing to admit) then there is a demonstration of this truth; if this demonstration can be appealed to in any demonstration, then there will be a demonstration, of any other possible world, that it is not actual. See previously for the distinction between '*in se*' and '*hypothetice*' possibility.

[32] This also means that '*in se*' necessity does not obey the K rule: if $\Box p$ & $\Box(p \supset q)$ then $\Box q$. The antecedent might be true, while the concepts contained in q are not sufficient to demonstrate q 'on their own.'

[33] *Dt.Met.* §10, *Meta.* §7, and AG 19.

[34] Cf. Adams (1994), 12–15; Mates (1986), 105–7; and Pichler (1910), 32–48. For a discussion of texts in which Leibniz suggests that there are contingent facts about possibles *qua* possible, i.e. facts about possibility not governed by the principle of non-contradiction, see Adams (1994), 30–4.

must be ultimately composed of logically incompatible components, although we may not possess sufficient insight to discover them.

Logicism should not be confused with the once widespread view that analyticity and necessity are co-extensive.[35] Although it is a matter of some controversy in contemporary philosophy what exactly analyticity is, and, indeed, whether there is an analytic/synthetic distinction at all, it is relatively uncontroversial that, if there is an analytic/synthetic distinction, analyticity is a semantic property of sentences.[36] Logicism, on the contrary, is a view about the modal status of *propositions*, not *sentences*; although these early modern thinkers do not draw this distinction in those terms, they would acknowledge that logicism is not a view about bits of human language.[37] Instead, logicism holds that a proposition is necessary just in case its truth is grounded in the *natures* or *essences* of the objects the proposition is about. Logicism is not a view about the *meanings* of propositions, if meanings are understood as something that we are in a position to grasp in virtue of semantic competence with our language.

1.3. Logicist Epistemology

Having briefly outlined the logicist modal metaphysics, we now turn to the account given by Leibniz, Wolff, and Baumgarten of how we attain knowledge of possibility and necessity. In their modal epistemologies, Wolff and Baumgarten closely follow Leibniz, especially the theory presented in "Meditations on Knowledge, Truth, and Ideas," first published in 1684 and recommended to Wolff by Leibniz in their correspondence.[38] In this essay, Leibniz classifies concepts according to four properties: clarity, distinctness, adequacy, and whether one's grasp of them is intuitive or symbolic. Concepts have these properties relative to minds; a concept that is distinctly understood by one mind might be indistinctly understood by another.[39] I clearly understand a concept just in case I know whether a given object falls under the concept or not; for instance, I clearly understand the concept <*red*> because I can tell, of a visible object, whether it is red or not.[40] I distinctly understand a clear concept if, in addition, I can give necessary and sufficient conditions for an object to fall under it; I understand the concept <*red*> only indistinctly, or obscurely,

[35] E.g., Ayer (1936).

[36] For an overview of these issues, see Boghossian (1996).

[37] Cf. the discussion of the ontological status of logical entities like propositions in Mates (1986), 47–68.

[38] Cf. Wolff's very similar epistemology of ideas in *Dt.Log.* §§1.1–23 and *Log.* §§78–95. Leibniz also refers to "Meditations" in other writings available to eighteenth-century readers, e.g., *NE* 297. Cf. Wundt (1945), 140, 153–8; and Honnefelder (1990), 304–5.

[39] More perspicuously: token concepts (modes of individual minds) have these properties *simpliciter*. When we talk about 'the' concept of X, though, we are talking about what all token concepts of X have in common in virtue of being token concepts of X. Since the clarity, distinctness, etc. of token concepts differ from mind to mind, we need to specify which mind, and thus which token concept, we are talking about. For instance, the concept <*red*> is indistinct for me because my token concept of <*red*> is indistinct.

[40] Cf. *Dt.Log.* §§1.9–12.

because I cannot give a set of marks that are necessary and sufficient for something to be red.[41]

To understand the notion of adequacy we need to understand more about the structure of concepts. A concept is either simple (if it has no parts) or complex (if it has parts).[42] A complex concept is composed of other concepts, called the 'marks' of that concept, which it is said to 'contain.' The marks of one concept can themselves contain other concepts, and concept containment is transitive: if mark C_1 is contained in (is a mark of) concept C_2, and C_2 is contained in (is a mark of) concept C_3, then C_1 is contained in (is a mark of) C_3. Since concept containment is transitive, the marks of a concept may not be immediately or directly contained in it; finding its marks may require unpacking the immediate constituents of the concept, and their constituents, etc.

Leibniz defines the adequate understanding of concepts as follows:

> One has distinct knowledge of an indefinable notion, since it is primitive, or its own mark, that is, since it is irresolvable and is understood through itself and therefore lacks requisites [...] [W]hen everything that enters into a distinct notion is, again, distinctly known, or when analysis has been carried to completion, then knowledge is adequate.[43]

The ultimate constituents are atomic concepts, which are "understood through themselves," that is, they are not composed of further concepts. I adequately understand a complex concept just in case I can decompose it into its constituent concepts, and can resolve those concepts into their constituent concepts, and so on, to the atomic marks that compose the original concept, which are by definition adequate.[44]

When thinking with complex concepts, we typically do not hold in mind simultaneously all of the primitive constituents of the concept and their complex manner of arrangement. Instead, we mentally assign some symbol to one of the concepts that compose the larger concept. In such cases, our grasp of the concept is *symbolic*. If we mentally distinguish all of the constituents of the concept, down to its primitive constituents and the complex manner in which they compose the concept, our grasp of the concept is *intuitive*. All thought about primitive concepts is intuitive, because they have no conceptual constituents.[45]

Given the logicist conception of possibility from section 2, a concept is possibly instantiated if and only if it contains no contradictory marks. When we have an intuitive and adequate grasp of a concept, we have insight into its logical structure and we can see whether there are any contradictions present in it. Since primitive concepts are logically atomic, they are also logically independent; primitive concepts stand in no relations of logical entailment with one another. Therefore, the only

[41] See '*DM*', §8 (AG 41); *NE* 262, 293–5, and 346; as well as *Dt.Log.* §1.15.
[42] See *Dt.Log.* §1.17. [43] AG 24.
[44] Cf. *Mon.* §35; *NE* 120; L 160, 167; and *Dt.Log.* §1.16. [45] AG 25.

source of contradiction within concepts is their complex composition out of primitive parts.

This epistemology of concepts is related to Leibniz's distinction between nominal definitions and real definitions. A nominal definition provides a set of marks that are individually necessary and jointly sufficient for an object to fall under it. I distinctly understand a concept just in case I can give a nominal definition of it. A real definition, by contrast, shows why it is possible for objects to fall under it; a real definition decomposes the concept into its logically atomic constituents and shows that there is no contradiction among them. As Leibniz writes in "Meditations": "we also have a distinction between *nominal* definitions, which contain only marks of a thing to be distinguished from other things, and *real definitions*, from which one establishes that a thing is possible" (AG 26).[46] Only if we have an intuitive and adequate understanding of a concept can we give a real definition of it and have insight into the possibility of an object falling under it.

This, in turn, is related to Leibniz's distinction between *a posteriori* and *a priori* knowledge.[47] In *a posteriori* knowledge, gained paradigmatically through experience, we know only that something is the case. In *a priori* knowledge we know the reasons *why*.[48] If we experience an object that instantiates it, then we can know that a given concept is possibly instantiated.[49] However, this can only explain our knowledge of *actualized* possibilities, and it does not give us *a priori* knowledge of possibility: we know only that the concept is possibly instantiated, because actually instantiated, but not *why* this is the case. To know *why* a concept is possibly instantiated we need a *real definition* of that concept; we need to analyze the concept into its primitive constituents and see that there is no logical inconsistency among its marks. In this case, we have adequate intuitive understanding of the concept.[50] A real definition of a concept also gives us knowledge of non-actual possibilities; whether or not the concept is actually instantiated, if we know that it is logically consistent then we know that it is possibly instantiated.[51]

This conception of *a priori* knowledge as knowledge 'from the grounds' should be distinguished from the more familiar 'justificatory' definition of *a priori* knowledge, as follows:

(*A Priori*$_G$) S knows *a priori*$_G$ that *p* if and only if G is a sufficient ground of the fact that *p* and S derives *p* from G.
(*A Priori*$_J$) S knows *a priori*$_J$ that *p* if and only if (i) *S* knows that *p* and (ii) the justification in virtue of which S knows that *p* does not come from experience.

[46] Wolff draws a similar distinction in *Dt.Log.* §§1.41–9; see also *Log.* §§141, 152, 179, and 191.
[47] Cf. *Dt.Met.* §§77, 368, and 372.
[48] For more on Leibniz's use of the terms '*a priori*' and '*a posteriori*' see Adams (1994), 109–10.
[49] Cf. *Dt.Log.* §1.34. [50] AG 26 and *NE* 294. [51] See *NE* 162, 266, 254, 311, 352, and 346.

This distinction will be a continuing theme in this book, for, as I will argue, Kant often has the 'from the grounds' conception in mind when he discusses *a priori* cognition (*Erkenntnis*) and knowledge (*Wissen*).[52] These are only preliminary characterizations; the details of the definition of *a priori*$_{\mathrm{G}}$ knowledge will depend on what kind of ground is in question (e.g., efficient-causal grounds, or the formal grounds contained in a thing's essence). Leibnizian real definitions provide *a priori*$_{\mathrm{G}}$ and *a priori*$_{\mathrm{J}}$ knowledge; since our grasp of a real definition is purely intellectual, it does not depend on sensory experience (*a priori*$_{\mathrm{J}}$) and it acquaints us with the grounds of the possibility of the concept's object (*a priori*$_{\mathrm{G}}$). From now on, unless otherwise noted, by '*a priori*' knowledge I mean *a priori*$_{\mathrm{G}}$ knowledge.[53]

Leibniz makes this elaborate classification of concepts in "Meditations" in order to diagnose, and recommend a cure for, a major obstacle to cognition: errors about what is possible. The paradigm instance of a modal error, for Leibniz, is taking a covertly contradictory concept to be a concept of a possible object. Leibniz's example of a concept with a concealed contradiction is that of 'the fastest motion,' an apparently consistent concept, hence apparently a concept of a possible object.[54] Leibniz puts such importance on rooting out modal error because he holds that we can derive no knowledge from self-contradictory concepts. He writes: "For we cannot safely use definitions for drawing conclusions unless we know first that they are real definitions, that is, that they include no contradictions, because we can draw contradictory conclusions from notions that include contradictions, which is absurd."[55] Leibniz's reasoning appears to be as follows. If S is an impossibly instantiated concept, then it contains a contradiction between two of its constituents, call them A and ~A. If any proposition involving S as the subject concept were true, then the proposition *S is A* would be true, because in this judgment we merely predicate one of S's components of S. But, by parity of reasoning, *S is ~A* would be true. Hence, a contradiction arises if we admit as true any propositions with contradictory subject concepts. Because propositions with self-contradictory concepts cannot be true, propositions with self-contradictory concepts cannot be known. In the case of *a priori*$_{\mathrm{J}}$ knowledge, where we cannot rely on experience to show that our concepts are actually instantiated (hence possibly instantiated), we need real definitions of our concepts: proofs that they are logically consistent, hence possibly instantiated. Since metaphysical knowledge is paradigmatically *a priori*$_{\mathrm{J}}$, knowledge in metaphysics requires *a priori*$_{\mathrm{J}}$ knowledge of the possibility of the instantiation of its most basic concepts (e.g., substance, force, etc.)[56]

[52] See Smit (2009).

[53] For the Wolffian version of these doctrines, see *Dt.Met.* §372 and *Log.* §498.

[54] Leibniz uses the same example, for the same purpose, in his correspondence with Elizabeth (AG 238) and *DM* §23.

[55] AG 25.

[56] For instance, in the case of the ontological argument (see the rest of this chapter), we must first prove that the concept <*most perfect being*> is possibly instantiated. As Leibniz writes in "Meditations," "the fact that we think about a most perfect being is not sufficient for us to assert that we have an idea of it" (AG 25).

In fact, according to the view Leibniz defends in "Meditations," contradictory concepts are not genuine concepts. He writes:

> However, at first glance we might seem to have the idea of a fastest motion, for we certainly understand what we say; yet we certainly have no idea of impossible things. And in the same way, the fact that we think about a most perfect being is not sufficient for us to assert that we have an idea of it.[57]

When we think about a concept that contains a hidden contradiction, e.g., 'the fastest possible motion,' we take ourselves to be entertaining a thought with content. But we are mistaken; for Leibniz, concepts with concealed contradictions are not really concepts at all. To show that we have a concept of the most perfect being, we must show that this concept is a concept of a possible being.[58]

We saw earlier that the logicists hold that all necessary truths can be demonstrated from identities and definitions:

> (*Necessity*$_{\text{Dem}}$) For any proposition *p* it is necessary that *p* if and only if there is a (finite) demonstration of *p* from identities and definitions of the constituents of *p*.

The definitions in question are real definitions. If we allowed nominal definitions, like 'Julius Caesar is the individual that figures in most of the examples in this book,'[59] this principle would entail that it is necessarily the case that Julius Caesar figures in most of the examples in this book, which is absurd. So the principle should be formulated as:

> (*Necessity*$_{RealDef}$) For any proposition *p*, it is necessary that *p* if and only if there is a (finite) demonstration of *p* from identities and real definitions of the constituents of *p*.

The domain of the possible, for Leibniz, Wolff, and Baumgarten, is a logically structured domain of logically consistent concepts of possible objects, propositions about those possible objects, and their logical relations. By analyzing concepts into their more fundamental constituents, we gain insight into the logical structure of modal reality and attain *a priori* knowledge of possibility. A domain of objects is intelligible, for the logicists, to the extent that *a priori* knowledge of that domain is possible. Whether some domain is *a priori* knowable is a metaphysical matter: is there a sufficient ground for all facts about objects in this domain?[60] Since the domain of possibility is a logical domain, for the logicists, possibility is fully *a priori*

[57] AG 25.

[58] Leibniz's views on contradictory concepts, and the truth conditions of propositions involving them, are less fixed than he presents them in "Meditations." See Mates (1986), 67–8.

[59] This is a nominal definition because it provides necessary and sufficient conditions for being Julius Caesar. It does not give the meaning of 'Julius Caesar' (so descriptivism about proper names is not intended here).

[60] *Mon.* §28 (AG 216–17) and AG 209; cf. *Dt.Met.* §77.

knowable. Although we may not have the cognitive resources to analyze all concepts and all objects and determine their essences, the *a priori* intelligibility of the space of possibility consists in the existence of logically sufficient explanations of why concepts are possibly instantiated and why propositions are possibly true.

Readers may have noticed that, whereas in section 2 I phrased everything in terms of 'beings' (*Dinge*, *entia*), Wolff's and Baumgarten's term for the highest genus of ontology (the possible as such), I have switched in this section to talking about 'objects.' The reason is that in this section we have been primarily concerned with concepts of beings and the relation between beings and the concepts they fall under. I am using the term 'object' to refer to beings insofar as they fall under concepts, that is, insofar as they constitute the extensions of concepts. In contemporary terms, they are admissible values of bound first-order variables. We have seen that the principle of non-contradiction holds for all beings (*Dinge*, *entia*). Consequently, our quantifier expressions should be restricted to beings, for otherwise the principle of contradiction would have counter-examples (i.e. there would be objects that are not beings, *Undinge*, and for which the PNC would not hold). Conversely, the science of ontology (the science of *all* beings insofar as they are beings) presupposes that we can quantify over *all* beings, so every being is an object (of quantification). Henceforth, I will often refer to beings as 'objects'; by calling them 'objects' I mean to remind the reader that they are objects in the Quinean sense: admissible values of bound first-order variables. I will sometimes refer to them as q-objects.[61]

This also explains an awkwardness in my characterization of 'being' from section 2. I defined a 'being' as 'any possible entity whatsoever,' but if the reasoning of the previous paragraph is correct, then no definition of being by genus (e.g., entity) and differentia (e.g., possible) is possible; there is no wider genus of which 'being' is a species. So my characterization of beings as possible entities should be not understood as a *definition* but as an indication of how wide the class of beings is. Whether a non-circular definition of 'being' (as a count noun) is possible is obviously a difficult question in its own right; I will not attempt to address it here.

1.4. Ontotheism

All of the logicists, and at least the pre-Critical Kant, share the orthodox view that God exists necessarily.[62] Julius Caesar exists only contingently, like all finite created substances, but God exists necessarily. What accounts for this difference? Logicism, recall, is the view that:

> (*Necessity$_{RealDef}$*) For any proposition *p*, it is necessary that *p* if and only if there is a (finite) demonstration that *p* from identities and real definitions of the constituents of *p*.

[61] See Quine (1948). Whether concepts themselves are beings and/or objects is a complicated matter; see previous notes for a discussion of the ontological status of concepts.

[62] For Kant's pre-Critical view of divine necessity, see Ch. 5; for his Critical view, see Ch. 9.

Since God exists necessarily, there is a demonstration of his existence from his real definition (and perhaps from the real definition of <*existence*>). God's essence either contains existence as a component, or contains predicates that logically entail his existence. I will abbreviate this by saying that God's existence is *logically grounded* in his essence.

Since this point will be crucial for the next two chapters, it warrants further development. As we have seen, the logicists hold the natural view that objects necessarily have certain predicates in virtue of their essences (essential properties, and attributes), while their contingent predicates are merely compatible with their essences (accidents). The logicists extend this model of the modal status of predicates to the case of existence. God exists necessarily in virtue of the fact that <*existence*> is logically grounded in his essence (either by being contained in it, or logically entailed by it). God exists necessarily because his essence alone makes it the case that he exists. By contrast, Julius Caesar exists contingently, if he exists at all, because <*existence*> is not logically grounded in his essence. Caesar exists contingently because his essence alone does not make it the case that he exists.

This is a commitment that each of our logicists endorsed:

> Leibniz: For if there is reality in essences or possibles, or indeed, in eternal truths, this reality must be grounded in something existent and actual, and consequently, it must be grounded in the existence of the necessary being, in whom essence involves existence, that is, in whom possible being is sufficient for actual being.[63]
>
> Wolff: God exists through his essence, or his existence is essential.[64]
>
> Baumgarten: From God's possibility, it is valid to draw the conclusion that he exists, i.e. his existence is sufficiently determined by his very essence.[65]

Descartes and Spinoza also agreed that God exists in virtue of his essence.[66]

Kant introduced the term *ontotheology* to describe the part of theology that attempts to determine the predicates God has merely as a possible being (*ens*). Ontotheology, he says in his theology lectures, "considers God merely through concepts of possible beings in general [*mögliche Dinge überhaupt*]" (*Volck.RT*, Ak. 28: 1142).[67] The divine predicates discussed in ontotheology include his possibility, essence, necessity, substance, simplicity, immateriality, etc.[68] Consequently, an

[63] *Mon.* §45. This text is important because Leibniz asserts that God exists necessarily in virtue of his essence "involving" existence, even though he does not give the ontological argument in the *Monadology*; even where Leibniz does not use the ontological argument to *prove* God's existence, he still uses the underlying metaphysics to explain God's necessary existence.

[64] *TN* §27. Cf. *Dt.Met.* §§929–30 and *Ont.* §308.

[65] *Meta.* §820.

[66] See CSM II: 46–7 and *Ethics* Ip7d (a substance's "essence necessarily involves existence, or [*sive*] it pertains to its nature to exist"). See also Eberhard: "a being is necessarily actual if the ground of its actuality is contained in its essence" (Ak. 28: 561). I discuss Descartes in Chapter 2.2, but I forgo further discussion of Spinoza in this book.

[67] Cf. the virtually identical formulations in *Pöl.RT* (Ak. 28: 1002 f.). Cf. A632/B660, *MV* (Ak. 28: 451), MK_2 (Ak. 28: 777), and MK_3 (Ak. 28: 824).

[68] *Danz.RT* (Ak. 28: 1252).

ontotheological argument for God's existence is one that purports to prove the existence of God merely using modal concepts. Kant refers to ontotheological arguments using his more familiar coinage: *ontological* arguments.[69] Anselm, the discoverer of onto(theo)logical arguments, is also described as the founder of ontotheology.[70]

It is often overlooked that Kant sometimes uses 'ontological argument' in a broad sense[71] to include two very different kinds of onto(theo)logical argument: (i) *a priori*$_G$ ontological arguments that infer *from* God's possibility (his essence) *to* his existence, and (ii) *a posteriori*$_G$ ones, which infer from the possibility *of other things* to God's existence.[72] Both kinds of onto(theo)logical argument are *a priori*$_J$ because they do not depend upon experience. This broad sense of 'ontological argument' corresponds to the broad sense of 'ontotheology' defined above: theology using concepts of possible beings as such. In this broad sense of 'ontotheology' Kant himself is engaged in ontotheology in *Beweisgrund* and gives an *a posteriori*$_G$ argument he himself calls "ontological."[73] Throughout his philosophical writings he rejects type (i) arguments because they are based on a false metaphysical view (God exists in virtue of his essence). His attitude to type (ii) arguments is more complex, and it will take me the rest of this book to explore it completely.

The term *ontological argument* has had a significant career after Kant, and has come to refer specifically to *a priori*$_G$ (type (i)) arguments.[74] In fairness to the tradition, Kant often uses that term (and 'ontotheology') in the more specific sense.[75] To avoid confusion, I will follow this convention; henceforth, *ontological argument* will refer only to *a priori*$_G$ arguments that purport to derive God's existence from his essence. Likewise, I will use *ontotheism* to refer to the specific ontotheological doctrine that God exists in virtue of his essence. Ontotheism, as I will use that term, is the view that there is an *a priori*$_G$ ontological argument for the existence of God. It is not a commitment of ontotheism, however, that, given the limitations of our intellect, we can ever *give* an ontological argument for the existence of God, only that there is one to be given.

Kant rejected ontotheism and (*a priori*$_G$) ontological arguments almost from the moment he began publishing philosophy. In his 1755 *Nova dilucidatio* he writes:

> To say that something has the ground of its existence in itself is absurd [...] I find, indeed, the view repeatedly expressed in the teachings of modern philosophers that God has the ground of His existence in Himself. For my part, I find myself unable to support this view [...] I know

[69] *MV* (Ak. 28: 454) and *ML*$_2$ (Ak. 28: 599). [70] *Volck.RT* (Ak. 28: 1143).

[71] Logan (2007) is a notable exception. [72] *OPG*, Ak. 2: 155.

[73] *OPG*, Ak. 2: 160, 162. See also *Refl.* 6027 (Ak. 18: 427).

[74] Though it is not clear that it applies to the arguments discussed in Malcolm (1960), Plantinga (1967), 82–94, and Plantinga (1974), 197–221.

[75] *MV* (Ak. 28: 454), *ML*$_2$ (Ak. 28: 598–9), *Pöl.RT* (Ak. 28: 1027), *Volck.RT* (Ak. 28: 1173), and *Danz.RT* (Ak. 28: 1260).

that appeal is made to the concept of God; and the claim is made that the existence of God is determined by that concept. (*ND*, Ak. 1: 394)[76]

Variants of this claim—that no being can exist in virtue of its essence or its possibility—are found throughout the pre-Critical and Critical writings.[77] I will argue in sections 5 and 6 of this chapter that Kant's famous objection that "existence is not a predicate or a determination of a thing" (Ak. 2: 72) entails that nothing whatsoever can exist in virtue of its essence. This same claim is repeated almost word for word in *CPR* and throughout Kant's lectures on metaphysics and theology.[78] Kant's rejection of ontotheism is one of his most consistent doctrines. Given the historical continuity of this Kantian doctrine, it will be appropriate, in reconstructing Kant's arguments for it, to consider both Critical and pre-Critical texts, although the historical context of this chapter is Kant's pre-Critical *Auseinandersetzung* with logicism.

Throughout these writings, Kant levels many objections to specific ontological arguments, but my aim in this chapter (and Chapter 2) is to reconstruct his reasons for rejecting (*a priori*$_{\mathrm{G}}$) ontological arguments *as such*, that is, his reasons for rejecting ontotheism *tout court*. Getting the level of generality of Kant's rejection of ontological arguments right is important for several reasons. First, it affects what will count as success on Kant's part. He does not merely claim that particular ontological arguments of Leibniz, Baumgarten, etc. are invalid or unsound (though he does also claim this); he claims that ontological proofs *as such* are impossible. Kant raises various objections to specific ontological arguments, but those are not my concern here;[79] my concern is reconstructing Kant's objection to ontological arguments as such. It is also important because it shows that Kant's ultimate objection to ontological arguments is not directed to them as *arguments*. Kant's claim is not that the premises of an ontological argument do not validly entail its conclusion, although his view does entail that. Nor is Kant's point the epistemic point that ontological arguments fail to give sufficient reasons to accept that God exists, although his view does entail that. Kant's claim is that ontological arguments presuppose a false metaphysical view about the source or ground of necessary existence, what I am calling 'ontotheism.'[80] As Kant puts it succinctly in his lectures on metaphysics, "God himself cannot know his own existence through concepts" (*MK*$_2$, Ak. 28: 784). Even

[76] Cf. *MV*: "no intellect can have insight into the absolute necessity of a being" (Ak. 28: 455).

[77] *MH* (Ak. 28: 13, 55, 131, 133–4), *MV* (Ak. 28: 455), *MK*$_3$ (Ak. 28: 824), *MM* (Ak. 29: 814), and *Pöl.RT* (Ak. 28: 1027).

[78] *ML*$_1$ (Ak. 28: 313), *MV* (Ak. 28: 413, 455), *ML*$_2$ (Ak. 28: 598), *MK*$_2$ (Ak. 28: 783), and *MM* (Ak. 29: 822).

[79] For instance, Kant points out at A598/B624 that Leibniz's (and, following him, Wolff's and Baumgarten's) proof that <*ens realissimum*> is logically consistent does not suffice to prove that it is possibly instantiated. Cf. Ak. 28: 455.

[80] Kant introduces his discussion of existence in *Beweisgrund* by writing that the concept of existence must be treated with special precision because in the case of *necessary existence* "hat eine subtilere Nachforschung aus einem unglücklich gekünstelten, sonst sehr reinen Begriff irrige Schlüsse gezogen, die sich über einen der erhabensten Theile der Weltweisheit verbreitet haben" (Ak. 2: 71).

God cannot know his own existence through concepts, because God's existence is neither entailed by nor contained in his essence.[81]

That Kant's fundamental target is the ontotheist explanation of divine necessity is often overlooked, with the result that Kant's objection to ontological arguments is misunderstood.[82] In *Beweisgrund* Kant writes:

> That of which the opposite is impossible in itself is absolutely necessary. This is certainly a correct nominal definition. But if I ask: upon what does the absolute impossibility of the non-being of a thing depend? then what I am looking for is the real definition; this alone can serve our purpose. (*OPG*, Ak. 2: 81)

An object exists necessarily if and only if the non-existence of that object is impossible. Kant is not rejecting this principle, but pointing out that it is not informative. It is a nominal definition—it provides necessary and sufficient conditions for necessary existence—but it is not a real definition.[83] Kant's point is that this definition is not *explanatory*: it does not tell us *what it is* for a necessary being to exist necessarily, nor does it tell us, if there is a necessary being, *why* that necessary being is necessary (the ground of its necessary existence). Immediately after this passage, Kant summarizes his objection to the ontological argument, and then writes:

> In the final reflection of this work all of this will be made more convincing, by clearly explaining the untenability of the mistaken view that absolutely necessary existence could be explained by means of the law of contradiction. (*OPG*, Ak. 2: 82)

"The mistaken view" is the ontotheist view that God exists necessarily in virtue of his essence containing or entailing his existence. This view explains God's necessary existence via the law of non-contradiction because, in its logicist version, God's existence is a logical consequence of his essence; if God did not exist, this would be a contradiction.[84] That Kant's fundamental target in this work is a view about what *explains* God's necessary existence is further evidenced by the fact that he goes on to give an account of *what it is* for God to exist necessarily (his non-existence would cancel all real possibility) and to argue that there is a ground of God's necessary existence: he grounds all real possibility.[85] I explore both points in detail in Chapters 4 and 5.

Likewise, Kant's discussion of the ontological argument in the *CPR* does not begin as a criticism of them as arguments *that* God exists, but as explanations of *why* God

[81] This is clear in *ND* (Ak. 1: 394, 396), as well as the Herder metaphysics lectures (Ak. 28: 13, 14).

[82] For instance, Proops (2015)'s reconstruction of Kant leaves him without an objection to ontological arguments as such, but only individual objections against individual arguments. Kant's claim, however, is more sweeping: ontological proofs as such are impossible (A592/B620).

[83] For the distinction between nominal and real definitions, see *JL* §106 (Ak. 9: 143–4) and Ch. 9.2, where it is discussed in more detail.

[84] Cf. Baumgarten: "If God were not actual, the principle of [non-]contradiction would be false" (*Meta.* §824).

[85] Cf. *OPG* (Ak. 2: 82–3).

exists necessarily. In the immediately preceding section of the Ideal of Pure Reason, "The grounds of proof of speculative reason for inferring the existence of a highest being," Kant describes reason as assuming the existence of an absolutely necessary being and then casting about for an explanation of *why* that being necessarily exists. He first argues that reason has no grounds for regarding the concept of an unlimited being—one possessed of every reality—as a necessary being, and, conversely, no grounds for rejecting limited beings as candidates for necessary existence. His discussion of the ontological argument in section 4 ("On the impossibility of an ontological proof of the existence of God"), therefore, begins with reason already having formed the concept of a necessary being, and inferred (illegitimately) its existence.[86]

For now, I am identifying logicism as the view that:

> ($Necessity_{RealDef}$) For any proposition p, it is necessary that p if and only if there is a (finite) demonstration that p from identities and real definitions of the constituents of p.

When combined with the assumption that God necessarily exists, this entails that:

> (1) There is a finite demonstration of the proposition *God exists* from the real definition of God (which states his essence) and existence.

This means that logicism is committed to a specific form of ontotheism: the relation between the real definition of God (which states his essence) and the proposition that *God exists* is logical entailment. Ontotheism is a more general position than logicism. For instance, Descartes is an ontotheist (he accepts that God exists in virtue of his essence) but he is not a logicist.[87] In this chapter and the next I am focusing on Kant's objection to ontotheism in general; if successful, this would also refute logicism.

1.5. Possibilism

Ontotheists hold that God exists necessarily in virtue of the fact that his existence is grounded in his essence. They are committed to the following in-virtue-of claim:

> (1) ($\Box$God exists) in virtue of the fact that (God's existence is grounded in his essence).

This is an instance of the general principle that objects have predicates necessarily in virtue of those properties being logically grounded in their essences. That general principle entails that:

[86] A592–3/B620–1. Kant connects the ontological argument to the issue of what explains necessary existence in a number of different texts. Cf. the metaphysics lectures: *MH* (Ak. 28: 131), *MV* (Ak. 28: 418), *MvS* (Ak. 28: 498–500), ML_2 (Ak. 28: 556–8, 599), and MK_2 (Ak. 28: 724, 783).

[87] That is one reason Descartes will be important to my argument; the other is that Descartes' ontological argument was deeply influential on Leibniz.

(2) *a*'s essence grounds *a*'s being F ⊃ □(F*a*).

In other words, any predicate grounded in an object's essence is a predicate the object necessarily has. Furthermore, the consequent of this conditional obtains in virtue of the antecedent; the object in question necessarily has the predicate *because* (in virtue of the fact) that predicate is grounded in its essence.

However, this conception of the relation between necessary predication and essence runs into problems if we make the assumption that necessarily every object exists, i.e.

(3) □(*x*)(exists(*x*)).

Note that this is *not* the claim that every object *de re* necessarily exists; it is merely the claim that, necessarily, the quantifier 'for all' ranges only over existing objects. It is an immediate consequence of the most natural way of defining the predicate 'exists' in terms of the universal quantifier:

(4) exists(*x*) = ¬(*y*)(*y*≠*x*).[88]

Intuitively, (4) says that an object exists just in case it is not the case that no object is identical to it. If we define 'exists' in this fashion, then the sentence inside the scope of the necessity operator in (3) is a logical truth; since logical truths are necessary, (3) is true. If we define the predicate 'exists' in this fashion, (2) entails:

(5) *a*'s essence grounds *a*'s being F ⊃ □(exists(*a*)).[89]

In other words, any object with an essence sufficient to ground some of that object's predicates is an object that necessarily exists. This entails that either no object other than God has essential predicates, or that every object with essential predicates necessarily exists. Neither is an acceptable consequence.

The ontotheist might try to escape this consequence by modifying (2) as follows:

(6) *a*'s essence grounds *a*'s being F ⊃ □(exists(*a*) ⊃ F*a*).

But notice that the consequent of this conditional is trivial where 'exists' replaces 'F':

(6*) *a*'s essence grounds *a*'s existence ⊃ □(exists(*a*) ⊃ exists(*a*)).

[88] I define the predicate 'exists' using the universal quantifier rather than the so-called 'existential' quantifier (the two are interdefinable). I go on to argue that the ontotheist should understand that quantifier not as expressing 'there exists an F' but as expressing 'there is an F.' Thus, introducing '∃*x*F*x*' at this stage would only serve to confuse the issue.

[89] Because □(F*a*) entails that □(¬(*y*)(*y*≠*a*)), which, by (4), is equivalent to □(exists(*a*)). The intermediate step—from □(F*a*) to □(¬(*y*)(*y*≠*a*))—fails in a 'free logic' (Nolt 2014). However, I think that the ontotheist who adopts a free logic will require a theory structurally isomorphic to the one I argue for in the main text; for more, see the supplementary article "Free Logic and Ontotheism" on my website (see Notes on the Text). Thanks to Catharine Diehl, Andrew Stephenson, and Damien Melamedoff for pressing me on this point.

The triviality of this conditional undermines the ontotheist position. The ontotheist view, after all, is that the antecedent of this conditional explains the consequent. If the consequent of the conditional is trivial, then the antecedent of the conditional is doing no explanatory work. To accept (6) is to trivialize the doctrine that God necessarily exists in virtue of his existence being grounded in his essence. If (6) is taken to be the meaning of (2), then every object necessarily exists in precisely the same sense that God necessarily exists: necessarily, if it exists, it exists.

In order to counter these problems, the ontotheist needs to retreat to (2) and reject the principle that every object exists. Recall:

(2) a's essence grounds a's being F $\supset \Box(\mathrm{F}a)$.

and the example from section 1:

(7) Caesar's essence grounds Caesar's being human

from which it follows that:

(8) $\Box$(human(Caesar)).

The ontotheist needs to understand (8) in a way that does not entail that it is necessary that Caesar exists, that is, to interpret (8) so that it is consistent with:

(9) $\Diamond\neg$exists(Caesar).

Claim (8) is the claim that in any counterfactual situation (or 'possible world') Caesar is human; claim (9) is the claim that in some counterfactual situation (or 'possible world') Caesar does not exist. What the ontotheist needs in order to make these consistent is a distinction between an object having properties in a counterfactual situation and that object existing in that counterfactual situation. This will involve distinguishing, in each counterfactual situation, the objects that are merely the subject matter of true propositions in that counterfactual situation, and the objects that exist in that situation. Accepting (2) means accepting the consequence that every object has its essential properties in every counterfactual situation, but this does not require embracing the conclusion that every such object *exists* in every counterfactual situation. In other words, the ontotheist needs to accept that there are values of a and F such that:

(10) $\Diamond(\mathrm{F}a\ \&\ \neg\mathrm{exists}(a))$.

In general, where a is a being that does not necessarily exist, but has an essence that grounds its possession of property F, it will follow that a has F in every counterfactual situation, even ones where it does not exist. In other words, the ontotheist must accept:

(11) $\Diamond\neg(x)(\mathrm{exists}(x))$.[90]

[90] Cf. Fine (1994*a*), 3–4.

This requires rejecting the most natural definition of the 'existence' predicate:

(12) $\text{exists}(x) =_{\text{def}} \neg(y)(y{\neq}x)$

or, equivalently:

(13) $\text{exists}(x) =_{\text{def}} \exists y\ (y{=}x)$.[91]

What my discussion so far brings out is that the ontotheist needs to deny (12) and (13) and claim that existence is not equivalent to falling within the scope of the universal quantifier (more precisely, being the value of a variable bound by the universal quantifier). The ontotheist needs to understand 'exists' as a predicate that (potentially) applies to only a subset of the objects that fall within the domain of the universal quantifier. Consequently, the ontotheist should not think of the quantifier '∃'—interdefinable in the standard way with the universal quantifier—as the *existential* quantifier, but as a broader quantifier that includes not only existing objects but (potentially) non-existent objects as well. The ontotheist might think of this quantifier as expressing the natural language quantifier expression 'there is' [*es gibt*] (that is why I have refrained from using '∃' up to this point). It is important to point out that the ontotheist does not need to claim there *are* non-existent objects, but only that such objects are possible; however, for many of the ontotheists considered here, their other philosophical commitments may push them to accept non-existent possible objects (e.g., Leibniz's possible worlds).

It might be objected that the ontotheist can do the same work by distinguishing between the objects that are *actual* and the objects that are not, while maintaining that, necessarily, all objects (both actual and non-actual) exist.[92] On such a view, objects can have properties (e.g., those contained in their essences) in counterfactual situations in which they are not actual. However, many of the philosophers we are discussing use 'existence' and 'actuality' interchangeably. One notable instance is Kant, because he does not consistently distinguish existence [*Dasein*] and actuality [*Wirklichkeit*].[93] Kant may have been simply following standard usage in German philosophy at the time. In his *Vorbereitung zur natürlichen Theologie*, from which Kant sometimes lectured, Eberhard's list of recent proofs of the existence of God seems to use '*Dasein*' and '*Wirklichkeit*' interchangeably (Ak. 18: 563); Baumgarten also identifies '*actualitas*' and '*existentia*' in *Metaphysica* §55. Consequently, I will treat existence and actuality, insofar as they are predicates of objects, as equivalent.[94]

[91] These definitions are equivalent by stipulation: I am defining the quantifier '$\exists x$' in terms of the universal quantifier.

[92] E.g., David Lewis famously believed in non-*actual* objects, but had no truck with the neo-Meinongian view that there are non-*existent* objects. See Lewis (1986), 97–8.

[93] Cf. *ND* (Ak. 1: 396), *OPG* (Ak. 2: 72, 75), and *Danz.RT* (Ak. 28: 1151, 1256, 1291). See, however, MK_3 (Ak. 29: 986) as well as *Refl.* 6324, where Kant explicitly distinguishes actuality from existence. In Chs. 8.4 and 10.7 I argue that Kant's Critical view does include a distinction we could call the existence/actuality distinction, but that distinction will not affect the dialectic here.

[94] 'Actuality' here refers to a predicate of objects (actual(x)). Actuality as a propositional operator (actually p) is different.

Since the ontotheist needs to presuppose that there may be non-existent objects, and 'objects' here means (possible) beings, this is equivalent to the presupposition that there may be non-actual/non-existent possible objects/beings.[95]

This thesis, that there could be non-existent objects, I will call *possibilism*:

(*Possibilism*) There could be merely possible objects (beings, *entia*, *Dinge*) that do not exist.

The ontotheist, in order to account for God's necessary existence, must accept possibilism, or so I have argued. The negation of possibilism is what I will call *actualism*:

(*Actualism*) Necessarily, there are no non-existent merely possible objects (beings, *entia*, *Dinge*).

Earlier, I pointed out that one path to possibilism is to reject the most natural definition of existence in terms of the 'there is' quantifier:

(14) $exists(x) =_{\text{def}} \neg(y)(y{\neq}x)\ [=_{\text{def}} \exists y\ (y{=}x)]$.

On the other hand, if we adopt this as the definition of existence, the following expresses the claim that every object exists:

(15) $(x)\ \neg(y)(y{\neq}x)$.

Since this is a logical truth, and logical truths are necessary, this entails that necessarily every object exists, actualism. So actualism is a direct consequence of defining the existence predicate as in (14); possibilism entails the negation of that definition.

Some readers may feel that they began reading a book on eighteenth-century metaphysics and are now reading a book on contemporary analytic philosophy. What have possibilism and non-existent possibilia to do with Leibniz, Wolff, and Baumgarten, much less Kant? Recall that 'being' is the highest genus of ontology and all beings have essences.[96] For each being (*ens*) there is an answer to the question, what is it for that being *to be*? For creatures, an answer to the question 'what is it for that being to be?' does not determine whether that being exists. God is the unique being whose essence grounds his existence. The question of possibilism—could there be non-actual possibilia?—can be formulated in this terminology as the question: could there be a being that has an essence but lacks existence? This question, I take it,

[95] 'Object' here means object of quantification. I argued in §3 that objects of quantification (the objects there are) are all and only beings (*Dinge*).

[96] I am assuming, as Leibniz, Wolff, Baumgarten, and (I think) Descartes do, that every *ens* has an essence. This assumption may not be universally shared in the Scholastic tradition, though. Thanks to Kris McDaniel for drawing my attention to this point.

is not an anachronistic interjection, but one that these ontotheists could make perfect sense of.[97]

1.6. Real Predicates

In *Beweisgrund* Kant objects to the ontological argument that "existence is not a predicate or a determination of a thing" (Ak. 2: 72), and in the *CPR* he writes that "being is obviously not a real predicate" (A598/B626). In this section I explore what these claims mean.

First of all, 'real predicate' and 'determination' are synonymous, as Kant makes clear in a parthenthetical remark at A598/B626: "a real predicate (that is, a determination of a thing)." A logical predicate is any concept that can be a predicate in a judgment: "anything one wishes can serve as a logical predicate, even the subject can be predicated of itself" (A598/B626).[98] So from the fact that there are existential judgments, judgments whose predicate is <*exists*>, it follows immediately that <*exists*> is a logical predicate. So there is really one claim, expressed in two synonymous ways: <*exists*> is not a determination, and <*exists*> is not a real predicate. While Kant does not explicitly distinguish 'real' from merely 'logical' predicates in *Beweisgrund*, this distinction is implicit when he claims that, although existence is not a (real) predicate, it is admissible to use it as a (logical) predicate in ordinary speech as long as one does not thereby assume that it is a determination of objects:

> Nonetheless, the expression 'existence' is used as a predicate. And, indeed, one can do this safely and without troublesome errors, as long as one does not insist on trying to derive existence from merely possible concepts [...] (*OPG*, Ak. 2: 72)

However, it is not at all clear what Kant means by denying that existence is a "determination." On this point, his definition in the *Critique* is unhelpful: "the **determination** is a predicate, which goes beyond the concept of the subject and enlarges it. It must therefore not be contained in the subject concept" (A598/B626).[99] On the most literal reading of this passage, a determination of an object is a 'synthetic predicate,' a predicate of the object that is not contained in its concept. More precisely, since every object falls under indefinitely many concepts, a concept C is a synthetic predicate of concept C* just in case C is not contained in C* (the judgment

[97] Descartes explicitly discusses the essence/existence issue; see Chapter 2.2. For Wolff's views see *Dt. Met.* §544; for an analysis of Wolff's relation to the Scholastic debate about essence and existence see Honnefelder (1990), 320, and 367–70. See also the supplementary article "Essence and Existence" on my website (see Notes on the Text). Thanks to Uygar Abaci for pointing out the relevance of the essence-existence debate to my argument here.

[98] Since the predicate of one judgment can be made the subject of another, this means that all concepts that can be predicates of judgments—that is, all concepts whatsoever (A69/B94)—are logical predicates; see A94/B129. See Proops (2015), 11 for more evidence that predicates are concepts.

[99] For more on the notion of a 'synthetic predicate' see *JL* (Ak. 9: 59).

*All C*s are C* is synthetic). However, as other commentators have pointed out, this interpretation commits Kant to the following inconsistent triad:

(1) Existence is not a determination of any concept, i.e. the predicate <*exists*> is not synthetic with respect to any concept.
(2) All existential judgments are synthetic.
(3) If a judgment is synthetic, then its predicate is synthetic with respect to its subject.[100]

Since Kant repeatedly asserts (2) in this section of the *CPR* and (3) follows from the definition of a synthetic judgment, either the interpretation of determinations as 'synthetic predicates' is mistaken, or Kant contradicts himself within the space of a few paragraphs.

Another possibility is that Kant means 'determination' here in Baumgarten's technical sense: "what is either posited to be A, or posited not to be A, is DETERMINED."[101] However, this refers to any predicate whatsoever; that there are existential judgments (judgments in which the predicate is <*exists*>) entails that existence is a determination in Baumgarten's sense. In his own copy of *Metaphysica* Kant identifies Baumgarten's definition of 'determination' with that of a *logical* predicate (*Refl.* 3520, Ak. 17: 33); since <*exists*> is a logical predicate, it is a determination in Baumgarten's sense.

A more tempting possibility is to deny that <*exists*> is a predicate of objects at all, but a predicate of concepts, anticipating the Fregean theory of the existential quantifier as a second-order concept that applies to concepts that have a non-empty extension. This interpretation finds support in this oft-quoted passage from *Beweisgrund*:

> But when existence occurs as a predicate in common speech, it is a predicate not so much of the thing itself as the thought which one has of the thing. For example: existence belongs to the narwhal [*Seeeinhorn*] but not to the unicorn [*Einhorn*]. This simply means: the representation of a narwhal is an empirical concept; in other words, it is the representation of an existent thing. [...] The expression 'A narwhal is an existent animal' is not, therefore, entirely correct. The expression ought to be formulated the other way around to read 'The predicates, which I think collectively when I think of a narwhal, attach to a certain existent animal.' (*OPG*, Ak. 2: 72–3)[102]

Kant claims that the judgment *a narwhal is an existent animal* does not assert that some predicate is contained in the concept <*narwhal*>, and does not attribute further predicates to the objects that fall under <*narwhal*>; it asserts that there is at least one object that falls under <*narwhal*>, i.e. that the concept is instantiated.[103] I take this to

[100] Cf. Wood (1978), 105 and Shaffer (1969), 125. [101] *Meta.* §34. [102] Cf. A599/B627.

[103] Kant gives the mistaken impression that, on his analysis, 'narwhals exist' is equivalent to 'there is a narwhal *and I have experienced it*'. This would make the following judgment false as a matter of meaning: there are narwhals but no one has ever encountered one. But he drops the misleading suggestion that the

be clear evidence that the *fundamental* (though not the *only*) role of the predicate *<exists>* is to apply to concepts: it applies to a concept if and only if that concept is instantiated by an object. While this is not yet the complete Fregean theory of the existential quantifier, it does anticipate it.[104] Consequently, I will borrow from the contemporary symbolism and formalize *narwhals exist* as:

(4) $\exists x(\text{narwhal}(x))$

which is to be read as 'there is an object in the extension of *<narwhal>*.'[105]

This, however, cannot *exhaust* the content of Kant's claim that existence is not a determination or 'real predicate' because it is not something ontotheists *need* to deny and thus has no force *by itself* against the ontological argument. As we saw in section 5, ontotheists are committed to possibilism, the view that there could be non-existent objects. This is entirely compatible with the view that to make existence claims we need a quantifier expression, a second-order predicate that applies to a concept just in case it is instantiated; the ontotheist needs merely to add that in existential judgments like *narwhals exist* the quantifier is restricted to *existing* objects.[106] So the ontotheist can fully accept Kant's analysis of existential judgments as long as he interprets the quantifier-expression in (4) as implicitly restricted to *existing* objects, that is:

(4*) $\exists x{\in}\text{E}\ (\text{narwhal}(x))$

where E is the set of all existing objects. This is equivalent (assuming E is non-empty) to the following claim, using an unrestricted quantifier and an existence predicate for objects:

(5) $\exists x(\text{exists}(x)\ \&\ \text{narwhal}(x))$.[107]

In other words, the ontotheist can fully accept that existence is a second-order predicate (a quantifier) as long as it is a *restricted* quantifier; alternately, that

instance of the concept must be *experienced* when he analyzes 'God exists' as "an existing thing has those predicates, which we collectively designate with the expression: God" (Ak. 2: 74).

[104] For one thing, Kant has not yet developed the Fregean function-object analysis of judgment, and his neglect of relational predicates leaves him without the resources to develop a theory of polyadic quantification. Friedman (1992*a*), 96–135 discusses the limitations of Kant's logic.

[105] Rosenkoetter (2010) objects to what he calls the 'Frege-anticipation' thesis that it is incompatible with Kant's claim that the assertoric function of judgment corresponds to the category of existence. He claims that the Frege-anticipation thesis would entail that "Kant would need to hold, in parallel, that all assertoric judgments can be reduced to *q* is true" (552). However, it is unclear what Rosenkoetter's argument for this claim is, nor is it clear why the defender of the Frege-anticipation thesis cannot hold Rosenkoetter's own account of the assertoric function of judgment.

[106] One can (as e.g., Quine did) hold that existence is a quantifier without holding that it is a second-order predicate of concepts. For the purposes of this book I will be identifying the view that existence is a quantifier with the view that it is second-order.

[107] Where the extension of the predicate 'exists(x)' is E, the set of existing things used to restrict the quantifier in (4*).

existential judgments are made using an unrestricted quantifier ('there is,' which ranges over all objects) and an existence predicate that applies to only some objects. This interpretation may even be encouraged by Kant's own text, quoted above, because he interprets *narwhals exist* as the judgment that "the predicates, which I think collectively when I think of a narwhal, attach to a certain existent animal". This may give the (false, or so I will argue) impression that Kant himself thinks that existential judgments are made using an unrestricted quantifier ('there is') and an existence predicate for objects.

This brings out an important, and, I believe, too often neglected point: the real issue between the ontotheist and Kant over existence is not whether existence is a quantifier (second-order predicate) but whether it is a restricted or unrestricted quantifier. The real issue is whether there is an existence predicate *for objects* that applies to only a subset of them (equivalently, whether the existence quantifier is a restriction of the 'there is' quantifier); the ontotheist (I have argued) must maintain that the existence predicate *for objects* applies only to a subset of them.[108]

The 'synthetic predicate' interpretation, from above, assumes that, when Kant writes that the determination "enlarges" the subject concept, all he means is that the determination is not one of the marks analytically contained in the subject concept. But this is not the only sense in which a predicate might be said to "enlarge" the subject concept of a judgment. A predicate might also enlarge a concept by enlarging its *content* and rendering that contept more determinate by restricting the range of objects that can fall under it. I propose, then, the following interpretation of Kant's technical term 'determination':

> (*Defn.*) A concept *P determines* a concept *C* if and only if it is possible that there is an object that instantiates *C* and *P* and it is possible that there is an object that instantiates *C* but not *P*.[109, 110]
>
> (*Defn.*) A predicate *P* is a *determination* if and only if *P* determines at least one concept.[111]

One concept can determine another, in the sense of specifying the nature of the objects falling under the concept. <*Scalene*> determines the concept <*triangle*>, but <*having interior angles that sum to 180 degrees*> does not. It does not add any new specification to the concept <*triangle*>, even though it is not analytically contained in

[108] Forgie (2007) argues that Kant's claim that existence is a second-order predicate should not be conflated with Gassendi's claim that existence is not a property but the precondition for having properties in the first place. My point is that Kant's attempt to refute ontotheism requires him to make a version of Gassendi's claim: there cannot be objects that do not exist.

[109] Cf. *Pöl.RT* (Ak. 28: 1027), *Volck.RT* (Ak. 28: 1176), and *Danz.RT* (Ak. 28: 1258), as well as the texts from Kant's metaphysics lectures on '*determinieren*' cited below.

[110] This interpretation is similar to that of Van Cleve (1999), 188 and Hanna (2001), 133.

[111] I am assuming that existence, in virtue of being a logical predicate, is a concept.

that concept.[112] Thus, on my definition, all determinations are synthetic predicates, but not every synthetic predicate of an object is a determination.

This interpretation also preserves the natural reading of the passage: "the determination is a predicate, which goes beyond the concept of the subject and enlarges it [the concept]. It must therefore not be contained in the subject concept [*sie muß also nicht in ihm schon enthalten sein*]" (A598/B626—my emphasis). On my reading, the second sentence is a consequence of the first, but is not identical to it: because a determination enlarges the subject concept in the sense of adding further content to it, the determination cannot be analytically contained in the subject concept. On the 'synthetic predicate' reading, the second sentence merely restates the first sentence.[113]

In addition to respecting the natural reading of this passage, this interpretation has at least five other advantages. First, it allows us to escape the inconsistent triad of views; just because <*exists*> does not determine any subject concept, it does not follow that existence is not a synthetic predicate. This is because:

(1) For any concept C, necessarily anything that falls under C falls under <*exists*>; <*exists*> is not a determination

is compatible with:

(2) No judgments of the form *there exists a C* are analytic; existence is not a mark of any concept.[114]

The second advantage to this reading is that it allows us to explain *why* <*exists*> is not a determination: it does not 'enlarge' or further specify any concept, because it is the concept that every object falls under. If <*exists*> were a determination of some concept C, it would follow that it is possible for there to be objects that fall under C but not <*exists*>; it would follow that it is possible for there to be non-existent objects.

This point connects directly to our earlier discussion of the quantificational theory of existence. As a predicate of objects, <*exists*> is a determination just in case there is some concept C such that:

(3) $\Diamond \exists x(Cx \,\&\, \neg \text{exists}(x))$

from which it follows that:

(4) $\Diamond \exists x(\neg \text{exists}(x))$.

[112] Because for Kant, the judgment that every triangle has internal angles that sum to two right angles is synthetic, not analytic.

[113] Similarly, when Kant defines 'determination' in the Mrongovius metaphysics lectures, he says "the logical predicate can be analytic, but determination is always synthetic" (Ak. 29: 819). He notably does *not* claim that all and only synthetic predicates are determinations.

[114] The compatibility of these two claims follows from Kant's acceptance of synthetic *a priori* judgments, which, by definition, are necessarily true and not analytic. See B4 and Ak. 8: 235.

If the object-predicate <*exists*> is a determination, it follows that it is possible for there to be a non-existent object. But notice that, conversely, if (4) is true, it follows that <*exists*> is a determination; (4) entails that <*exists*> determines the trivial concept <*x*=*x*>, the concept under which all objects necessarily fall. So existence is a determination if and only if (4) is true.

The natural way of defining the existence predicate *for objects* is by using the quantifier '∃' (which is equivalent to the definition in terms of the universal quantifier, given in brackets):

(5) $\text{exists}(x) =_{\text{def}} \exists y\,(y=x)\ [\leftrightarrow \neg(y)(y\neq x)]$.

Applying this to 'exists' in (4), we get:

(6) $\Diamond\exists x\neg\exists y(y=x)$

which is a logical falsehood. There cannot be an object such that there is no object to which it is identical. The definition of existence using the quantifier '∃' (claim (5)) entails that <*exists*> is not a determination. If we interpret 'determination' as I have argued we should, Kant's claim that existence is not a determination, while not philosophically uncontroversial, is highly plausible. This is the third advantage of my reading. Kant's claim that existence is not a determination is equivalent to defining the object-level existence predicate in the natural way, using the quantifier '∃'. It is equivalent to claiming that the quantifier expression 'there is' ranges only over *existing objects*, that is, that there are no non-existent objects. On such a theory of existence, it is appropriate to call the quantifier '∃' the *existential* quantifier. This also shows that, although the *fundamental* sense of existence for Kant may be given by the existential quantifier '∃', he also has the resources to define an existence predicate for objects, as in (5). Furthermore, if the existence predicate for objects, <*exists*>, is defined in terms of the quantifier '∃' then we can translate unproblematically between them. Henceforth, when I talk about <*exists*>, I will always have the object-predicate in mind; when I mean the second-order predicate of concepts (what is now called a quantifier), I will use '∃'.

One might well wonder, then, if the fundamental meaning of existence is a quantifier, why does Kant bother discussing an existence predicate for objects? To refute the ontotheist, of course! As I argued earlier, the ontotheist can accept that existence is a quantifier, but then the real issue becomes: is the existential quantifier a restriction of the wider quantifier 'there is'? On my interpretation, by denying that existence *as a predicate of objects* is a determination, Kant is answering this question in the negative. This is equivalent to denying that there are (or could be) non-existent objects, which, as we saw in section 5, would be devastating to the metaphysics of ontotheism in general, and, I will argue in Chapter 2, the specific ontological arguments given by Kant's specific opponents (Descartes, Leibniz, etc.). So the fourth advantage of my reading is that Kant's claim that <*exists*> is not a determination, if true, spells serious trouble for ontotheism.

Fifthly, and finally, this interpretation of what it is to 'determine' a concept is well supported by Kant's lectures on metaphysics. For example:

> The principle that, of two contradictorily opposed predicates, one pertains to the object is called the 'principle of excluded middle among two contradictories.' This is called determining [*determinieren*] for various objects are left indeterminate by the concepts we have of them. For example, the concept of human [*der Begriff des Menschen*] is indeterminate. [A human] can be learned or not, can be a man or a woman, etc. If I then say that one of the two must pertain to [an object], and I posit one of them, [the object] is determinate; indeterminate means determinate only in respect of what we think through our concept. For example, the concept of an angle is indeterminate, for it can be obtuse or acute. (*MV*, 28: 410)[115]

To determine [*determinieren*][116] a concept is to predicate one of two contradictorily opposed concepts, each of which is a possible way that the original concept can be specified; e.g., <*acute*> determines <*angle*> because it is possible for angles to be acute, and possible for them to be obtuse. We can say that <*acute*> and <*obtuse*> possibly divide the extension of <*angle*>. To say that <*exists*> does not determine any concept means therefore that <*existence*> and <*non-existence*> do not possibly divide the extension of any concept. The extension of *no* concept divides into existing and non-existing instances of that concept because there are *no* non-existent objects.[117]

If Kant is right that existence is not a determination, and if I am right that ontotheism—the doctrine that God exists in virtue of his essence—entails that existence is a determination, then Kant has refuted ontotheism *as such*. The generality of Kant's objection entails that there can be no being, not even God, that exists in virtue of its essence. Since logicism, combined with the traditional doctrine that God necessarily exists, entails that God exists in virtue of his essence, this also refutes logicism. If existence is not a determination, then there is at least one necessary truth, that God exists, that is not a logically necessary consequence of the essences of possible beings. In the next chapter I reconstruct in detail Kant's arguments that existence is not a determination.

[115] Cf. *MH* (Ak. 28: 14, 19, 24–5, 843, 845), *MM* (Ak. 29: 818), *MvS* (Ak. 28: 491), and ML_2 (Ak. 28: 551–2).

[116] These passages typically use the Latinate phrase '*determinieren*' but we have seen that in *Beweisgrund* '*Determination*' is also Kant's term for what he will later refer to as a '*Bestimmung*' or 'real predicate' (Ak. 2: 72).

[117] Crusius's technical notion of *Determination* in *Ent.* §23 is quite similar; it is likely that Kant is using the Crusian notion of determination, rather than Baumgarten's.

2

Is Existence a Real Predicate?

2.1. Introduction

In the first half of this chapter I reconstruct in detail the ontological arguments of Descartes (section 2), Leibniz (section 3), and Baumgarten (section 4), to show that each of them is committed to *possibilism*, the view that there could be objects that do not exist. This confirms my argument, in section 1.5, that ontotheology as such is committed to possibilism. In the second half of the chapter I reconstruct Kant's arguments in *Beweisgrund* and in the *CPR* that existence is not a determination, which, I argued in Chapter 1.6, means that possibilism is false (there cannot be non-existent objects).

Although Wolff does give an ontological argument in Part II of his *Theologia naturalis* (1737), it introduces little that is not already found in Leibniz's arguments, so I will largely ignore Wolff in this chapter.[1] Descartes is a difficult figure in this context. First of all, he is not a logicist; on the contrary, he believes in brute necessary connections, knowable to us through clear and distinct perception, but not grounded in any relation of logical entailment. However, like the logicists, he is an ontotheist; he believes that God exists necessarily in virtue of the 'involvement' of existence in God's essence. Furthermore, Descartes in many ways set the pattern for early modern ontological arguments. Consequently, I begin with Descartes' version of the ontological argument and argue that it too is committed to possibilism.

2.2. Descartes' Ontological Argument

Descartes' ontological argument in the Fifth Meditation is the most famous version of the argument in modern philosophy. It deeply influenced Leibniz's thinking about the ontological argument, as well as that of Wolff, Baumgarten, and Kant.[2] Descartes

[1] *TN* was translated into German by Gottlieb Friedrich Sagen under the title *Natürliche Gottesgelahrheit nach beweisender Lehrart abgefasst* (*NG*); both versions are included in Wolff's *Werke* (see Note on Sources). Wolff first introduces the concept of compossibility (§1), defines the *ens perfectissimum* as the being in which all compossible determinations are present to the highest degree (§6), argues that the *ens perfectissimum* is possible because its concept contains no negations (because all of its determinations are unlimited, hence positive) and thus cannot contain a contradiction (§13), claims that existence is a determination in things (§20), and concludes that the *ens perfectissimum* has existence to an unlimited degree, hence it exists (§21).

[2] Cf. Wolff's summary of the recent history of the ontological argument in *NG* §13.

is widely credited with resuscitating the 'Anselmian' argument, although I think that Anselm's argument is different enough from the Cartesian argument that I forgo discussion of it here.[3] Although the Cartesian argument is originally given in the body of the Fifth Meditation,[4] I think the three-line, syllogistic version of the argument in the First Replies is clearer, so I will start there:

> [1] That which we clearly and distinctly understand to belong to the true and immutable nature, or essence or form of something, can truly be asserted of that thing. [2] But once we have made a sufficiently careful investigation of what God is, we clearly and distinctly understand that existence belongs to his true and immutable nature. [3] Hence we can now truly assert of God that he does exist.[5]

Premise [1], Descartes' argument for it, and the very notion of a clear and distinct perception on which Descartes bases much of his argument in *Meditations*, are highly problematic, as was forcefully pointed out by Descartes' contemporaries in the Objections, and as has been pointed out again and again by scholars since. I will not focus on those problems with the argument, though, because they are not problems with the ontological argument as such.

As the second set of Objections points out, since only possible beings have true and immutable natures, Descartes must assume that God is possible: "it does not follow from this that God in fact exists, but merely that he would have to exist if his nature is possible, or non-contradictory" (CSM II: 91/AT II: 127).[6] But Descartes had already anticipated, and answered, this objection in his response to Caterus in the First Replies: "possible existence is contained in the concept or idea of everything we clearly and distinctly understand."[7] The possibility of God is guaranteed by our clear and distinct understanding of his true and immutable nature. So expanding upon the original argument and filling in its enthymematic premises, we can reconstruct it as follows:

<u>Descartes' Ontological Argument:</u>

(1) If I clearly and distinctly understand that *p* then *p*.
(2) If I clearly and distinctly understand a true and immutable nature, then the object whose nature that is, is possible.
(3) There is a true and immutable nature of God and I clearly and distinctly understand it.

[3] Cf. Barnes (1972).

[4] Cf. First Objections and Replies (Objection: 99–100; Reply: 113–21); Second Objections and Replies (Objection: 127; Reply: 150–2, 163–7); and Fifth Objections and Replies (Objection: 322–6; Reply: 382). References give the page numbers in vol. 7 of AT; this pagination is also given in the margins of CSM. (See 'Note on Sources and Translations'.)

[5] CSM II: 83/AT VII: 115–16. I have inserted the premise numbers and the line spacing; Descartes presents the argument in continuous prose.

[6] Presumably this is because there are no true and immutable natures of *impossibilities* (non-things, *Undinge*).

[7] CSM II: 83/AT VII: 116.

(4) ∴ God is possible. [From (2) and (3).]
(5) I clearly and distinctly understand that every perfection is contained in God's true and immutable nature.
(6) ∴ Every perfection is contained in God's true and immutable nature. [From (1) and (5).]
(7) Any property contained in the true and immutable nature of a possible object is a property had by that object.
(8) God has every perfection. [From (6) and (7).]
(9) Existence is a perfection.
(10) ∴ God exists.[8] [From (8) and (9).]

Lines (1)–(4) establish the possibility of God; lines (5)–(10) spell out precisely how Descartes goes about showing that existence is contained in God's true and immutable nature and deriving the conclusion that God exists.

In the Replies, Descartes' interlocutors raise a number of objections to this argument, but few have been as influential as Gassendi, who writes:

> [...] existence is not a perfection either in God or in anything else; it is that without which no perfections can be present. For surely, what does not exist has no perfections or imperfections, and what does exist and has several perfections does not have existence as one of its individual perfections; rather its existence is that in virtue of which the thing itself and the perfections are existent, and that without which we cannot say that the perfections are existent, and that without which we cannot say that the thing possesses the perfections, or that the perfections are possessed by it. Hence we do not say that existence 'exists in a thing' in the way that perfections do; and if a thing lacks existence, we do not say it is imperfect, or deprived of a perfection, but say instead that it is nothing at all.[9]

Gassendi is claiming that existence, far from being one perfection among others, is the condition without which a thing has no perfections, indeed no properties.[10] Any bearer of properties must exist; if a thing does not exist, "we do not say that it is imperfect... but say instead that it is nothing at all," that is, that there is no such thing. Gassendi is assuming an actualist theory of existence: every object (being) exists. This means that (7), sometimes known as the 'Cartesian predication principle,' has to be understood as the conditional claim that *if* some property is contained in an

[8] In some presentations of the argument Descartes claims that it is *necessary* existence that is one of the divine perfections (Second Replies); in other texts he identifies existence as the perfection in question (e.g., the Fifth Meditation itself, as well as the First and Fifth Replies). In the second set of Replies he switches in the space of a few paragraphs from claiming that it is existence that is one of the divine perfections (AT VII: 151) to claiming that it is necessary existence (AT VII: 152) to deciding in favor of the latter in his 'geometric' presentation of the argument (AT VII: 166). I have focused on the former presentation of the argument (in terms of existence *simpliciter*) in order to facilitate comparison of Descartes with the logicists.

[9] CSM II: 224/AT VII: 323.

[10] Forgie (2007) cautions us against assimilating Gassendi's objection to the ontological argument (existence is not a property) to Kant's (existence is not a determination). On this point I agree; but I dissent from his reading of Kant. See Ch. 1.

essence and there is an object with that essence, *then* the object has the property.[11] The Cartesian predication principle cannot (if we are assuming actualism) be the claim that if some property is contained in an essence then the object of that essence contains the property, because that presupposes the essence has an object, i.e. some existing object instantiates that essence. Instead, it has to be understood as:

(7*) $(x)[(\text{E is the essence of } x \,\&\, \text{F is contained in E}) \supset Fx)]$, for any essence E and any property F.

But recall the work (7) is supposed to do in the argument. It is supposed to license the inference from the claim that every perfection is contained in the true and immutable nature of God (6) to the conclusion that God has every perfection (8). In order to do that, though, Descartes would have to show for some x and some E that E is the true and immutable nature of x, i.e. he would have to show of some x that x is God. In other words, he would have to first show that God exists in order to show that God exists! Descartes' ontological argument is pointless, according to Gassendi, because it requires its conclusion as one of its premises.

The logical form of Descartes' argument, as Gassendi understands it, is:

Gassendi's Reading of Descartes' Ontological Argument:

(1*) (I clear and distinctly understand that p) $\supset p$.
(2*) I clearly and distinctly understand an essence E $\supset \Diamond\exists x$(E is the essence of x).
(3*) There is an essence (true and immutable nature) of God and I clearly and distinctly understand it. Call this essence 'G.'
(4*) $\Diamond\exists x$(G is the essence of x). [From (2*) and (3*).]
(5*) I clearly and distinctly understand that every perfection is contained in G.
(6*) ∴ Every perfection is contained in G. [From (1) and (5).]
(7*) $(x)[(\text{E is the essence of } x \,\&\, \text{F is contained in E}) \supset Fx)]$, for any essence E and any property F.
(8*) ∴ For any perfection F, $(x)(\text{G is the essence of } x \supset Fx))$. [From (6*) and (7*).]
(9*) Existence is a perfection.
(10*) ∴ $\exists x$(G is the essence of x).

Thus, the argument is invalid on Gassendi's reading because (10*) does not follow from (8*) and (9*). Descartes never proves that *there is* a God (an entity that has the divine essence, my clear and distinct perception of which was the origin of the argument), only that it is possible that there is one. The most Descartes is warranted in concluding is the conditional claim that:

(10**) $(x)(\text{G is the essence of } x \supset \text{exists}(x))$.

[11] Caterus points out in the First Objections that the most Descartes is entitled to is this conditional claim (CSM II: 72/AT VII: 99). Compare this to Gassendi's remark that: "all that this means is that *if* anything is a man, it must resemble other things to we which we apply the same label 'man'" (CSM II: 222; AT VII: 320). Cf. A594–5/B622–3.

If we assume that bound variables only take *existing* objects as values, this is *trivial*; it means something like: if anything *existing* is God (has the divine essence), then it exists. As another insightful critic of Descartes' ontological argument, Caterus, put it in his Reply: "you cannot infer that the existence of God is anything actual unless you suppose that the supreme being actually exists; for then it will actually contain all perfections, including the perfection of real existence."[12] Note, though, that (10**) is trivial *only* if we are defining the predicate 'exists' in the usual way:

(Ex) exists $(x) =_{\text{def}} \neg(y)\ (y \neq x) = \exists y(y=x)$.

This, as discussed earlier, is tantamount to assuming actualism. Gassendi, and Caterus, are correct that the conclusion of this argument (10**), is trivial if we assume actualism.

Descartes' response to Gassendi is instructive:

> Here I do not see what sort of thing you want existence to be, nor why it cannot be said to be a property just like omnipotence—provided, of course, that we take 'property' to stand for any attribute, or for whatever can be predicated of a thing; and this is exactly how it should be taken in this context.[13]

Descartes is reiterating that God has the properties contained in his essence, regardless of whether God exists or not (because at this point in the argument, we have not yet proved that God exists). In other words, Descartes is thinking that he has already proved that *there is a God* whose true and immutable nature he understands, otherwise he would be unable to use the Cartesian predication principle (7*) to prove anything other than a conditional conclusion. But this means Descartes is thinking of the logical form of the initial steps of the argument as:

(1†) (I clear and distinctly understand that p) $\supset p$.
(2†) I clearly and distinctly understand an essence E $\supset \exists x$(E is the essence of x & possible(x)).[14]
(3†) There is a divine essence & I clearly and distinctly understand it. Call this essence G.
(4†) ∴ $\exists$x(G is the essence of x & possible (x)). Let g be such an object. [From (2†) and (3†).][15]

At this point, Descartes is licensed in claiming that there is a God, because he has shown there is an object with the divine essence. The rest of the argument shows that God, the being with the divine essence, exists:

[12] CSM II: 72/AT VII: 99. [13] CSM II: 263/AT VII: 383.

[14] The second disjunct of the consequent is redundant because I have stipulated that quantifiers range only over beings, and beings as such are possible (§1.3). However, I retain it so as to make explicit the difference between how the actualist (2*) and possibilist (2†) deal with possibility: the actualist uses a modal operator, while the possibilist uses a wider quantifier.

[15] For ease of exposition, I am ignoring the issue of whether g is the unique such object.

(5†) I clearly and distinctly understand that every perfection is contained in G.
(6†) ∴ For every perfection F, F is contained in G. [From (1†) and (5†).]
(7†) $(x)[(\text{E is the essence of } x \ \& \ \text{F is contained in E}) \supset Fx)]$, for any essence E and any property F.
(8†) For any perfection F, F*g*. [From (4†), (6†), and (7†).]
(9†) Existence is a perfection.
(10†) ∴ Existence(*g*).

Descartes is thinking of the argument as showing that there is a God, and thus introducing God as a subject for non-conditional predication, prior to proving that God exists. This only makes sense if Descartes rejects the claim that:

(*Actualism*) $\Box(x)(\text{exists}(x))$.

But Descartes has to reject actualism anyway, because it is in tension with (2†). Assuming an actualist theory of existence, (2†) would entail that every idea I clearly and distinctly understand is the idea of an actually existing object. This would have the absurd consequence that every mathematical idea is the idea of an actually existing object. From the mere fact that I can clearly and distinctly understand the idea of a chiliagon it would follow that there is an actually existing chiliagon. To a contemporary reader, this might appear unproblematic; chiliagons are abstract objects, so it might seem unproblematic that if we have a clear and distinct idea of them, they exist. However, mathematical objects are not, according to Descartes, 'abstract' in the contemporary sense; mathematical objects are bodies (*res extensa*), so an actualist ontology would entail that, for every geometrical idea we clearly and distinctly understand, there is a body with precisely that shape. I take it this is a consequence Descartes would want to avoid.

In some texts, Descartes appears to endorse explicitly this commitment to non-existent possibilia.[16] However, in other passages his attitude is more ambivalent; for instance, in *Principles of Philosophy* he endorses a merely 'rational' distinction between essence and existence:

> we do understand the essence of a thing in one way when we consider it in abstraction from whether it exists or not, and in a different way when we consider it as existing; but the thing cannot be outside our thoughts without its existence [...] Accordingly I say that... there is a lesser distinction between [them] [...] I call it a rational distinction.[17]

[16] AT VII: 64–6/CSM II: 45. Cf. Kenny (1968) and Griffin (2013), 9–33. More recently, some scholars have questioned whether Descartes accepts non-existent possibilia; see esp. Nolan (1997) and (2005), and Cunning (2008).

[17] CSM III: 349/AT IV: 349. Thanks to Nolan (2005) for drawing my attention to this passage. Note, though, that Descartes goes on immediately to write that, "if by essence we understand a thing as it is objectively in the intellect, and by existence the same thing insofar as it is outside the intellect, it is manifest that the two are really distinct" (CSM III: 281). This complicates matters, however; it is not clear this is compatible with a merely rational distinction between essence and existence. In *Principles*, §26 Descartes claims that the existence of a substance is one of its attributes, and since there is merely a rational

This doctrine means that every thing that has an essence also exists; the distinction between essence and existence is a distinction between two different ways of conceptualizing one and the same thing: through its primary attribute (essence) or as the fully concrete thing it is (existence).[18] If this is the case, then there cannot be non-existent possibilia with true and immutable essences; such an ontology would require a deeper distinction between essence and existence than is allowed by the 'rational' distinction doctrine.

My central interpretive claim is that Descartes' ontological argument commits him to a possibilist ontology; without it, the argument is invalid, as Gassendi and Caterus pointed out. In Kantian terms this means he is committed to regarding existence as a real predicate, a predicate that some objects have and some lack. That some of his other views may be incompatible with this ontology—for instance, the merely rational distinction between essence and existence—shows only that in one part of his philosophy he is committed to a thesis that he rejects in another part. This is not an uncommon situation, even for the greatest thinkers, like Descartes.

2.3. Leibniz's Ontological Argument

Leibniz developed his own ontological argument, and his views about existence, in attempting to overcome what he saw as the defects in Descartes' argument. His most fundamental objection to the Cartesian argument—repeated in numerous texts—is that Descartes fails to prove that God is possible. Descartes' argument proves, at most, that God's existence follows from his essence, his possibility. To conclude that God exists we must first discharge the assumption that God is possible, that there is a consistent essence of God.[19] Descartes' method of doing this is irreparably flawed, according to Leibniz, by its reliance on the method of clear and distinct perception.

However, Leibniz's theory of conceptual analysis and adequate ideas in "Meditations" provides a replacement for the rule of clear and distinct perception that does not suffer from the same problems as its predecessor. If we adequately understand a concept, i.e. can give a real definition of it, then the concept is logically consistent and therefore possibly instantiated. If our adequate understanding of one concept reveals another concept contained in it, then the latter concept pertains necessarily to the

distinction among the attributes of a substance, it follows that the distinction between the essence (principal attribute) and existence of a substance is merely rational; however, it is unclear what he means in saying that existence is an attribute of a (contingent) substance. See Cunning (2008) for further discussion.

[18] Cf. the discussion of the distinction between essence and existence in Question XXXI of Suárez's *Metaphysical Disputations* (Suárez 1983); for critical discussion, see Wippel (1982*a*), (1982*b*), and Witt (2010).

[19] Leibniz makes this same point—that the Cartesian argument establishes only that if God is possible then he exists—in a number of texts: "Meditations on Knowledge, Truth, and Ideas" (AG 25), *DM* §23, Leibniz's correspondence with Elizabeth (AG 238), and a wide range of letters and unpublished notes from throughout his career (L 165 f., 168, 211, 231, 286).

former concept. Clear and distinct perception, for Leibniz, consists in ascertaining the logical relations among concepts and their internal logical consistency.[20]

To return to the original objection—that Descartes fails to prove that a most real being is possible—Leibniz's logicist modal metaphysics is designed to dispense with this objection. As we saw in chapter 1.3, complex concepts are composed of simple, logically atomic concepts. The operations by which complex concepts are formed are conjunction, negation, and limitation. The primitive concepts, therefore, are logically simple (they are not conjunctions of other concepts), purely positive (they are not negations of other concepts), and they are concepts of unlimited perfections. Leibniz takes the unlimited perfections to be logically prior to limited perfections, i.e. limited perfections are 'understood' through unlimited ones.[21] This means that all concepts of finite things are logical constructs—through conjunction, negation, and limitation—of infinite and unlimited perfections. A finite mind possesses limited versions of, for instance, the unlimited perfections of understanding and will.[22]

The unlimited perfections are the logically atomic constituents of all other concepts. Leibniz's definition of logical entailment in terms of conceptual analysis means that an atomic concept, one with no internal logical structure, cannot stand in entailment relations with another atomic concept. The unlimited perfections, being logically atomic, are logically independent. Since the unlimited perfections are logically independent, they are logically compatible.[23] Therefore, the concept of a being possessed of all possible unlimited perfections, <*ens perfectissimum*>, is a logically consistent concept. Consequently, given the logicist analysis of possibility in terms of logical self-consistency, an *ens perfectissimum* is possible.[24] This is Leibniz's point in *Monadology* §45: "nothing can prevent the possibility of what is without limits, without negations, and consequently without contradiction" (AG 218).

Before continuing, I want to flag two features of Leibniz's ontological argument (and the ontological arguments of the 'Leibnizians', Wolff and Baumgarten) that I am going to ignore in what follows. The first is the *uniqueness* of the *ens perfectissimum*. All three logicists held the doctrine of the identity of indiscernibles, and since the concept of an <*ens perfectissimum*> completely determines the intrinsic properties of its object, it follows that there cannot be two distinct objects instantiating it. I will take it for granted that, if there is *at least* one *ens perfectissimum*, there is *only one*. Secondly, I am going to ignore the difference between *perfection* and *reality* and the

[20] It should be noted, however, that Leibniz's theory is subject to a problem analogous to one that afflicts Descartes: how do we know that a candidate analysis of a concept is the correct one? Furthermore, Leibniz's epistemology rests on a logicist metaphysics, which is problematic in its own right, as I will argue at length in Chs. 3–4.

[21] L 167/*DSR* 101. Cf. the correspondence with Elizabeth (AG 240).

[22] Cf. *Mon.* §48. For an extensive discussion of Leibniz's views on limited and unlimited perfections, see Adams (1994), 115–19.

[23] See *DSR* 69, though, for a slightly different picture of the relation among the divine attributes.

[24] Adams analyzes in great detail Leibniz's argument for the compatibility of the unlimited perfections in (1994), 142–8.

concomitant difference between conceiving of God as the *ens perfectissimum* and as the *ens realissimum*. In *Beweisgrund* Kant claims that "'perfection' [*Vollkommenheit*] always presupposes a relation to a being endowed with cognition and desire" (Ak. 2: 90) and substitutes the morally neutral term 'reality' in its place. While there are reasons to dispute whether Kant is right about the Leibnizian concept of perfection,[25] I am going to sidestep the issue and just treat 'perfection' as standing in for the more morally neutral metaphysical concept of 'reality,' which we can think of as a gradable degree of being.[26] Ignoring these complications is appropriate to my goal: not to reconstruct Kant's objections to particular ontological arguments (e.g., that the identity of indiscernibles might be false, or that perfection is distinct from reality) but to ontological arguments *as such*.

The modal independence of the unlimited perfections allows Leibniz to formulate his own version of the ontological argument in "Meditations." If we make explicit its suppressed premises, that argument is:

Leibniz's "Meditations" Argument:

(1) Any logically consistent concept is the concept of a possible thing.
(2) The concept <*ens perfectissimum*>—the concept of a being that possesses every unlimited perfection—is logically consistent.
(3) ∴ The *ens perfectissimum* is possible. [From (1) and (2).]
(4) Existence is a perfection.[27]
(5) ∴ Existence is contained in the concept <*ens perfectissimum*>. [From (2) and (4).]
(6) Whatever is contained in the concept of a possible thing can be predicated of that thing.
(7) ∴ The *ens perfectissimum* exists. [From (5) and (6).]

Note, though, that premise (1) is ambiguous in the same way that premise (2) in Descartes' argument was. It can be read as having either of two distinct logical forms:

(1*) C is a logically consistent concept $\supset \Diamond \exists x Cx$.
(1†) C is a logically consistent concept $\supset \exists x(\text{possible}(x) \,\&\, Cx)$.

(1*) would allow Leibniz to prove that *possibly* there is an *ens perfectissimum* or, at most, the conditional conclusion that *if* there is one *then* it exists; (1†) would allow

[25] Leibniz's view is that the concepts of cognition (perception of perfection) and desire (appetition of perfection) are to be understood through the concept of perfection, not the other way around, so Kant's claim is false if it is understood as a claim of conceptual *dependence*; see *DM* §1–6.

[26] For a particularly vivid Kantian illustration of the idea that there are degrees of being, see *Refl.* 4244, Ak. 17: 477–8.

[27] There are a number of texts in which Leibniz accepts the premise that existence is a perfection, e.g., "Meditations" ("the most perfect being includes all perfections, among which is existence", AG 25); the correspondence with Elizabeth (AG 237); *NE* 437–8; and L 167, 231. However, there are other texts in which Leibniz accepts that existence is "involved in" God's essence, but without straightforwardly identifying existence as one of the perfections.

him to discharge the antecedent of that conditional (that there is an *ens perfectissimum*) and prove that there is an existent *ens perfectissimum*, but at the price of committing him to a possibilist ontology of non-actual possibilia. (1†), plus Leibniz's doctrine that there are logically consistent concepts of non-actual possible worlds and the substances that compose them, entails that there are non-actual possible objects.

These are familiar points from the discussion of Descartes' argument. Interestingly, though, while Leibniz offers a Cartesian version of the ontological argument (along the lines of (1)–(7) above) in several texts,[28] in other texts he rejects this argument for the very reason that it proves only a conditional conclusion. In a note from 1678 he writes:

> Spinoza reasons thus, following Descartes: It is the same to say that something is contained in the nature or concept of some thing, as to say that very [predication] is true about that thing (as it is contained in the concept of a Triangle, or follows from its essence, that its three angles are equal to two right angles). But necessary existence is contained in the same way in the concept of God. Therefore it is true about God to say that necessary existence is in him, or that he exists. To this reasoning, and others like it, it can be objected that all those propositions are conditional, for to say that three angles equal to two right angles are involved in the nature or concept of a triangle is to say only that *if* a triangle should exist, it would have this property. So in the same way, even if it be granted that necessary existence belongs to the concept of God, still all that will be inferred from that is that *if* God should exist, then he would have this property (of necessary existence), or that if God should exist, he would [exist] necessarily.[29]

For our purposes, we can ignore the reference to Spinoza and focus on Leibniz's critique of the Cartesian argument (which, he claims, Spinoza takes over into his own philosophy).[30] Leibniz's claim here is interesting for a number of reasons. First, it is precisely the objection that Gassendi and Caterus raised. Secondly, as we saw previously, Descartes was not bothered by this objection, because, as he writes in the Fifth Meditation, "everything which I clearly and distinctly perceive to belong to that thing really does belong to it."[31] That Leibniz does not adopt the Cartesian response shows that, in this text at least, he is hesitant to embrace the ontology of non-actual possibilia to which this would commit him. It is not clear, though, that this is Leibniz's settled view on the matter.

This means that Leibniz understood that the ontological argument, as conceived by Descartes and, in some texts, himself, is committed to a possibilist ontology. In the rest of this section I want to explore Leibniz's work on the ontological argument with this question in mind: does Leibniz succeed in developing a theory of existence to support the ontological argument, while avoiding this commitment to possibilism?

[28] E.g., in several notes from 1676 (*DSR* 47, 102).

[29] A II.i.393. Cf. the discussions of this passage in Adams (1994), 161–2 and Griffin (2013), 40.

[30] In the *Ethics*, a draft of which Spinoza showed Leibniz when they met in 1676 in Amsterdam, Leibniz presumably has in mind Ip7 and Ip11Dem. For recent work on the complex Leibniz–Spinoza dialectic, see Newlands (2010), Lin (2012), and Griffin (2013), 58–82; for a comprehensive study, see Laerke (2008).

[31] AT VII: 64–6/CSM II: 45.

Or are Leibniz's ontological arguments inescapably committed to an ontology of non-actual possibilia?

The problem that exercised Leibniz during his most intensive period of work on the ontological argument, his stay in Paris in the 1670s, is how to give an account of what a 'necessary being' is and why such a being is possible.[32] Abstracting for a moment from the details of these various arguments, the general idea behind Descartes' original argument was that God is a necessary being because existence is 'involved' in his essence, though Descartes may not have provided a satisfactory account of that 'involvement.' Leibniz's main task in amending the Cartesian argument—aside from finding a way around appealing to the rule of clear and distinct perception—was to clarify what it means for existence to be 'involved' in God's essence. The "Meditations" expresses the most straightforward way in which this can be understood: existence is one of the perfections that make up God's essence as the *ens perfectissimum.*

That Leibniz continued to develop more sophisticated models of why an *ens perfectissimum* would be an *ens necessarium* suggests that he found something deeply wrong with this model, and it is not hard to determine what that might be. The "Meditations" picture of how God's essence 'involves' existence (existence is simply one of the perfections) leads to insuperable difficulties, as Gaunilo[33] and others had pointed out. If existence is a 'first-order' perfection like others, i.e. the kind of property that can be a constituent of an essence, this very quickly leads to a multiplication of necessary entities. The perfection *existence* must be compatible with the concept <*sandwich*>; otherwise, it would be a necessary truth that no sandwiches exist (which, fortunately, is not true). So the concept <*existent sandwich*> is a consistent concept. It follows that a sandwich exists, by parity of reasoning with the ontological argument. So there is an ontological argument for the existence of at least one sandwich. The straightforward explanation of how existence is contained in God's essence—it is just one more perfection like all of the others—leads to absurdity.[34]

An alternate explanation would be that necessary existence is the infinite unlimited perfection possessed by God, rather than the limited, contingent existence possessed by finite things; on such a view, necessary existence is a primitive concept, and contingent existence is a logically complex concept involving a limitation of necessary existence. In some passages Descartes himself appears to favor this model for understanding how existence is involved in the true and immutable nature of God.[35] However, as Adams (1994) points out, this view is subject to an analogue of the Gaunilo problem. If necessary existence is one of God's perfections, then it is a logically atomic property, containing no negations. Consequently, it is compatible with any other property. Take the essence E of any finite thing and conjoin necessary existence to it, forming <*E + necessary existence*>. Since necessary existence is

[32] See Griffin (2013) and Adams (1994), 157–76.

[33] Gaunilo's objection to Anselm's argument can be found in Plantinga (1965).

[34] Cf. Adams (1994), 150–1.

[35] E.g., AT VII: 152, 156.

logically atomic, the resulting concept is logically consistent, hence there is something that necessarily exists and has essence *E*. So now we have a profusion of ontological arguments, just as we had with Gaunilo. Simply making necessary existence a perfection in its own right will not help Leibniz understand the involvement of existence in the divine essence.[36]

For these reasons, during his period of intense work on the ontological argument in the 1670s Leibniz tried to develop an account of the 'involvement' of existence in the essence of the *ens perfectissimum*, without assuming that it is a perfection itself. "For perfections seem to be qualities," he writes, "as existence is not."[37] He developed several such accounts, but I will discuss only one of them here,[38] for it helps to bring out the desiderata on any theory of existence that might repair the ontological argument. Leibniz, while denying that existence is a perfection itself, retains the connection between existence and the perfections that constitute a thing's essence. His idea is that existing is not possessing a certain degree of reality, but a comparative degree of reality: being more perfect than the alternative possibilities (what Leibniz calls "mutually incompatible things").[39] The alternative possibilities are not merely individuals; if they were, Leibniz would be claiming that all of the most perfect individuals exist, something he consistently denies. God's choice is, in the first instance, a choice of which possible *world* to create (the most perfect one) and so the "mutually incompatible things" are possible worlds. His idea is that, for a finite being, to exist is to be part of the most perfect possible world.[40] Since God is not part of a *world*, but still exists, we can generalize this notion of existence as follows: to exist is to be part of the most perfect possible maximal state of affairs, where a state of affairs is maximal just in case any state compatible with it is part of it. This definition of existence will apply univocally to God and his creatures.[41]

[36] Cf. Adams's objection to the idea that necessary existence is a primitive perfection (Adams (1994), 151). A possible Leibnizian response to this problem would be to claim that although <*existent sandwich*> (or <*necessarily existent sandwich*>) are consistent concepts they are not *essences*; the ontological argument only works for essences, as I have reconstructed it. This is a promising Leibnizian response, but to be dialectically successful it would need to be backed up with an account of which concepts are essences; without such an account, it reduces merely to the (unobjectionable but trivial) claim that no sandwich can essentially exist because no sandwich can have existence in its essence. See *Refl.* 3706 (Ak. 17: 240–2) for some evidence that Kant appreciated this point. Thanks to Colin Marshall for pressing me on this point.

[37] A II.i.313. There are texts, however, in which Leibniz asserts that existence is a perfection: L 177, L 231, and *NE* 358.

[38] E.g., I forgo discussion of Leibniz's idea, discussed in Adams (1994), 151–6, of arguing that the *ens perfectissimum* would exist necessarily because it would be conceptually and hence ontologically independent of every other being. Given that the ultimate purpose of this book is to understand Kant, ignoring this Leibnizian strategy is appropriate because Kant would think of any such argument for the necessary existence of the *ens perfectissimum* as a cosmological, rather than ontological argument, because it derives existence from a version of the principle of sufficient reason.

[39] A VI.iv, 1354. See the discussion of this passage in Adams (1994), 165 and Griffin (2013), 38.

[40] *DSR* 21, 67, and Adams (1994), 165–7. Cf. "On the Ultimate Origination of Things" (AG 149–55), as well as Blumenfeld (1973) and Look (2005).

[41] That Leibniz has a 'univocal' conception of existence is denied by Nachtomy (2012). I agree with Adams (1994), 170, though, that this would be highly problematic for Leibniz.

However, this view faces a very serious problem. Whether or not a given series of finite beings (for instance, the actual ones) is part of the most perfect maximal state of affairs is a fact determined by their essences. God understands that they are so compatible and, on the orthodox theological picture, creates them; his creation does not *make* them compatible with the most perfect world.[42] But this means the envisaged theory of existence entails that actually existing creatures exist in virtue of their essences, *not* in virtue of God's creative activity, his will. This is unacceptable because Leibniz, following theological orthodoxy, holds that God is the *only* being who exists in virtue of his essence. One of Leibniz's consistent objections to Spinoza is that Spinoza denies God's will, and thus his goodness and wisdom, any role in explaining the existence of finite creatures (modes of the one Spinozistic substance); for Spinoza, finite modes are a necessary consequence of the eternal nature of God, the one substance. The theory of existence as 'compatibility with the most perfect possible maximal state of affairs' has the same unacceptable Spinozistic consequence.

For our purposes, though, the most important point is that on this conception, existence is not a 'first-order' property that can be contained in an essence (namely, God's), it is the 'higher-order' property of having perfections (properties) that make an object part of the most perfect possible maximal state of affairs.[43] Adams (1994) stresses the importance of this 'higher-order' conception of existence and interprets Leibniz as anticipating, in his Paris notes, what is usually taken (in my view, correctly) as a Kantian insight: that existence is a predicate that applies in the first instance to concepts *themselves*. Existence is the predicate that applies to a concept if and only if it is instantiated by an object. As we saw above, there are texts from this period that suggest that Leibniz was at least hesitant to adopt a view on which existence is a property possessed by some possible objects but not by others. However, there are really three theses about existence that need to be distinguished here:

(*Higher-order*) Existence is higher-order, either by being a predicate of concepts (the predicate of being instantiated) or a higher-order property of objects (e.g., of being comparatively more perfect than other objects).

(*Quantifier*) Existence is fundamentally a predicate of concepts (e.g., the predicate of being instantiated), not of objects. In contemporary terms, it is a quantifier.[44]

(*Actualism*) There are no non-existent objects.

[42] For more on Leibniz's view that possibility depends upon God's understanding, while existence depends upon his will, see Newlands (2013). The grounding of possibility and actuality in divine faculties is also an important theme in Kant's modal theory; see Ch. 4.

[43] By 'higher-order' property I will mean a property of having certain specified properties; e.g., redness is a first-order property while the property 'having at least one color property' is a higher-order property.

[44] One can (as e.g., Quine did) hold that existence is a quantifier without holding that it is a second-order predicate of concepts. For the purposes of this book I will be identifying the view that existence is a quantifier with the view that it is second-order.

As I argued in Chapter 1, Kant holds the second two doctrines; since (*Quantifier*) entails (*Higher-order*), he is also committed to the first. Adams convincingly shows that Leibniz must hold that existence is higher-order to avoid the problems that arise when existence is treated like any other predicate (e.g., the Gaunilo problem), but this does not by itself show that he took it to be a predicate *of concepts* (much less that he endorsed actualism). While there is some textual evidence that Leibniz experimented with thinking of existence as a second-order predicate (a quantifier), and, what is more, the 'widest' quantifier there is (actualism), this does not show it was ever his settled view.[45]

Looking back over Leibniz's various attempts to repair the ontological argument, we can see him trying to develop a theory of existence that meets certain desiderata, among which are:

(*Univocal*) There is a univocal sense of 'exists' in which both God and finite creatures exist. There may be more specific senses of 'exist' (e.g., necessary existence) that apply only to one or the other, but there must be a common sense of 'exists.'

(*Difference*) In this univocal sense of 'exists,' God exists essentially, but none of his creatures do. It follows from God's essence that he exists, and God is the only being of whom this is true.

(*Higher-order*) Existence is higher-order, either by being a predicate of concepts (e.g., the property of being instantiated) or a higher-order property of objects (e.g., of being comparatively more perfect than other objects).

However, Leibniz was unable to develop a theory of existence that met these desiderata and, as a result, was unable to repair the ontological argument satisfactorily. In the next section I will argue that, on one reading, Baumgarten succeeded where Leibniz did not.

2.4. Baumgarten's Ontological Argument

The natural next step would be to consider Christian Wolff's ontological argument and whether it commits him to possibilism. However, as I noted in section 1, Wolff's own ontological argument in *Theologia naturalis* is an uninteresting derivative of Leibniz's "Meditations" argument. Furthermore, as Kant himself points out, Wolff's definition of existence as the "completion of possibility" [*complementum possibilitatis*] is too indeterminate to ascertain what, precisely, his theory was.[46] I therefore pass

[45] Cf. Adams (1994), 158–64 and Mates (1986), 170–88.

[46] "The Wolffian definition of existence, that it is a completion [*Ergänzung*] of possibility, is obviously very indeterminate. If one does not already know what over and above possibility can be thought in a thing [*über Möglichkeit in einem Dinge kann gedacht werden*], one will not learn it from this definition" (*OPG*, Ak. 2: 76). For more on Wolff's view of existence, see *Ont.* §§173–4. Honnefelder (1990), 363–71 concurs that Wolff adds little to what is found in Leibniz and Baumgarten.

over Wolff and consider Baumgarten, whose theory of existence is more developed than Wolff's and is the target of constant criticism by Kant.[47]

Baumgarten gives an ontological argument for the existence of God in *Metaphysica* §§803–9 that is largely derivative of Leibniz. This may explain why Baumgarten has not received the attention of scholars of the ontological argument that, I think, he deserves.[48] In addition to this relatively uninteresting variant of the Leibnizian argument, though, Baumgarten also developed an innovative theory of existence as complete determinacy:

> Apart from essence, (§53) the possible is either determinate in respect of all affections which are compossible in it, or not (§34, 10). In the first case, it is ACTUAL [*actuale*]; in the second case it is called PRIVATIVE (merely possible) NON-BEING (*nihil*, cf. §7).[49]
>
> EXISTENCE [*EXISTENTIA*] (*actus*, cf. §210, *actualitas*) is the totality [*complexus*] of all affections that are compossible in a being, i.e. the complement of the essence, or internal possibility, insofar as these are considered only as a totality of determinations (§40).[50]

The parentheses are Baumgarten's own references to previous definitions and principles within *Metaphysica*. As we have seen, 'determination' is a perfectly general term for Baumgarten: it refers to any predicate that can be attributed to a being (§34).[51] Every possible being (*ens*) is fully determinate with respect to its essential properties and attributes. In addition to essential properties and attributes, beings also have accidental properties, which Baumgarten divides into intrinsic accidents (modes) and relational accidents (§37). Each being is either fully determinate in respect of its intrinsic accidents (modes), or it is not. In the former case it is actual/exists; in the latter, it is merely possible, non-actual/non-existent. Following Baumgarten in *Metaphysica* §55 (see previous quotation), I will henceforth identify existence and actuality. For a being to exist is for it to possess a fully determinate set of modes (intrinsic accidents) in addition to its essential properties and attributes. Baumgarten is committed, therefore, to at least the following principle:

> (*Ex-Determinacy*) For all x, x exists if and only if, for any intrinsic predicate A, either x is A or x is ~A.[52]

[47] Baumgarten's theory of existence is (implicitly or explicitly) criticized by Kant in *Refl.* 5760, 5783, 5784, 5786, 6245, 6255, 6290, and 6322; in the metaphysics lectures, *MV* (Ak. 28: 413), *MvS* (Ak. 28: 503), *ML*$_2$ (Ak. 28: 554), *MD* (Ak. 28: 630), *MK*$_2$ (Ak. 28: 722–4, 779), and *Volck.RT* (Ak. 28: 1156), as well as the more familiar discussions in the *CPR* and *OPG* (see the next chapter). It is even alluded to in the *CJ* (Ak. 5: 475–6).

[48] Symptomatic of this is Proops (2015), 13–15, who does not consider the obvious rejoinders Baumgarten could make to (Proops's reconstruction of) Kant's objection.

[49] *Meta.* §54.

[50] *Meta.* §55.

[51] Thus it does not have the specifically Kantian meaning of "a predicate which goes beyond the concept of the subject and enlarges it" (A598/B626).

[52] The second disjunct means that x has the negative property ~A; I discuss the difference between having the negative property ~A, and merely lacking A, and the related distinction between negations of predicates and negations of propositions, subsequently.

By itself, this is a very minimal claim. It says only that there are no existents that are indeterminate with respect to one or more intrinsic predicates. Since Baumgarten identifies existence with complete *intrinsic* determination, henceforth I will drop the explicit restriction to intrinsic predicates; when I talk about accidental determinations in Baumgarten, I mean only accidental intrinsic determinations (modes).

At various places in *Metaphysica* Baumgarten hints at an argument for God's existence, based on this theory of existence as complete determination. For instance:

> [...] God is also determined in regard to the rest of his internal perfections as much as anything whatsoever can be determined in regard to an internal perfection. Therefore, God is actual (§54).[53]

> A non-actual God would be a being enjoying all realities, and yet missing some reality. In regard to all internal perfections, he would be determined as greatly as a being can be internally determined, and yet, in regard to some of these perfections he would not be so determined (§54).[54]

For the sake of readability, I have eliminated many of Baumgarten's in-text citations to other parts of the *Metaphysica*, except for the final references in both passages to §54, quoted above, where he introduces his complete determination theory of existence. Very roughly, Baumgarten's idea in the first passage is that God is essentially completely determinate with respect to every intrinsic predicate; if *A* is an intrinsic predicate, God either essentially has *A* (e.g., if *A* is an infinite unlimited perfection or a conjunction thereof) or God essentially lacks *A* (e.g., if *A* contains or entails any limitation of perfection). God's essence has no 'room' to be 'filled out' with intrinsic accidents. In the second passage, Baumgarten puts the point negatively: if God did not exist, he would be not completely determinate (by the definition of existence), but since God is *essentially* completely determinate, this is absurd. I will focus on the direct version of the argument, which we can provisionally reconstruct as follows:

Baumgarten's Complete Determination Argument, Preliminary Reconstruction:

(1) The concept <*ens perfectissimum*> is logically consistent (contains no contradictions). [This was proven earlier, on essentially Leibnizian grounds—see *Meta.* §§806–9.]
(2) Anything that is logically consistent (contains no contradictions) is possible.
(3) ∴ The *ens perfectissimum* is possible. [From (1) and (2).]
(4) Every possible being has all of the properties contained in, and entailed by, its essence (essential properties and attributes).
(5) The essence of the *ens perfectissimum* entails that it is completely determinate with respect to every intrinsic predicate.
(6) Complete intrinsic determinacy is existence.
(7) ∴ The essence of the *ens perfectissimum* entails that it exists.

[53] *Meta.* §819. [54] *Meta.* §823.

(8) ∴ The *ens perfectissimum* exists. [From (4) and (7).]
(9) Anything whose essence entails its existence is a necessary being.
(10) ∴ The *ens perfectissimum* is a necessary being. [From (8) and (9).]

One potential source of skepticism about this argument is that, from a Kantian perspective, the definition of existence on which it rests (premise (6)) fails to distinguish between concepts and objects. Is this a specification of which concept <*existence*> is, or does it specify necessary and sufficient conditions on existent *objects*, or both? Consequently, there are two ways of understanding Baumgarten's definition of existence/actuality and two ways of understanding this argument. On the first reading—which I will call the 'intensional' reading—Baumgarten is distinguishing between two different ways of conceptualizing one and the same set of objects: we can think of those objects as merely possible, or we can think of them as actually existing. On the 'extensional' reading, by contrast, the distinction between the possible and the actual is not merely a conceptual (or 'rational') one. On this reading, Baumgarten is distinguishing 'possible' and 'actual' in respect of their extensions; he is distinguishing between two kinds of objects, those that are merely possible and those that actually exist.

If the intensional reading is correct, then the complete determination argument is quite weak. Consider, again, the problem that marred many earlier ontological arguments. Premise (2) can be understood in either of two ways:

(2*) Concept C contains no contradictions $\supset \Diamond\exists x Cx$.
(2†) Concept C contains no contradictions $\supset \exists x(Cx \;\&\; \text{possible}(x))$.

For what should now be familiar reasons, the argument is invalid if we assume only (2*), while it is committed to a possibilist ontology if we assume (2†). However, the intensional reading of Baumgarten's theory of existence is incompatible with a possibilist ontology, for it allows only a conceptual, rather than extensional, distinction between possible and existent objects. This leaves Baumgarten with (2*), which will allow him to prove only the conditional conclusion that *if* there is a God, he exists.

One way for Baumgarten to overcome this defect in his argument would be to appeal to a principle like:

(*DI*) Concept C is a fully determinate concept of an object $\supset \exists x Cx$.

This principle is fully compatible with the intensional reading, and hence with actualism, and would allow Baumgarten to prove almost trivially that there exists an *ens perfectissimum*. However, this principle is subject to the fatal objection that, as Kant never tired of pointing out when lecturing from the *Metaphysica*, there are in principle completely determinate concepts that are not instantiated, for instance, completely determinate concepts of non-actual possible worlds.[55] I take this to be a

[55] *MV* (Ak. 28: 413), *MvS* (Ak. 28: 503), *ML*$_2$ (Ak. 28: 554), *MD* (Ak. 28: 630), *MK*$_2$ (Ak. 28: 722–4, 779), and *Volck.RT* (Ak. 28: 1156); cf. *Refl.* 5760, 5783, 5784, 5786, 6245, 6255, 6290, and 6322.

devastating objection to DI and to Baumgarten's complete determination argument for God *if the 'intensional' reconstruction is correct* (i.e. if there is only a conceptual distinction between possible and actual objects).

The obvious weakness of Baumgarten's argument on the intensional reading suggests that another reading is possible. Baumgarten, after all, was no fool; Kant even refers to him as "the excellent analyst Baumgarten" (A22/B36). On the *extensional* reading of Baumgarten's theory of existence, the distinction between the *actual* and the *possible* is an extensional distinction between two sets of objects, one of which (the actual) is a subset of the other (the possible). On this way of understanding Baumgarten's theory of existence, there is a domain of possible objects and each object in the domain has an essence. For objects other than God, these essences are incompletely determinate; for some intrinsic predicate A, they do not contain (or entail) either A or ~A. Merely possible objects (possibilia) have only the properties contained in or entailed by their essences, and the essences of finite beings are incompletely determinate. Consequently, finite possibilia are incompletely determinate. They have their essential properties and their attributes, but no accidents. Existing finite beings have their essential properties and their attributes, but are also equipped with a fully determinate set of accidents. The extensional reading, therefore, attributes to Baumgarten the familiar principle:

(*Possibilism*) $\Diamond\exists x(\neg \text{exists}(x))$.

In fact, it attributes to Baumgarten the view that there *are* non-existent objects (beings, *entia*).

The extensional reading allows us to reconstruct Baumgarten's argument in such a way that it is not subject to the obvious Kantian rejoinder that there are completely determinate but uninstantiated concepts. To do so, it must distinguish between concepts in general and *essences*: concepts that contain only the essential properties of a being. It must appeal, not to:

(2†) Concept C contains no contradictions $\supset \exists x(Cx \;\&\; \text{possible}(x))$

from above, but to a principle of the form:

(2††) $\Diamond\exists x Cx \;\&\;$ C is an essence $\supset \exists x(Cx \;\&\; \text{possible}(x))$.

While not uncontroversial, this principle is a theorem in many modal systems.[56] Intuitively, it says that if it is possible for there to be some being with a certain essence, then there *is* such a being, although it may not actually exist. It entails that there could not *be* 'more' possible beings that there *are*, although 'more' of the merely possible beings could have the predicate of existence.

[56] It is a consequence of the Barcan Formula: $(x)\Box Ax \supset \Box(x)Ax$. For an influential recent defense of the Barcan Formula and its converse, see Williamson (2013).

With these pieces in position, we can now give an extensional reconstruction of Baumgarten's argument. To bring out its complete logical structure, I have expanded the preliminary reconstruction from earlier and made explicit some enthymematic premises (letting EP abbreviate the concept <*ens perfectissimum*>):

The Complete Determination Argument (Extensional reconstruction):

(1) The concept EP is logically consistent.

(2) Concept C is logically consistent ⊃ ◇∃*x*C*x*. (*Note*: this principle is the unproblematic one from above, which is not committed to possibilism. So far, this argument has not made use of any possibilist ontology.)

(3) ∴ ◇∃*x*EP*x*. [From (1) and (2).]

(4) The concept EP is an essence; anything that falls under the concept EP is essentially an *ens perfectissimum.*

(5) ◇∃*x*C*x* & C is an essence ⊃ ∃*x*(C*x* & possible(*x*)). [*Note*: this is (2††) from above.]

(6) ∴ There is a being whose essence is EP and this being is possible. [From (4) and (5).] Call this being *g*.

(7) For any object *x* and any concepts E and F, if E is the essence of *x*, necessarily F*x* if and only if the proposition *Ex* entails the proposition *Fx*. [*Note*: this is just the logicist theory of necessary predicates; see *Metaphysica* §§34–8.]

(8) For any intrinsic predicate A, if A*g* is logically compatible with EP*g*, EP*g* entails A*g*.

(9) For any intrinsic predicate A, if A*g* is not logically compatible with EP*g*, EP*g* entails (~A)*g*.

(10) ∴ For any intrinsic predicate A, EP*g* entails A*g* or EP*g* entails (~A)*g*. [From (9) and (10).]

(11) ∴ For every intrinsic predicate A, E*g* entails A*g* or (~A)*g*. [From (10); if E*x* entails either of the disjuncts, it entails the disjunction.]

(12) ∴ E*g* entails, for any intrinsic predicate A, A*g* or (~A)*g*. [From (11).]

(13) For all *x*, *x* exists $=_{\text{def}}$ for any intrinsic predicate A, A*x* or (~A)*x*. [*Note*: this is Baumgarten's theory of existence as complete determinacy.]

(14) ∴ E*g* entails *g* exists. [From (12) and (13).]

(15) ∴ Necessarily *g* (the *ens perfectissimum*) exists. [From (7) and (14).]

Intuitively, this argument shows that the *ens perfectissimum* essentially exists because the *ens perfectissimum* has a completely determinate essence, and being completely determinate is equivalent to existence, according to Baumgarten. The *ens perfectissimum* has no 'gaps' in its essence to be filled in with accidents.[57]

[57] Proops (2015) is thus incorrect when he claims that, "according to Baumgarten, existence is something external to the concept of a thing" (14). Existence is external to the essences of *finite* things (it is a mode); it is contained in the essence of God, the unique *ens necessarium*. Proops's reconstruction of

By contrast, finite creatures have incompletely determinate essences. When they are actualized or brought into existence this incompletely determinate set of essential properties is supplemented with a set of accidents, making them completely determinate. Finite creatures do not exist necessarily and there are no sound ontological arguments for their existence. To see this, observe that a crucial premise of the complete determination argument—(8)—does not hold where *x* is a finite object and *E* is its essence:

> (8*) For any intrinsic predicate A, if A*x* is logically compatible with E*x*, E*x* entails A*x*.

This is false where *x* is a finite creature and E is its essence, because for some intrinsic predicates *A*, E*x* is logically compatible with Ax but does not entail it; for some A, E*x* entails neither A*x* nor (~A)*x*.

The validity of this ontological argument shows that Baumgarten's theory of existence satisfies the three Leibnizian desiderata on a theory of existence:

> (*Univocal*) There is a univocal sense of 'exists' in which both God and finite creatures exist. There may be more specific senses of 'exist' (e.g., necessary existence) that apply only to one or the other, but there must be a common sense of 'exists.'
>
> (*Difference*) In this univocal sense of 'exists,' God exists essentially, but none of his creatures do. It follows from God's essence that he exists, and God is the only being of whom this is true.
>
> (*Higher-order*) Existence is higher-order, either by being a predicate of concepts (e.g., the property of being instantiated) or a higher-order property of objects (e.g., of being comparatively more perfect than other objects).

It supplies a *univocal* theory of existence because there is a common sense in which God and finite creatures exist: they are fully determinate. I explained in the previous paragraph why it satisfies the *difference* requirement. By taking existence to be a 'higher-order' property of objects (the property of having a fully determinate set of accidents), it avoids the problems that plagued Descartes and Leibniz when they took existence to be one perfection among others that can partly constitute an essence (Gaunilo's objection and variants thereof). For instance, there is no corresponding ontological argument for the existence of a 'completely determinate island' because a completely determinate concept of an island is not an *essence*; no completely

Kant's objection to Baumgarten—existence cannot be contained in the concept of the *ens realissimum* because it is not contained in the concept of finite things—is ineffective. Further, Proops is too hasty in claiming that Baumgarten's ontological argument is subject to Caterus's objection that the ontological argument proves only a connection between concepts in thought, not the existence of a thing outside thought. Kant's own response to the Caterus objection could be made by Baumgarten himself; see *Refl.* 3706, Ak. 17: 240–1. Although I have left out the details of Wolff's theory, I do not see why Wolff could not make the same response as well.

determinate island *essentially* has the completely determinate set of accidents contained in that concept. It is no surprise, then, that Kant devotes so much time and effort to refuting Baumgarten's theory of existence as complete determination; the ontological argument based on this theory is, of all of the ontological arguments we have canvassed so far, the strongest.

This argument relies on the extensional reading of Baumgarten's theory of existence. We might then wonder whether this is the *correct* reading. But recall that the distinction between the extensional and the intensional reading of Baumgarten's theory arose when we imposed a sharp distinction between *concepts* and *objects*, or, equivalently, between distinguishing concepts in virtue of their *intensions* (conceptual content) and in virtue of their *extensions* (the objects that fall under them). Since Baumgarten does not rigorously maintain this distinction (nor, arguably, does anyone before Kant), there is no reason to think that his theory is determinately either extensional or intensional. It can be read in either way.

In fact, there is evidence that Kant was aware of this ambiguity in Baumgarten's theory. For instance, in *Beweisgrund*, he writes:

> [...] the proposition that a possible thing, regarded as such, is indeterminate with respect to many of its predicates, could, if taken literally, lead to serious error. For such indeterminacy is forbidden by the law of excluded middle, which maintains that there is no intermediate between two predicates which contradict one another. It is for example impossible that a man should not have a certain stature, position in time, age, location in space, and so forth. Our proposition [that existence is complete determination] must rather be taken in the following sense: the predicates which are thought together in a thing in no way determine the many other predicates of that thing. Thus, for example, that which is collected together in the concept of a human being as such specifies nothing with respect to the special characteristics of age, place, and so forth. But then this kind of indeterminacy is to be found as much in an existing thing as it is in a merely possible thing. (Ak. 2: 76).

Kant's point in the first half of this passage is that the distinction between the incompletely and the completely determinate cannot be a distinction between two kinds of objects, because incompletely determinate objects would violate the principle of excluded middle. Instead, Kant argues, the distinction is between objects and concepts: objects are fully determinate but concepts of them can be incompletely determinate. I read this passage as Kant remarking on exactly the ambiguity in Baumgarten's theory that I have noted: read extensionally, it runs afoul of the principle of excluded middle, and read intensionally it provides no basis for a sound ontological argument.

One might think that the extensional version of Baumgarten's theory of existence is decisively refuted by Kant's point: if an object x is not determinately A and not determinately ~A, this violates the principle of excluded middle. But how to apply the principle of excluded middle to *predicates* is a controversial matter in eighteenth-century logic, and at least one prominent philosopher of this period denies that the principle of excluded middle *for propositions* ($p \vee \neg p$, for any p) entails the corresponding principle for *predicates* ($Ax \vee (\sim A)x$, for any x and any A)—Kant himself! As I have argued elsewhere,[58] the Critical Kant does not hold that the principle of excluded middle as applied to predicates—what Kant calls, tellingly, the "principle of complete determination"—is a logical principle at all.

In this book, I do not have the space to explain why, in the Critical period, Kant denied that the principle of complete determination is a logical principle. I want merely to explain *why* it is coherent to deny that the principle of excluded middle entails the principle of complete determination. The principle of excluded middle (in the case of simple subject-object predications) is:

(*PEM*) For any predicate A and object x, $Ax \vee \neg(Ax)$

while the principle of complete determination (in the case of such predications) is:

(*PCD*) For any predicate A and object x, $Ax \vee (\sim A)x$.

The difference is in their second disjunct. The second disjunct of the PEM is what Kant calls a 'negative judgment.' Applied to the judgment Ax it says that it is *not* true that x falls within the extension of the predicate A. The second disjunct of the PCD is what Kant calls an 'infinite judgment.' It says that x falls within the extension of the predicate $\sim$A. The negation in a negative judgment applies to a whole judgment, while the negation in an infinite judgment applies to a predicate; I have used two different negation strokes and positioned the parentheses to bring out these differences in scope. We might assume that the negative judgment $\neg(Ax)$ entails the infinite judgment $(\sim A)x$. We might, in other words, assume that if x is *not* in the extension of A, then x is in the extension of $\sim$A. But that is equivalent to assuming that the extension of $\sim$A is the *complement* of the extension of A, which is equivalent to assuming that the principle of complete determination is true. So the PEM alone does entail the PCD. Kant denies that, as a matter of logic, the extension of a predicate A and its negation $\sim$A must be exhaustive; no logical laws are violated if an object x belongs neither to the extension of A *nor* to the extension of $\sim$A. This means that the following represents a logical possibility, according to the Critical Kant:

(1) $\neg Ax \ \& \ \neg((\sim A)x)$.

In this case, object x is *indeterminate* with respect to predicate A.

In its extensional version, Baumgarten's theory of existence entails that where x is a merely possible object and A is an intrinsic accidental predicate, x is indeterminate with respect to A. I am not going to argue that Baumgarten would analyze this as (1) or that he would follow Kant in distinguishing between infinite and negative

[58] Stang (2012).

judgments, and deny that the principle of complete determination follows from the principle of excluded middle. Whether or not the historical individual Alexander Gottlieb Baumgarten ever held the extensional version of his theory of existence, and whether he would defend it by distinguishing PCD from PEM as I have done above, the philosophical fact remains: such a theory and such an argument are positions within logical space, and, what is more, may be stronger than any ontological argument given by Leibniz or Descartes (or Wolff). We have already seen that Kant offers objection *both* to the extensional *and* to the intensional versions of Baumgarten's theory of existence. Given that Kant's ambition is to prove that no ontological proof whatsoever is possible, this is entirely the correct procedure. However, this entails that he must be able to refute whatever the best version of Baumgarten's theory is, *including* the version that adopts Kant's own strategy for maintaining PEM while abandoning PCD.[59] In the remainder of this chapter I will reconstruct in some detail Kant's arguments that existence is not a 'real predicate' with the following question in mind: do they carry any force against the most defensible version of his theory of existence that Baumgarten could have given (whether or not he actually did)? If the answer to that question is no, the most we will be able to claim on Kant's behalf is that he refuted several historical ontological arguments, but that he did not show (as he claimed to have) that ontological proofs as such are impossible.

2.5. The Argument in *Beweisgrund*: Leibniz

Having argued in §1.5 and §1.6 that ontotheism as such is committed to the claim that <*exists*>[60] is a real predicate, and in the first half of this chapter that the ontological arguments actually given by Descartes, Leibniz, and Baumgarten assume that it is, I will now reconstruct Kant's arguments that existence is not a real predicate, starting with *Beweisgrund*.

After a brief introduction, the first 'Reflection' of the first Division of *Beweisgrund*[61] begins with Kant's argument that existence is not a real predicate, which I here divide to clarify its internal structure:

> [A] This proposition [that existence is not a predicate or a determination of a thing] seems strange and absurd, but it is indubitably certain. Take any subject you please, for example, Julius Caesar. Draw up a list of all the

[59] Note that my point holds even if the argument of Stang (2012) is wrong and Kant does not deny that PCD is a logical principle; the availability of that view in logical space means that Kant must refute it to refute the in-principle possibility of ontological proofs.

[60] In what follows I will sometimes drop the angled brackets, for ease of comprehension.

[61] *Beweisgrund* consists of three Divisions (*Abteilungen*), which are themselves divided into a series of 'Reflections,' which consist in one or more numbered sections. For ease of reference I will refer to parts of *Beweisgrund* by the Division number, followed by the Reflection number, followed by the section number; e.g., II.3.2 will refer to the second Division, third Reflection, second section.

predicates which may be thought to belong to him, not excepting even those of space and time. You will quickly see that, with all of these determinations, he can either exist, or not exist. The Being who gave existence to the world and to our hero within that world, could know every single one of these predicates without exception, and yet still be able to regard him as a merely possible thing which, in the absence of that Being's decision to create him, would not exist.

[B] Who can deny that millions of things which do not actually exist are merely possible from the point of view of all the predicates they would contain if they were to exist? Or who can deny that in the representation which the Supreme Being has of them there is not a single determination missing, although existence is not among them, for the Supreme Being cognizes them only as possible things? It cannot happen, therefore, if they were to exist, they would contain an extra predicate; for, in the case of the possibility of a thing in its complete determination, no predicate at all can be missing.

[C] And if it had pleased God to create a different series of things, to create a different world, that would have existed with all the determinations, and no additional ones, which he cognizes in it [*die er an ihr doch erkennt*], although that world is merely possible. (Ak. 2: 72)

As I read this passage, [A] contains Kant's primary argument that existence is not a determination. He first claims that existence is not contained in the complete concept of an object, the concept that contains every (real) predicate that applies to the object ("draw up a list of all the predicates which may be thought to belong to him").[62] He concludes from this that existence is not a (real) predicate, a determination. [B] might be thought to add something, because Kant there explicitly considers merely possible objects, but in my reconstruction of [A] below I will show that this thought is already implicit in [A]. In [C] Kant shifts gears slightly, and rephrases the argument in [A] in terms of possible worlds, rather than possible individuals (e.g., Caesar). In the next two sections Kant expands upon and explicates this claim (e.g., his discussion of absolute and relative positing) but does not give any substantial further argument for it. If we want to understand Kant's argument that existence is not a determination we must examine [A] closely.[63]

That Kant focuses on complete concepts of finite substances (e.g., Caesar) is puzzling. He ultimately needs this argument to refute the ontotheist view that the existence of the unique infinite substance, God, follows logically from its essence. On the ontotheist picture, essences of finite substances are compatible with existence, but do not contain or entail existence; this is why finite substances exist

[62] This may not refer to the predicates that actually belong to Caesar, but to all of the predicates that could possibly apply to Caesar. I consider that possibility below.

[63] The last sentences of the paragraph quoted above expand slightly upon this argument, but do not contain additional arguments.

only contingently, while God is the unique necessary being. So complete concepts of finite substances are, strictly speaking, irrelevant to the ontotheist view of the necessary existence of God. Ontotheists are not *as such* committed to affirming, or denying, that the complete concept of a finite substance determines whether it exists or not; they are committed, at most, to denying that the *essence* of a finite substance determines whether it exists or not.[64] This passage is especially puzzling in light of my interpretation; if the point of Kant's objection is to deny that there are non-actual possible objects, how are complete concepts of actual objects at all relevant to his argument?

One explanation for this puzzling fact would be that Kant identifies an object's essence and its complete concept. If this were Kant's view, he could argue that existence is never contained in a complete concept, and then validly conclude that it is never contained in an essence. However, to interpret Kant as identifying essences and complete concepts would be to attribute to him a form of essentialism even stronger than Leibniz's, for Leibniz, who holds that no individual could exist with a different complete individual concept, distinguishes between essences and complete individual concepts. Since it is questionable whether Kant accepted even that weaker Leibnizian doctrine, it is unlikely that he identifies essences of finite substances with their complete concepts.

Alternatively, it might also be thought that Kant is simply arguing from a stronger claim to a weaker one: since a complete concept contains all the predicates of an object, and an essence contains a subset of them, if its complete concept does not contain existence, then neither does its essence. But the ontotheists in question (Descartes, Leibniz, Wolff, Baumgarten) already agree that the existence of a finite substance does not follow from its essence! On this issue, Kant and the ontotheist are in full agreement, so it cannot be the point of his argument. The ultimate aim of Kant's argument is to show that not even God's existence follows from his essence. In order to derive that conclusion, Kant would need to argue first that God's complete concept does not contain (or entail) existence, which he notably does not do, focusing instead on the complete concepts *God has* of finite substances.

Finally, it might be thought that Kant is arguing that Leibniz's view that the complete concept of any individual (e.g., Julius Caesar) contains all of his (real) predicates entails that existence is not a real predicate (not a determination). Kant could be reasoning that the predicates contained in a complete individual concept (CIC) are the real predicates (determinations) of the individual and that these are the predicates God thinks of the object as having when he thinks of that individual as a denizen of some possible world, as the member of a candidate maximal plan for

[64] Leibniz holds that, in some sense, existence is 'contained' or 'involved' in the complete individual concepts of actually existing substances: an infinite analysis of such a concept and an infinite comparison of those concepts with concepts of other possible worlds God could have created would reveal why these individual substances, and not others, are actualized. See Ch. 1.2.

creation.[65] If existence is a real predicate, then it is contained in God's CIC of Julius Caesar as a possible being. So God already thinks of him as existing in thinking of him as possible, which entails, absurdly, that God cannot think of any individual as merely possible. Spelled out fully, this reconstruction of Kant's reasoning would go as follows:

(1) All of the real predicates Julius Caesar has if he exists are contained in God's CIC of him. ["The Being who gave existence to the world and to our hero within that world, could know every single one of these predicates without exception [. . .]."]
(2) If existence is a real predicate, then if Julius Caesar exists he has the real predicate of existence.[66]
(3) ∴ If existence is a real predicate, then existence is contained in God's CIC of Julius Caesar. [From (1) and (2).]
(4) If God's CIC of Julius Caesar contains existence, then God's CIC of Julius Caesar represents him as existing.
(5) ∴ If existence is a real predicate, then God's CIC of Julius Caesar represents him as existing. [From (3) and (4).]
(6) But God's CIC of Julius Caesar does not represent him as existing; it represents him as possible, as existing *only if* God wills to create him. ["[. . .] and yet still be able to regard him as a merely possible thing which, in the absence of that Being's decision to create him, would not exist."]
(7) ∴ Existence is not a real predicate. [From (5) and (6) by *modus tollens*.][67]

This is a very weak argument, which has no dialectical force against Leibniz or any other ontotheist. First of all, why would Leibniz admit (1)? Why cannot Leibniz hold that the CIC of an individual contains all of the real predicates it would have if it were to exist *except* existence itself (and predicates that logically entail existence)? On such a modified Leibnizian view, God's CICs of possible individuals contain all of their predicates except the predicate they obtain in virtue of being created by God: existence. That Leibniz *himself* fails to make this qualification to his doctrine of CICs is dialectically irrelevant; Kant would have succeeded at best in showing that

[65] What I mean by calling possible worlds 'maximal plans for creation' is, first of all, that Leibnizian possible worlds have their being in the content of God's thoughts; unlike Lewis worlds, they are not concrete entities. Secondly, they are maximal: if w is a world and p is logically consistent with w (w does not entail $\neg p$) then w contains p (w entails p). See L 661–2. Though my interpretation of Kant's views on existence and his arguments for those views are diametrically opposed to his Meinongian reading, I have benefited greatly from reading the work of Tobias Rosefeldt and from discussing these issues with him. See Rosefeldt (2008) and (2011).

[66] Strictly speaking, this should read: if <*exists*> is a real predicate, etc. For the sake of readability, though, I leave out the angled brackets here.

[67] The premises of this argument are unaffected if we substitute for 'Julius Caesar' the name of some non-actual individual substance, hence the [B] portion of the text does not really add anything not already found in [A] (see previous discussion).

Leibniz himself is not a particularly effective advocate for his own ontological argument. Even the ontotheist can admit that, while existence is a real predicate, it is a real predicate with a very different status than the other real predicates, something of which Leibniz was, in fact, quite aware.[68]

Even more problematic, though, is claim (6). Why must Leibniz think that God's CIC of Julius Caesar does not represent him as existing? God's CIC of any individual substance represents what properties that substance would have *if* God were to create it. Presumably, existence is one of those properties. But how then, one might ask, can God also think of the individual substance as being merely possible, as being a candidate for creation, rather than already actually created? Compare this to an ordinary case: I can think of how my friend Joe would look *if he were bald* while, at the same time, thinking that he is not bald. My thought can represent the non-actual but possible situation of Joe's being bald while also representing Joe as being actually not bald. Likewise, God's concept of an individual can represent that individual as existing *in that it represents the individual as it would be if it were to exist* but does not represent that individual as actually existing. The error here is thinking that from:

(a) God's concept of *X* represents *X* as *F* and God is infallible

it follows that:

(b) *X* actually is *F*.

This does not follow. For infallibility does not preclude God from representing possibilities that he does not thereby represent as actual. On the contrary, God's omniscience requires that he *does* represent non-actual possibilities and represents them *as* non-actual possibilities!

It is possible that Kant had this anti-Leibnizian argument in mind. If so, he failed to make any advance against ontotheism or the possibilist ontology. However, as I have reconstructed the argument, he is arguing against an opponent who is committed to the Leibnizian view that God's concepts of individual substances *qua possible* are fully determinate, i.e. that God thinks of each possible substance as belonging to at most one fully determinate possible world. This assumption is encoded in the argument in premise (1). If God's CIC of Julius Caesar does not restrict him to the actual world, then his CIC must include 'world-indexed' information: it must include the information that JC has one set of real predicates in one possible world, and a distinct set in another world.[69] For instance, if Julius Caesar is bald in world w but not bald in world w^*, God's CIC of Caesar does not represent Caesar as bald *simpliciter*; it represents him as bald in w and not-bald in w^*.[70] So the

[68] See Ch. 2.3 and Adams (1994), 157–76. [69] Cf. Adams (1994), 71–4.

[70] Recall Kant's precise words: "draw up a list of all the predicates which may be thought to belong to him [*alle seine erdenkliche Prädicate*], not excepting even those of space and time." This could mean: think of all of the complete sets of predicates that Caesar could have in different worlds.

argument as stated applies only against someone who accepts Leibniz's doctrine that an individual exists in at most one possible world. But this is not a view that Kant has any reason to expect would be shared by all ontotheists, much less by all philosophers who think that existence is a real predicate. It would have the effect of making his argument apply only against Leibniz himself. However, a modified version of Kant's argument can be given that does not encode this Leibnizian assumption:

(1*) For any world *w* in which Caesar would exist and any real predicate F Caesar would have in *w*, God's CIC of Julius Caesar contains *F in w*.
(2*) If existence is a real predicate, then, for any world *w* in which Caesar would exist, Caesar would have the real predicate *exists in w*.
(3*) If existence is a real predicate, then for any world *w* in which Caesar would exist, God's CIC of Caesar contains *exists in w*. [From (1*) and (2*).]
(4*) If God's CIC of Julius Caesar contains *exists in w*, then God's CIC of Julius Caesar represents him as existing in *w*.
(5*) ∴ If existence is a real predicate and *w* is a world where Caesar would exist, then God's CIC of Julius Caesar represents him as existing in *w*. [From (3*) and (4*).]
(6*) God's CIC of Julius Caesar does not represent him as existing in *w*, for any world *w*; it represents him as possible, as existing only if God wills to create him.
(7*) ∴ Existence is not a real predicate. [From (5*) and (6*) by *modus tollens*.]

But, as with the previous argument, premise (6*) is false. By representing JC as existing in *w*, God is not thereby representing JC as existing *simpliciter*; he is representing JC as being such that he would exist if he (God) were to actualize *w*.

But, it will be pointed out, one of the worlds at which God's CIC represents Caesar as existing is the actual world. So we can replace (6*) and (7*) with the following:

(6†) The actual world @ is a world where Caesar would exist.
(7†) ∴ If existence is a real predicate then God's CIC of JC represents him as existing in @. [From (5*) and (6†).]
(8†) If God's CIC of Julius Caesar represents him as existing in @ then God's CIC of JC represents him as actually existing.
(9†) But God's CIC of Julius Caesar does not represent him as actually existing; it represents him as possible, as existing only if God wills to create him.
(10†) ∴ Existence is not a real predicate. [From (7†)–(9†), by *modus tollens*.]

However, this argument is invalid. Its appearance of validity rests on conflating two subtly different senses of 'actual', the first in (8†), and the second in (9†). It is worth going into detail about them, because the distinction will be important for understanding Kant's critique of Baumgarten as well. In the first sense—call it the 'rigid' sense of 'actual' (actual_R)—'the actual world' is just a name for *this* world, the world that God in fact created. Earlier, I introduced the name '@.' In any counterfactual context, '@' just refers to this world. In the second sense—call it the non-rigid sense (actual$_{NR}$)—'actual' refers to whatever is created by God. This is the sense in which, if God had created a different world, things that are actual would not have been actual (which is false for actual$_{R}$). I am not sure I can give a satisfactory analysis of the meaning of actuality$_{NR}$, but perhaps it will help to understand these notions if we note that they are materially equivalent to one another and to p itself, since this is the actual world (and the truth of p is evaluated with respect to the world in which it is asserted):

(*) Actually$_{R}(p) \leftrightarrow$ actually$_{NR}(p) \leftrightarrow p$

However, this equivalence only holds contingently, because the two operators behave differently in modal contexts. In particular, actuality$_{R}(p)$ does not change in truth value in different possible worlds:

(*Actual*$_{R}$) Actually$_{R}(p)$ is true in w if and only if p is true in @.

No matter which world God created, it would actually$_{R}$ be the case that, for instance, Caesar exists. Consequently, actually$_{R}(p)$ is not necessarily equivalent in truth value to p itself. If God had not created Caesar, *Caesar exists* would be false, but *actually*$_{R}$ *Caesar exists* would be true. By contrast, the truth value of *actually*$_{NR}(p)$ changes in different possible worlds:

(*Actual*$_{NR}$) For any world w, actually$_{NR}$ (p) is true in w if and only if p is true in w.

Consequently, in any possible world, p is true if and only if actually$_{NR}(p)$. If God had created a different possible world, some propositions that are actually$_{R}$ the case would not be actually$_{NR}$ the case. When we consider the possibility of God creating some world other than @, we are considering the possibility that @ might not have been actual$_{NR}$.

To return to the argument, we can now see that (8†) is true only if it involves the rigid notion of actuality:

(8†$_{R}$) If God's CIC of Julius Caesar represents him as existing in @ then God's CIC of Caesar represents him as actually$_{R}$ existing.

while (9†) is true only if it involves the non-rigid notion of actuality:

(9†$_{NR}$) But God's CIC of Julius Caesar does not represent him as actually$_{NR}$ existing; it represents him as merely possible, as existing only if God wills to create him.

Intuitively, (9†$_{NR}$) says that God's CIC of Julius Caesar does not prejudge whether God creates @ or not; it does not represent him as actually existing in the non-rigid sense. Since it is consistent for God to represent Julius Caesar as actually$_{R}$ existing (existing in @) without representing him as actually$_{NR}$ existing, there is no conflict here and (10†) does not follow by *modus tollens*.

2.6. The Argument in *Beweisgrund*: Baumgarten

A better explanation of Kant's focus on completely determinate concepts of possibilities is that his unnamed target is Baumgarten. But this raises the question of why Kant would direct his objections at Baumgarten specifically, rather than to all ontological arguments as such. In section 4 I argued that there are reasons to think that Baumgarten's 'complete determination' argument is the strongest ontological argument fielded by any of Kant's ontotheist opponents (or at least, among those known to Kant).[71] Furthermore, there is good evidence that Kant thought of Baumgarten in these terms, for Kant returns repeatedly, in published and unpublished writings, to the complete determination theory of existence, in order to refute it. One possible explanation of this is that he agrees that Baumgarten's complete determination argument is 'the argument to beat.'

If we read the 'Julius Caesar' passage quoted previously with Baumgarten in mind, a natural Kantian objection emerges: the completely determinate concept of Julius Caesar does not represent him as existing. One can think of an object as completely determinate as one wishes without thereby thinking of it as an existing object. This is the reason he introduces God and God's thoughts about possibilia into the argument: our finite minds cannot form a genuinely completely determinate concept.[72] The argument is that God can have a completely determinate thought about some possible object, without representing it *as* an existing object. If existence were equivalent to complete determination, then God's thought of that object would represent it as existing. Therefore, existence is not equivalent to complete determination, contra Baumgarten. This is why Kant emphasizes that, although God has fully determinate concepts of non-actual possible individuals (he "could know every single one of these predicates without exception, and yet still be able to regard him as a merely possible thing"), he nonetheless does not thereby represent those individuals as existing.

While this argument may initially appear damning for Baumgarten's position, in fact Baumgarten can easily reply. Baumgarten needs to explain how God can have a thought that represents its object as being completely determinate without representing it as existing.[73] The expression 'a thought that represents the object as being so-and-so' could mean either a thought that represents the object as actually$_{NR}$ being so-and-so, or one that represents the object as it would be, if it were so-and-so. To use an example from earlier, I can think, of Joe, that he is actually$_{NR}$ bald, or I can think

[71] It may be that one of the versions of the ontological argument Leibniz worked on but never published is stronger than Baumgarten's, though the critical discussion in Adams (1994) and Griffin (2013) does not make that seem very likely.

[72] Without talking about God, Kant would have to formulate his argument in an awkwardly counterfactual form: *per impossibile*, were I to form a fully determinate concept, etc.

[73] In this section, I always mean 'actually' in the non-rigid sense. See previous section for more on the rigid and non-rigid senses of 'actually.'

of Joe as he would be if he were bald. Both thoughts could be described as the thought 'of Joe as bald,' but this description conceals an important difference: the first thought is a thought about (non-rigid) actuality, and the second is a thought about possibility. The thought of Joe as he would be if he were bald is not merely the thought that Joe is possibly bald, or that Joe could be bald. It is a thought about how Joe would be in respect of appearance and hairline if that possibility were (non-rigidly) actual.

Similarly, God can think about Caesar either as actually$_{NR}$ being fully determinate, or God can think about Caesar as he would be *if* he were fully determinate. In fact, he can do the latter in infinitely many ways, because there are infinitely many ways Caesar could be completely determinate because there are infinitely many fully determinate sets of accidents he could have (contra Leibniz). Baumgarten can accommodate the claim that God can have existentially neutral thoughts that represent their objects as completely determinate by interpreting this to mean that God can have thoughts about merely possible and hence indeterminate objects that represent those objects as they would be *if* they were completely determinate, i.e. if they were actual$_{NR}$. Nothing in the contents of these thoughts requires God to think that these objects are actually$_{NR}$ completely determinate. Given that God is omniscient, he can only think of actual$_{R}$ things as actually$_{NR}$ being completely determinate; if he were to think of a merely possible object as actually$_{NR}$ being completely determinate, he would be misrepresenting it. Existentially neutral but completely determinate thoughts about possibilia pose no problem for Baumgarten. This takes the wind out of the sails of Kant's objection to Baumgarten.

Over the course of this chapter and the previous one we have seen three Kantian objections to Baumgarten's 'complete determination' ontological argument:

1. There are completely determinate but uninstantiated concepts.
2. By denying the principle of complete determination, the extensional version of Baumgarten's theory violates the principle of excluded middle.
3. God can have completely determinate concepts of non-actual individuals without representing them as existing.

In section 4 I explained why the first objection carries little or no weight against Baumgarten. I argued further that, contra the second objection, Kant gives us no reason why Baumgarten cannot follow *Kant himself* in denying that the principle of excluded middle entails the principle of complete determination. I have just argued that the third objection fails as well. I conclude, therefore, that in *Beweisgrund* Kant fails to establish that Baumgarten's complete determination theory of existence is false and thereby fails to refute his 'complete determination' version of the ontological argument. This means that, at least in that early text, Kant fails to refute Baumgarten's version of the doctrine that existence is a real predicate.

2.7. The Argument in the *Critique of Pure Reason*

In the *CPR* Kant also argues that existence is not a real predicate, in the famous section "On the impossibility of an ontological proof of the existence of God." Within that section, the arguments that have received the lion's share of scholarly attention are the 'hundred dollars' argument and the other arguments in that paragraph (A599/B627) and the next paragraph on the following page (e.g., the 'missing reality' argument). I think that those arguments add little that was not already present in *Beweisgrund*. For reasons of space, therefore, I forgo further discussion of them here.[74]

However, that section also contains an apparently distinct objection, which I will call the 'no contradiction' objection, that some readers may think is a stronger objection, at least to *logicist* ontological arguments. I will argue, on the contrary, that the independence of this objection from the claim that existence is not a real predicate is merely apparent, and that it does not carry any additional force against the logicist.

Kant states the non-contradiction objection in a relatively straightforward passage:

> If I cancel the predicate in an identical judgment and retain the subject, then a contradiction results; hence I say that the former necessarily pertains to the latter. But if I cancel the subject together with the predicate, then no contradiction results; for there is no longer anything that could be contradicted. To posit a triangle and cancel its three angles is contradictory; but to cancel the triangle together with its angles is not a contradiction. It is exactly the same with the concept of an absolutely necessary being. If you cancel its existence, you cancel the thing itself with all its predicates; where then is the contradiction supposed to come from? [...] But if you say, **God is not**, then neither omnipotence nor any other of his predicates is given; for they are all canceled together with the subject, and in this thought not the slightest contradiction shows itself. (A594–5/B622–3)

Kant's point is that if we take an analytic judgment like *all triangles have three angles* and negate it, we get:

(1) Some triangles do not have three sides

which, given the definition of <*triangle*>, is equivalent to:

(2) Some (three-sided plane figures) do not have three sides

and this is clearly a violation of the PNC. Thus, we have a very natural account of why analytic judgments are necessary: their negations violate the PNC. However, if we take an existential judgment like *God exists* and negate it, we never get a violation of

[74] For readers interested in a reconstruction of those arguments, see the supplementary article on the arguments of the *CPR* on my website (see Notes on the Text). Notice that both the 'hundred dollars' argument and at least one of the arguments in the subsequent paragraph on A600/B628 ("when I think a thing, through whichever and however many predicates I like (even in its thoroughgoing determination)") are both arguments about completely determinate concepts. So it is prima facie plausible that they restate the basic argument of *Beweisgrund*, reconstructed above.

the PNC. So it is never logically impossible that something fail to exist, not even God. Although Kant does not make this objection in *Beweisgrund* it is implicitly contained in his observation that the proposition that *nothing whatsoever exists* does not entail a contradiction;[75] it is not a point wholly original to the *CPR*.

This Kantian point can seem especially strong if we interpret it as the logical point that a negated existential statement of the form:

(3) $\neg\exists xFx$

does not entail a contradiction, and, what is more, can be added to any consistent set of propositions that lack existential import without producing an inconsistency. If we take Kant's target to be the logicist view that God exists with logical necessity then this might appear to be a knock-down objection: no claim of the form $\exists xFx$ is logically necessary because its negation (3) does not entail a contradiction. Since this objection does not prima facie depend on controversial claims about whether existence is a determination or real predicate it might seem to be Kant's most straightforward and best objection to the ontological argument.

Kant claims that judgments of the form *there are no Fs* do not entail contradictions. Without going into detail about the nature of the entailment relation, I take it that it is uncontroversial that p entails q if and only if there is a proof of q from p using only logical principles, and everything here depends upon what counts as a *logical* principle. While Kant does not yet, in 1763, have his developed Critical theory of the formality of 'pure general logic', his claim in *Beweisgrund* that the proposition *nothing exists* is not a violation of the PNC (the highest principle of logic) is an anticipation of one important aspect of it.[76] For if *nothing exists* does not violate the PNC, and the PNC is the highest principle of logic (the principle from which all other logical principles flow, either directly or indirectly), then *logic alone* does not guarantee that anything exists, that is, that there are any objects in the extensions of any concepts. In contemporary logic, this means that empty domains are not excluded on purely logical grounds. The same result holds in Kant's Critical logic. (Pure general) logic, writes Kant in the *CPR*, abstracts entirely from the relation of cognition to objects;[77] this means not only that logic abstracts from the differences *among* objects, but logic abstracts from whether concepts have objects at all, that is, whether they have null extensions. So logic, for both the pre-Critical and Critical Kant, does not guarantee that there are any *objects*; no existentially committal propositions can be among the principles of (pure general) *logic* properly so called.

[75] Ak. 2: 78. [76] Ak. 2: 78.

[77] A55/B79. I am not claiming that pure general logic does not contain a certain concept of an object; I am claiming that pure general logic does not assume that *there are* objects in the sense under discussion here: objects of quantification (see next section).

This might be thought devastating to the logicist, for each of the ontological arguments I reconstructed in the previous chapter crucially relies on what is now called a 'comprehension' principle, a principle that allows us to infer that a certain concept has a non-empty extension:

Descartes: I clearly and distinctly understand essence E $\supset \exists x$(E is the essence of x & possible(x))
Leibniz: C is a logically consistent concept $\supset \exists x$(possible(x) & Cx)[78]
Baumgarten: C is a logically consistent concept & C is an essence $\supset \exists x$(possible(x) & Cx)

These are crucial components of their ontological arguments because they allow one to infer from a claim about the essence or concept of God (left-hand) to the claim that *there is* an object that falls under that concept (right-hand). However, from a Kantian point of view, these are not *logical* principles. They concern the conditions under which a certain concept has a non-empty extension (e.g., when it is logically consistent) and logic abstracts entirely from whether concepts have non-empty extensions.

But the logicists need not agree with Kant about what logic is. As other scholars have argued,[79] Kant's thesis that logic is formal (that it abstracts entirely from objects) is an innovation in the rationalist logical tradition; there is no reason to expect that Descartes, Leibniz, or Baumgarten would share it. The logicists can plausibly claim that certain comprehension principles (like those above), or certain principles about the domain of merely possible being, are logical principles, and propositions that are incompatible with them are logically impossible. The logicists, after all, conceive of logic as describing the metaphysical space of possible beings and their essences. They can plausibly claim that it is a *logical* principle that there is a domain of possible beings and the widest quantifier 'there is' ($\exists x$) ranges over (takes as the admissible values of its bound variables elements in the set of) possible beings, including the merely possible ones; consequently, $\sim\exists x(x=x)$ is logically impossible and there may be other 'existentially' quantified propositions whose negations are logically impossible (notably, $\exists x$God(x)). The logicist are not bound to agree with Kant that logic abstracts entirely from objects.

But notice that the crucial logicist move is to claim that logic does not abstract from objects but considers all possible objects as such and their principles (which are logical principles). This view rests on the assumption that there is a distinction between objects in general (possible objects) and existent objects; in other words, it rests on the assumption that existence is a determination, a particular way objects can be. For instance, the comprehension principles listed above become highly implausible if we interpret the quantifier expression $\exists x$ as expressing existence; in

[78] I suspect that the true Leibnizian comprehension principle is something like: C is a complete concept $\supset \exists x$(possible(x) & Cx), but that did not play an explicit role in my reconstruction of Leibniz's ontological arguments.

[79] McFarlane (2002) and Lapointe (2012).

Leibniz's case they would entail that every consistent concept is instantiated by an existent object.

We are led here to a conclusion very similar to the one I drew in the previous chapter when we considered whether the claim that existence is not a real predicate means merely that existence is a quantifier expression that applies in the first instance to concepts (and at best derivatively to objects): these alternate interpretations of Kant's objection deliver decisive objections to ontotheism *only if* they are supplemented with the further claim that there are no non-existent objects (on my interpretation, the meaning of the claim that existence is not a determination). That claim is Kant's fundamental objection to ontotheism. In fact, the very organization of the corresponding section in the *CPR* supports this. Kant begins with the "no contradiction" objection at A594–5/B622–3 and, aware that this *alone* will not quiet the resolute ontotheist, states rather huffily: "I would have hoped to annihilate this over-subtle argumentation without any digressions through a precise determination of the concept of existence, if I had not found the illusion consisting in the confusion of a logical predicate with a real one (i.e. the determination of a thing) nearly precludes all instruction" (A598/B622). Kant only then makes his famous claim that "being is obviously not a real predicate." I interpret the structure of this section, and Kant's huffiness, to mean that the decisive objection, the objection that will finally eliminate ontological arguments, is the claim that existence is not a real predicate. Kant's other objections (that existence is in the first place predicated of concepts, not of objects; that no negative existential is logically contradictory) are parasitic on *that* claim.

2.8. Absolute Positing

Having discussed in so much detail Kant's *negative* thesis about existence (it is not a real predicate), it is worth briefly discussing his *positive* theory of existence. Since Kant's theory of existence in the *CPR* is tied up with distinctively Critical doctrines (most importantly, the distinction between concepts and intuitions), I will explore the positive theory of existence in *Beweisgrund*, specifically, where Kant makes a distinction between two kinds of positing:

> The concept of positing [*Position*] or setting [*Setzung*] is perfectly simple: it is identical with the concept of being in general. Now something can be thought as posited merely relatively, or to express the matter better, it can be thought merely as the relation (*respectus logicus*) of something as a characteristic mark of a thing. In this case, being, that is to say, the positing of this relation, is nothing other than the copula in a judgment. If what is considered is not merely the relation but the thing posited in and for itself, then this being is the same as existence. (*OPG*, Ak. 2: 73)

Kant here distinguishes two forms of positing, which I take to be two ways in which *being* can be asserted: relative positing (in which the relation of two predicates is asserted in a judgment) and absolute positing (in which an object is posited as an instance of a predicate). Relative or 'copulative' positing asserts that whatever falls under one predicate falls under another, without asserting that any object instantiates either predicate. To take Kant's example a few paragraphs later:

> If I say 'God is omnipotent' all that is being thought is the relation between God and omnipotence, for the latter is a characteristic mark of the former. Nothing further is being posited here. Whether God is, that is to say, whether God is posited absolutely or exists, is not contained in the original assertion. (*OPG*, Ak. 2: 74)

The relata in this case are the predicates <*God*> and <*omnipotent*>; the former relatum cannot be God himself because the judgment *God is omnipotent* can be made whether or not the concept <*God*> is instantiated, i.e. whether or not there is a God to stand in such a relation. In fact, Kant goes on to say, we can make judgments involving concepts that cannot possibly be instantiated, like <*the God of Spinoza*>.

In absolute positing, by contrast, we assert that there is an object instantiating a concept (a set of predicates):

> If I say 'God is an existent thing' it looks as if I am expressing the relation of a predicate to a subject. But there is an impropriety in this expression. Strictly speaking, the matter ought to be formulated like this: 'Something existent is God.' In other words, there belongs to an existent thing those predicates which, taken together, we designate by means of the expression 'God.' These predicates are posited relative to the subject, whereas the thing itself, together with all its predicates, is posited absolutely. (Ak. 2: 74)

When we posit absolutely we assert that some concept, e.g., <*God*>, is instantiated by an object, and in so doing we also copulatively assert that all of the predicates contained in the concept pertain to that object as well: anything that instantiates <*God*> also instantiates <*omnipotent*>, <*omniscient*>, etc. Every absolute positing brings with it a relative positing ("these predicates are posited relative to the subject [concept], whereas the thing itself, together with all its predicates, is posited absolutely").[80]

This means we need to distinguish and account for all of the following cognitive acts:

(i) *Absolute positing*: there is an object that instantiates the concept F.
(ii) *Judging*: anything that falls under concept F falls under concept G.

Absolute positing, as Kant characterizes it and as I have defined it, is *general*: it says that some object or other instantiates a concept. But we also need to be able to judge of particular objects that they instantiate concepts (subsume them under concepts). So we need to add to our inventory of cognitive acts:

(iii) *Subsumption*: object x instantiates concept F.

[80] Cf. Abaci (2008).

Otherwise, we would be able to judge (to use another Kantian example) *there are narwhals* and *narwhals have horns* but not *this object is a narwhal* (in the presence of a narwhal). But the ability to perform the cognitive act I have called 'subsumption' involves the ability to think of particular objects (e.g., using the demonstrative 'this object'), which I will call:

(iv) *Acquaintance*: being cognitively in contact with an object x so that it can figure in an act of subsumption.

Kant's 1760 works do not contain a developed theory of these different cognitive acts and their relations; for that we must wait until the *Critique of Pure Reason* in 1781. For instance, one question we might have is: can we understand an act of absolute positing (there is an F) without understanding what acquaintance with such an object (with an x that is an F) would be? I will explore Kant's Critical answers to these questions in the second half of this study. But before we continue we can use the notion of 'absolute positing' to answer a question that some readers may have: what does 'object' mean? The concept of object I have been working with is (in Kantian terms) the concept of an object of absolute positing.[81] My interpretation of the doctrine that existence is a real predicate—there are no objects that do not exist—can be restated in these terms as: there are no objects of absolute positing that do not exist. To posit an object and to assert that it exists are *one and the same* act. Hence Kant's claim that: "if what is considered is not merely this relation [between predicates] but the thing posited in and for itself, then this being is the same as existence" (*OPG*, Ak. 2: 73).[82]

[81] A forerunner of the quantificational notion of 'object.' See the discussion of Rosenkoetter and Forgie and the 'Frege-anticipation' thesis in Ch. 1.6.

[82] Cf. A593–601/B621–9.

3

Real Conflict, Real Grounds, Real Possibility

3.1. Introduction

In the previous chapter we examined Kant's reasons for rejecting the logicist explanation of divine necessity. If no existential proposition can be logically necessary, as Kant argues, there is at least one proposition—that God exists—that is necessarily true, but not logically necessary. Although this entails that logicism is, strictly speaking, false, it leaves much of logicism intact. For instance, the existence of God might be the only exception to the logicist theory of necessity; all non-existential necessarily true propositions might, nonetheless, be logically necessary. That would leave the logicists without a perfectly general account of possibility and necessity, but it might leave them with an elegant account of modality for finite beings. To a contemporary audience less concerned with vindicating the traditional doctrine of divine necessity, this may not seem like much of a problem at all, and Kant's attack on logicism as a whole might appear quite weak. In this chapter I examine the second strand of Kant's rejection of logicism: his rejection of the logicist doctrine that logical contradiction is the only source of metaphysical incompatibility between predicates.

The logicist view of possibility is:

($Logicism_P$) It is possible that p if and only if p does not entail a contradiction.

In the case of predicates of one and the same subject, this entails that:

($Logicism_{PRED}$) For predicates P and Q, it is possible that there is an object x with P and Q if and only if the proposition $\exists x(Px \;\&\; Qx)$ does not entail a contradiction.

I will refer to the relation of being possibly co-instantiated as being *metaphysically compatible* and the relation of being logically possibly co-instantiated as being *logically compatible*; logicism entails that these relations are co-extensive. Sometimes I will refer to the former relation simply as *compatibility*.

In the pre-Critical works of the 1760s Kant claims that there are logically compatible predicates that are not *metaphysically* compatible, or what I will call *really* incompatible predicates. For instance, he writes in *Beweisgrund* that: "the impenetrability of bodies, extension, and such like, cannot be attributes of that which has understanding and will" (*OPG*, Ak. 2: 85). However, Kant does not offer a direct argument for this claim; elsewhere he appeals to an "immediate judgment of the understanding" that "forces one to admit" that understanding and will "together with the greatest possible reality can co-exist in one and the same being" (*OPG*, Ak. 2: 88). Presumably, another "immediate judgment of the understanding" is supposed to provide our access to the incompatibility of thought and extensions. But these bald assertions of real compatibility and incompatibility are not likely to satisfy his logicist opponent, for whom there must be some hidden logical incompatibility between the concepts of thought and extension (and a lack of such logical incompatibility between the concepts of understanding and will).

In this chapter I reconstruct Kant's argument that there is real incompatibility, focusing on *Beweisgrund* and Kant's 1763 essay, *Attempt to Introduce the Concept of Negative Magnitudes into Philosophy*.[1] In *Negative Magnitudes* he distinguishes between the 'logical conflict' (when two predicates are logically incompatible) and the 'real conflict' of predicates. Two predicates stand in 'real conflict' just in case they are both realities (they are not mere absences of positive determination) and, when they are co-instantiated by a single object, each of them causes the effect the other would have on its own to be canceled. Kant's primary example is that of two forces of equal magnitude, but opposite direction, acting on the same body. Each force cancels the motion that would be caused by the other force, if that force were the only force acting on the body. But realities stand in real conflict only if they are *possibly* co-instantiated by the same object; realities in real conflict must be metaphysically compatible. So even if Kant were to prove that there is real conflict between predicates, this would not show that there is real incompatibility between predicates. Far from entailing metaphysical *in*compatibility, real conflict between predicates entails that they are *compatible*.[2]

In this chapter I explore how Kant uses the phenomenon of real conflict to show that there are really incompatible predicates. The key idea is that real conflict is a causal relation; predicates stand in real conflict when they mutually *causally* cancel one another's *effects*. *Cause*, in eighteenth-century German metaphysics, is usually taken to be one species of the more general notion of *ground*, and I begin, in section 2, by exploring the logicist theory of grounding, and in particular, of causation. In section 3 I explain one of the key elements of Kant's pre-Critical metaphysics: the distinction between logical grounds and 'real' grounds and his argument that causes are *real* grounds of their effects. In section 4 I show how this distinction between logical and real grounds is the missing premise in Kant's argument that logicism cannot allow for real conflict between predicates, and in section 5 I reconstruct his argument that, if there is real conflict between predicates, then there are really

[1] The topic is also broached in *Beweisgrund*, but not discussed as thoroughly as in *NG*, so it is on that work I will focus here; see Ak. 2: 86.

[2] As Abaci (2014) correctly points out.

incompatible predicates (though not the predicates that are in real conflict—see previous discussion). I conclude with a brief discussion of 'real' possibility.

Before continuing, I want to make a note about Kant's notion of a 'predicate' [*Prädicat*], which, from a contemporary perspective, plays two distinct conceptual roles in Kant's pre-Critical thought: (a) a metaphysical or 'worldly' role, in which a predicate is something in the world (e.g., a force in a substance), rather than the content of a representation, akin to a *property* in contemporary usage;[3] and (b) a 'semantic' role, in which predicates are concepts and can be combined in judgments.[4] There are at least two ways to interpret Kant on this point: (i) he is conflating two kinds of things that are (correctly) distinguished in contemporary philosophy, properties and concepts; or (ii) he is making the ambitious and interesting claim that the contents of our thoughts are also worldly items, i.e. that we think 'with' properties, the very items that metaphysically constitute and characterize objects.[5] While I find (ii) intriguing, I do not have the space here to articulate or develop this idea.[6] It would be premature, though, to simply adopt (i) and convict Kant of a conflation. Rather than try to resolve this problem, which would require answering some difficult philosophical questions (What exactly are concepts? Why should we assume that properties do not make up the contents of thoughts?), I will simply talk about predicates as Kant does: as 'amphibious' entities that are both worldly—they include physical forces and can be causally efficacious—and semantic—they have an internal logical structure and can be combined in judgments. In some contexts the semantic role will be primary; in others, the worldly role. In both contexts I will refer to the relation between an object and a predicate that object has as *instantiation*.

3.2. Logical Grounds

'Ground' is a very general notion in the pre-Kantian German rationalist tradition. The definition that perhaps best captures its wide variety of meanings is that given by

[3] *NM*, Ak. 2: 171, 172, 176, 193, and 200. Things Kant describes as predicates include: forces, feelings of displeasure, and tendencies. These are not mere representational contents in our (or the divine) mind: they are real determinations in things.

[4] See *OPG*, Ak. 2: 76, 77, 78, 80 (where he identifies the predicates *fiery* and *body* as concepts), 82, 83, and 156, and Prize, Ak. 2: 284. It might be objected that 'predicate' in logical contexts just means the second term of the judgment, but any concept can be the predicate (including the subject itself). See Ch. 1.3 and A69/B94.

[5] One way to resolve this difference in Kant's uses of the term 'predicate' would be to claim that predicates-as-properties *are* semantic-representational items in the mind of God; to say that some object has predicate/property F is just to say that it is subsumed under God's concept of F. However, I will argue in the next chapter that this cannot be right: God grounds all possible *predicates* but does not do so through his intellect; his intellectual representation of predicates presupposes their real possibility, and so I do not think we can analyze predicates as contents of divine thoughts. Yong (2014) defends such an 'intellectual' grounding of possibility.

[6] The idea that the space of thinkable content and the space of possibility are one and the same might be found in Hegel's *Science of Logic*; for an interpretation of Hegel, and Kant's *Beweisgrund*, along these lines see Yong (forthcoming), ch. 4.

Christian Wolff: "the ground is that through which one can understand why something is."[7] Kant argues that this definition is tacitly circular, because "why a thing is" means the ground of that thing.[8] Nonetheless, this definition, unlike more specific definitions given by individual thinkers, has the generality (indeed, possible emptiness) appropriate to this concept. A ground is something that explains why a thing is, or why it has the predicates it does.[9] 'Thing' here should be understood very generally as any possible being whatsoever: there are grounds of the truth of propositions, of predicates had by objects, alterations in substances, of the existence of substances, etc.

A standard distinction in eighteenth-century theories of grounds is between what I will call 'epistemic grounds' and 'explanatory grounds' (what Kant calls 'antecedently determining grounds').[10] Wolff, Crusius, and Kant all draw this distinction, but use different terminology which, in some cases, is then borrowed by other philosophers to mean something different, which creates ample opportunities for confusion;[11] I have introduced my own terms to avoid that confusion. The distinction concerns the grounding relation and what is grounded, not the ground itself; one and the same thing can be an epistemic ground and an explanatory ground. An epistemic ground provides justification for believing some proposition to be true; when the proposition is believed on the basis of that reason, this belief is justified. Most of these thinkers focus on the sub-class of epistemic grounds that are grounds of *knowledge*: when a belief is formed on the basis of these grounds, the belief counts as knowledge (assuming some background conditions are met).[12] For instance, smoke is an epistemic ground for knowing that there is fire. I will follow them and temporarily leave out of consideration epistemic grounds that confer an epistemic status short of knowledge.[13] An explanatory ground is simply something that explains *why* something is (or is the case) where the *explanandum* (the consequence, that which is grounded) is not the epistemic status of a belief. An explanatory ground of fire, for instance, is something that explains *why* there is fire (e.g., the cause of the fire),

[7] *Dt.Met.* §29; cf. *Meta.* §14 and Crusius's discussions of Wolff's and Leibniz's definitions of 'ground' in the footnotes to *De Usu* §1.

[8] *ND*, Ak. 1: 393; cf. *MH*, Ak. 28: 11.

[9] Since there are also grounds of the existence of objects, and existence is not a real predicate, we cannot restrict 'predicate' here to real predicates. See Chapter 1. However, in this chapter I will be concerned almost exclusively with real predicates; henceforth, unless otherwise noted, 'predicate' refers only to real predicates (determinations).

[10] This distinction is not exclusive; it is possible for the explanatory ground also to be an epistemic ground.

[11] *ND*, Ak. 1: 392–3. The most notable instance of this is that, as Kant himself points out, his distinction between 'real' and 'logical' grounds is very different from Crusius's distinction between 'real' and 'ideal' grounds (Ak. 28: 12, 37); in fact, Crusius himself borrows the 'real' and 'ideal' distinction from Leibniz and Baumgarten; see *Weg* §139, 142; *Ent.* §§34–35; *De Usu* §37; *Th.* §66; and *Meta.* §212. For critical discussion see Watkins (2005), 74–8, 162–5.

[12] In fact, it is common in this period to refer to what I am calling 'epistemic grounds' as *Erkenntnisgründe*.

[13] Until Chapter 9, where Kant's theory of epistemic grounds that are not *Erkenntnisgründe* will play an important role in reconstructing his Critical attitude to his pre-Critical theistic proof.

rather than why I (or anyone else) would be justified in believing there is a fire. There are different kinds of explanatory grounds, corresponding to different kinds of explanation, including logical grounds, causal grounds, and essential grounds. The very same thing can be both the ground of knowledge and an explanatory ground; for instance, the existence of fire both causally grounds smoke and is an epistemic ground of knowing that there is smoke.[14] It is important to note that the notion of explanation here is not an epistemic one: that *x* is an explanatory ground of *y* does not entail that *x* is a ground to believe that *y* exists (is an epistemic ground of belief that *y* exists) and it does not entail that *x* is an explanation of *y* that we are in a position to grasp or to understand. It does not, in other words, entail that *we* could explain *y* by citing *x*. The explanatory (ground-consequence) structure of reality is a metaphysical structure that potentially outstrips human capacities for grasping these explanations (finding grounds for given consequences).

In this chapter, though, my interest will be mainly in what I have called 'explanatory grounds',[15] so, unless otherwise noted, 'ground' means explanatory ground, or in Kant's terminology, an *antecedently determining* ground. Various kind of explanatory grounds are distinguished by different thinkers in this period, including (efficient) causal grounds, logical grounds, as well as what Crusius calls 'real-existential grounds.'[16] The first two I will discuss here, focusing on causal grounds; I discuss the third (real-existential grounds) briefly in Chapter 7.

I defined *logicism* in Chapter 1 as set of views about modality shared (I argued) by Leibniz, Wolff, and Baumgarten. However, it is not clear that these three thinkers shared a common set of views about *grounds* in general; Wolff and Baumgarten define *ground* in very similar ways, but by itself this is not decisive.[17] They differed over crucial issues in the metaphysics of causation (e.g., the nature of pre-established harmony), and this may indicate a deeper disagreement over the nature of grounds as such.[18]

Fortunately, we have abundant evidence, from his lecture transcripts, of how *Kant himself* read Baumgarten's general theory of grounds, and there are reasons to think he would have extended this interpretation to Wolff (and perhaps even Leibniz, though I will not consider Leibniz here). Baumgarten defines *ground* in *Metaphysica* §14 as "that from which it is knowable why something is," and, in his own copy of

[14] This is what Crusius calls 'real-ideal grounds *a priori*': a real ground because it is explanatory, and an ideal ground *a priori* because it is a ground of knowing not merely what is the case (*a posteriori*) but *why* (*a priori*). See *Weg* §142 and *Ent.* §35; for a similar idea in Baumgarten, see *Meta.* §311.

[15] Crusius and Baumgarten: real grounds; Kant: antecedently determining grounds.

[16] For Crusius's notion of 'real-existential grounds' see *Ent.* §§37, 79.

[17] Wolff: "the ground [*Grund*] is that by which one can understand why something is" (*Dt.Met.* §29). Baumgarten: "a ground [*ratio*] is that from which it is knowable why something is" (*Meta.* §14).

[18] Wolff upheld pre-established harmony only between souls and bodies (as a solution to the mind-body problem), while Baumgarten, in this respect the more orthodox Leibnizian, extended it to all substances. See Watkins (2005), 45–50, 74–8, and the translator's Introduction to Baumgarten (2004), xiii–xiv.

Metaphysica, Kant adds the marginal comment: "a ground is either analytic (logical) or synthetic (real ground)" (*Refl.* 3504, Ak. 17: 28), a comment which he expands upon at length in the Herder transcripts of his metaphysics lectures from the early 1760s (roughly simultaneous with the composition of *Negative Magnitudes* and *Beweisgrund*): "all grounds are either *logical*, in which the consequence, which is [partly—NS] identical to [the ground] as a predicate, is posited through the law of identity; or *real*, in which the consequence is not posited by the law of identity and is not [partly—NS] identical with the ground" (*MH*, Ak. 28: 11).[19] The distinction Kant draws here between logical and real grounds is a distinction in the 'connection' (*nexus*) between ground and consequence, the relation by which the ground posits the consequence (*MH*, Ak. 28: 12). A logical ground posits its consequence by the 'law of identity' (which I read as: the principle of non-contradiction) because the consequence is a predicate contained (immediately or mediately) in the concept of the ground; this means that if the ground were present, and the consequence were not, a contradiction would result. In the case of a real ground, the consequence is not contained (even mediately) as a predicate in the concept of the ground, and the reason the ground 'posits' the consequence cannot be explained logically, that is, through the principle of non-contradiction. In such a case, the presence of the ground without the consequence does not entail a contradiction. As Kant says a few pages later: "insight into the *nexus* of a real ground with its real consequence cannot be had through the law of identity" (*MH*, Ak. 28: 24).[20] This is not merely the epistemic point that *we* lack the cognitive resources to analyze the concept of the ground and find the consequence. Kant's point is a metaphysical one: the fact that X grounds Y is not a fact about concept containment or logical entailment. As he puts it in an unpublished *Reflexion*: "logical grounds are not distinguished from real grounds merely by the limits of my cognition [*Erkenntnis*], but in themselves" (*Refl.* 3719, Ak. 17: 266).[21]

Kant draws this distinction but does not make explicit its critical relevance to Baumgarten. Since Wolff's definition of *ground* is sufficiently similar to Baumgarten's,[22] and I am not aware of anything in the work of either thinker that would make a relevant difference here, I am going to treat them equivalently. Leibniz's views on grounding and causation are sufficiently complicated that I will leave them aside for now. It is not immediately clear from the text whether the distinction between logical and real grounds is supposed to be a *criticism* of Wolff and Baumgarten. It is not clear whether, for instance, Kant is claiming that Baumgarten (and Wolff) held that all grounds are *logical* grounds and *not* real grounds, or whether he is pointing out that Baumgarten (and by extension Wolff) overlooked this distinction entirely. It is clear enough, though, that they do hold that *essential* grounds are

[19] Cf. *MH*, Ak. 28: 12, 13, 18, and 25. Kant repeatedly return to this distinction in his metaphysics lectures: *MV* (Ak. 28: 402–4), *MvS* (Ak. 28: 486), *ML*$_2$ (Ak. 28: 549), *MD* (Ak. 28: 625), and *MK*$_3$ (Ak. 28: 807–10).

[20] Cf. *MV*, Ak. 28: 403.

[21] Cf. *Refl.* 3756, Ak. 17: 284.

[22] See previous discussion.

logical grounds. The essence of a being contains various predicates (its essential predicates) and the reason the essence 'posits' those predicates is logical: if the being lacked those predicates, a contradiction would result. The more difficult question, however, is whether they extend this view to (efficient) causal grounds: is the *nexus* between a cause and its effect a logical relation, for Baumgarten or Wolff? While it might seem obvious that this is the view of Wolff and Baumgarten (and even Leibniz), the textual evidence is not clear-cut.[23]

However, the transcripts from Kant's metaphysics lectures in the 1760s (the 'Herder' metaphysics) contain an interesting argument that Baumgarten (and by extension Wolff) is *committed* to the view that all grounds as such are logical grounds. The jumping-off point is the standard logicist distinction between '*in se*' necessity and *hypothetical* necessity.[24] *In se* necessity is what I have been referring to as necessity *simpliciter* since Chapter 1: that, the negation of which entails a contradiction. Something is *hypothetically* necessary when it is necessary *given* the assumption of something else (paradigmatically, its ground). If X is the ground of Y, and Y is not necessary *in se*, Y is hypothetically necessary, namely, on the hypothesis of X. Kant claims, though, that there is a connection between *in se* and hypothetical necessity: "every hypothetical necessity can be transformed into an absolute [*in se*—NS] necessity, if the hypothesis is added to the concept of the thing, e.g., all fallen humans sin" (Ak. 28: 18).[25] The idea of Kant's example is that *humans sin* is not necessary *in se* (its negation is not contradictory), but only necessary on the hypothesis that humanity is fallen; it becomes *in se* necessary if we add that hypothesis to the concept of the subject: *all fallen humans sin*. Translating this point back to the case of grounding: if X is the ground of Y and Y is not necessary *in se*, it is necessary on the hypothesis that X exists. This means, by Kant's conversion rule, that the conditional *if X exists then Y exists* is itself *in se* necessary ("every hypothetical necessity can be transformed into an absolute necessity"). Recall the logicist theory of *in se* necessity: it is *in se* necessary that *p* if and only if $\neg p$ entails a contradiction. But this means that the *nexus* between a ground and its consequence, and, more specifically, between an

[23] Eric Watkins follows Kant in reading Wolff and Baumgarten as positing a logical relation between ground and consequence. He points out that Wolff derives the PSR, the principle that every being has a ground, from the PNC, a logical principle, and reasons that the PSR must be a logical principle as well (Watkins 2005, 119). But even if the PSR follows from the PNC, this does not mean that the ground–consequence relation is a logical one. Watkins also points to the ubiquitous use of 'posit' [*setzen*, *poni*], an originally logical term, to describe the relation of ground to consequence (Watkins 2005, 119 and n. 17 on that page); but Kant himself talks of the 'absolute' positing of an object for a concept, while his point is *precisely* that the object is *not* 'contained in' the concept (Ak. 2: 74). Hogan (2005) moves from Kant's very early defense of real inter-substantial causation (in *Living Forces* and *Nova Dilucidatio*) to his upholding of real causal relation in the 1760s without making explicit that 'real' means something different in the two contexts: in the first context it means *non-ideal* (i.e. not merely an epistemic ground of knowing), and in the second context it means *non-logical*. While I agree with Watkins and Hogan that Kant is reading Baumgarten and Wolff (and perhaps Leibniz) correctly, I am not sure the textual case is so clear-cut.

[24] See *Meta.* §§102–5, *Dt.Met.* §§574–5, and *AzDM* §§195, 197. See also Leibniz, *Th.* §235.

[25] *MV*, Ak. 28: 426.

(efficient) cause and its effect, is a logical relation and can be explained through the principle of contradiction.

It is not completely clear that Wolff and Baumgarten would accept this argument; in particular, it is not clear that they accept that if B is necessary on the hypothesis of A, then *if A exists then B exists* is necessary *in se*. However, I am going to follow Kant in assuming that they do, and thus that they are committed to the principle that all grounds as such are what Kant calls *logical grounds*; I will attempt to reconstruct a plausible theory of grounding on this basis. However, a problem immediately arises for such a view: if A is the causal ground of B (taking A and B to be events, for the sake of simplicity), the logicists cannot claim that the proposition *A occurs but B does not*[26] is impossible *in se*, or impossible 'in its own nature', for truths impossible or necessary 'in their own nature' do not depend upon God's will; God's will makes things actual, but not possible or impossible 'in themselves.' So even God could not make it the case that the cause occurs without the effect occurring. This is not only implausible, but theologically unacceptable.[27] To take an example, if my striking a match causes it to ignite, the logicists would be committed to claiming that even God could not cause that match-striking to occur without the match igniting. The logicists need the relation between a causal ground and its effect to be weaker than strict logical entailment, *without* giving up on the more general idea that all grounding can be assimilated to logical grounding (assuming Kant's argument from the previous paragraph is correct).

The logicists can do this relatively easily by distinguishing, in the case of causal grounds, between the ground, the consequence, and the *conditions*. The basic idea is that there are some additional conceptual conditions that, in combination with the ground, are logically sufficient for the consequence.[28] So, in the case of causal grounding, Kant's 'conversion rule' from hypothetical to necessity *in se* should be reformulated: if Y is necessary on the hypothesis of its cause, X, then (assuming that causes and effects are events) there is some conjunction S of propositions about the background conditions such that if *if X occurs and S is true then Y occurs* is necessary *in se*. In the case of causal grounds, the natural candidates to include in the conditions are the facts about which natural laws obtain and background information about the substances involved (perhaps even the PSR itself). The proposition that I strike the match, plus propositions about the laws that govern the world (e.g., if I strike a match, the match is dry, there is sufficient oxygen, it is not in a Faraday cage, etc. then the match ignites), the background information (the match is dry, there is sufficient oxygen, it is not in a Faraday cage, etc.), and perhaps even the principle of sufficient reason itself, will suffice to entail the proposition that the match ignites. The logicists

[26] I am assuming that the existence of an event is equivalent to its occurring.

[27] Unless one individuates events very finely, i.e. builds the conditions into the essence of the event.

[28] Cf. Longuenesse (1998), 95–7, which offers textual evidence that this was in fact Wolff's view; see *Log.* §§505, 513, 516, and 523.

thus need to distinguish, in the case of causal grounding, between the *ground*, which explains why the effect occurs, and the *conditions*, with which the ground must be supplemented in order to entail the effect. To make such a view work, they would need to draw this distinction in a principled way, and there are significant philosophical difficulties in doing so, but I am not going to go into those details here.[29]

If Leibniz and Baumgarten do hold the 'logicist' view of (specifically, causal) grounding that Kant attributes to them, this would be a promising strategy for them to resolve an obvious problem about the (*in se*) contingency of causal connections. What is more, there is some indirect evidence that this is their strategy. In response to Kant's claim in the *CPR* that in a synthetic judgment the predicate is not contained in the subject (A7–10/B11–14)—a doctrine anticipated by the pre-Critical distinction between logical and real grounds—the rearguard Wolffians Eberhard and Maaß objected that, although the predicate may not be immediately contained in the subject, it may be 'mediately' or 'virtually' contained when the subject is supplemented with some third concept or conceptually articulated 'law of the understanding.'[30] Applying this to the cause–effect case, while the concept of the effect is not directly contained in the concept of the cause, it may be contained in the concept of the cause when that concept is supplemented with some additional conceptual conditions, like the background information and laws, or perhaps the PSR itself. While Maaß and Eberhard are somewhat less than crystal-clear on how this is supposed to work, I think my distinction between the ground and the conditions (articulated above) retains the core of their idea. It retains the original (at least according to Kant) logicist idea that the relation between cause and effect is itself explained by the PNC while allowing that the modal strength of the connection between cause and effect is something less than strict logical (*in se*) necessity (because without the condition the entailment does not go through).

3.3. Real Grounds

In *Negative Magnitudes* Kant presents a short but powerful argument that a cause cannot be the logical ground of its effect:

As for this real ground and its relation to its consequence my question presents itself in this simple form: how am I to understand the fact that, **because something is, something else is** [*weil Etwas ist, etwas anders sei*]? A logical consequence is only really posited because it is identical with the ground. Human beings can err: the ground of this fallibility is to be found in the finitude of man's nature, for if I analyze the concept of a finite mind [*Geist*], I see that fallibility is to be found in it. In other words, I recognize that fallibility is identical with

[29] They are parallel, in many respects, to the difficulties attendant to Nelson Goodman's 'cotenability' theory of counterfactuals; see Goodman (1954), ch.1.

[30] Maaß (1789), 223; see also Eberhard (1789). For critical discussion of the Eberhard–Kant dialectic on the analytic/synthetic distinction, see the Introduction to Allison (1973), as well as Hogan (2013).

> what is contained in the concept of a mind [*Geist*]. The will of God alone contains the real ground of the existence of the world. The will of God is something. The world which exists is *something completely different*. Nonetheless, the one is posited by the other. (*NM*, Ak. 2: 202)

Kant's point is that, while human finitude is a logical ground of fallibility, God's will is not a logical ground of the existence of the world. The relation of God's will to the world is an instance of *causation*; Kant's larger point is that causes in general are not logical grounds of their effects. At first, this argument may appear to beg the question against the logicist theory of grounds: "you may subject the concept of divine willing to as much analysis as you please: you will never encounter in that concept an existent world as something which is contained within the concept of God's willing, or as something posited by that concept through identity" (Ak. 2: 202). Kant's opponent could simply claim that God is the logical ground of the existence of the world in the sense that logical analysis of the concept of God's will (perhaps supplemented with the principle of sufficient reason, or the principle that God creates the most perfect possible world) reveals that God creates this world.

The key to seeing how this argument does not beg the question is the term 'identity.' Kant's point earlier in the paragraph is that if concept A is the logical ground of concept B, it follows that if an object falls under A, then *that very* object falls under B. Logical grounds explain why an object that falls under one concept also falls under another concept. As he puts it in the Herder metaphysics lectures: in the case of logical grounds, "the consequence, which is identical to [the ground] as a predicate, is posited through the law of identity" (Ak. 28: 11). The consequence is (partly) identical [*einerlei*] to the ground in that the consequence (the predicate) is *part* of the ground; the 'law of identity' means that if some object falls under concept A and if B is part of (a mark of) A, then the object instantiates B. In the case he discusses, humanity is a logical ground of fallibility because fallibility is partly *identical* to humanity—fallibility is a constituent mark of the concept of humanity—and so if any object instantiates humanity *that very same* object thereby instantiates fallibility. This model of logical grounds appears to provide no way of understanding how the positing of one object under a concept entails the existence of a distinct object under another concept. "The will of God is something," he writes; "The world which exists is *something completely different*. Nonetheless, the one is posited by the other."

It might be objected, though, that there are pairs of concepts such that, if one object falls under the first concept, it follows that a distinct object falls under the second concept. David instantiates the concept *<father>*; this entails that there is a distinct object instantiating the concept *<child of David>*. So it appears, contra Kant, that logical grounds can explain how the existence of one object falling under a concept (*<father>*) entails the existence of another object falling under another concept (*<child>*).

But further reflection on the father/child example supports Kant's conclusion. David's falling under the concept *<father>* entails that he has a child, but it does not *explain* why this other person exists and is his child. Intuitively, the explanation works the other way around: because this person exists and is David's child, David is a father. Nor does the fact that this other person exists and is a child of David's explain why he (David) exists. This leads to two conclusions. First, the father/child example does not give us a model for thinking about how one object falling under a concept explains the existence of a distinct object. Secondly, it gives us reasons to think that logical entailment can never issue in such an explanation. Causal explanation is asymmetric (at least in the case of creation), but logical entailment is not. David is a father if and only if there is a child of David's, but, intuitively, the relation of explanation works in only one direction: David is a father *in virtue of* the fact that there is a child of David's (and this is *because* of an act of procreation between David and the child's mother).

The analogue of the father/child strategy in the example of God's creation of the world would be to claim that it is built into the concept of God's will that God creates a world. Therefore, God's will is a logical ground of the existence of the world. But that will not work, for reasons analogous to the father/child case. If God creates the world then it may well be (if we follow e.g., Leibniz) that logical analysis of the complete concept of God's will reveals the predicate <*creates the world*>. But this only shows that the proposition that the complete concept of God's will is instantiated *entails* that the world exists; it does not show that the existence of God's will *explains* why the world exists.[31]

That Kant has anticipated this move is clear from the way his argument continues:

> Nor am I willing to be fobbed off by the words 'cause' and 'effect,' 'force' and 'action.' For if I already regard something as a cause of something else, or if I attach the concept of force to it, then I am already thinking of the cause as containing the relation of the real ground to its consequence, and then it is easy to understand that the consequence is posited in accordance with the rule of identity. For example, the existence of the world can be understood with complete distinctness in terms of the omnipotent will of God. But here 'power' signifies something in God, in virtue of which other things are posited. But this word already designates the relation of a real ground to its consequence, but it is this relation which I wish to have explained. (*NM*, Ak. 2: 203).

The move Kant has anticipated by his opponent is to build the causal connection between God and the world directly into the concept of God's will, so that logical analysis will reveal that one of the marks of that concept is <*creates the world*>. The objection Kant here levels against this move is that by building the causal connection into the content of the concept of the ground (God's will), his opponent has abandoned the original project of giving an account of what the causal connection consists in. The original logicist idea was that A causes B in virtue of the logical connection between A and B. If the logicist now says that <*causes B*> is built into the concept of A, and that is why A is a ground of B, then he has failed to give any

[31] The logicists might claim that only an infinite analysis would uncover the concept of the world within the concept of God's will, in order to defend the contingency of God's creative act; see Ch. 1.2 for more on the infinite analysis view of contingency.

account of the content of that logical mark, and has thus failed to actually account for the causal connection. Thus, Kant concludes, the causal connection between cause and effect cannot be assimilated to a logical relation of entailment or of concept containment.

3.4. Real Conflict

This principle—that causal grounds are *real* grounds of their effects—is the crucial premise in Kant's argument that logicists cannot admit the existence of 'real conflict'; without it, the argument does not go through. His preferred example of real conflict is that of two equal but opposite forces acting on the same body:

> The motive force of a body in one direction and an equal tendency of the same body in the opposite direction do not contradict each other; as predicates, they are simultaneously possible in one body. The consequence of such an opposition is rest, which is something (*repraesentibile*). It is, nonetheless, a true opposition. For that which is posited by the one tendency, construed as existing on its own, is canceled by the other tendency, and the two tendencies are true predicates of one and the same thing, to which they belong simultaneously. The consequence of the opposition is also nothing, but nothing in another sense to that in which it occurs in a contradiction (*nihil privativum, repraesentabile*). We shall, in the future, call this nothing: zero=0. Its meaning is the same as that of negation (*negatio*), lack, absence. (*NM*, Ak. 2: 171–2)[32]

A body, X, is acted upon by two forces, A and B, which are equal in magnitude, but opposite in direction. The features of this example that make it a case of real conflict are: (i) it is possible for one thing to be acted upon by both A and B; (ii) the two forces are positive determinations in themselves, not mere absences of positive determination; (iii) force A, by itself, has an effect, motion *a*, and force B, by itself, has an effect, motion *b*; and (iv) if one body is acted upon by both A and B, rather than the combined effect *a*+*b*, the effect is rest, a lack of motion. The rest of body X is what Kant calls a 'deprivation,' an absence of overall positive determination that is the result of a real conflict of forces. Forces A and B cancel one another's effects.[33]

On its own, though, the phenomenon of real conflict has little or no dialectical force against the logicist. On the logicist picture, grounds are logically intelligible grounds: the concept of force A contains the concept of motion *a*, and the concept of force B contains the concept of motion *b*. This may make it seem that the logicist cannot escape the conclusion that logically compatible realities cannot mutually cancel one another's effects; if motion *a* is a logical consequence of the concept of force A, it might seem that nothing logically compatible with A (e.g., B) can cancel this logical entailment.[34] However, as we saw in the previous section, logicist theories

[32] See *Weg* §160 for a Crusian anticipation of the Kantian concept of real conflict.

[33] For the distinction between 'deprivation' (*Beraubung, privatio*) and 'lack' (*Mangel, defectus, absentia*) see *NM* (Ak. 2: 177–8); cf. *OPG*, Ak. 2: 87.

[34] In contemporary terms, logical entailment is said to be 'monotonic.'

of causal grounds can distinguish between the cause and the conditions that make possible the grounding relationship. The logicist can claim that the concept of the force *plus* the conditions *entails* the effect. For instance, in the case where A is the sole force acting on the body, this will be included in the conditions, and in the case where the body is also acted on by B, this will be included.[35] Thus, the concept of force A can contain an infinite series of conditionals of the form: if a body is acted upon by force A and force *f*, then the body will experience motion *y*. One of these conditionals will be: if a body is acted upon by force A and by no other force, the body will undergo motion *a*. Another will be: if a body is acted upon by force A and by force B and no other forces, it will be at rest. Likewise for the concept of force B and motion *b*. The ground will be a set of forces mentioned in the antecedent of one and only one of these conditionals (e.g., A and B); the proposition that the ground exists, together with the propositions that make up the condition (e.g., that no other forces act on the body), will entail that the body is not moved, in Kant's example. By means of this strategy, the logicist can satisfy all four criteria given above:

(i) A and B are both positive qualities, i.e. neither is a mere lack of positive determination.
(ii) It is possible for something to be acted on by both A and B.
(iii) If a body were acted on by A alone, it would undergo motion *a*, and if a body were subject to force B alone, it would undergo motion *b*.
(iv) If a body were subject to force A and force B, the result would be a lack of motion.

By itself, then, Kant's argument that logicism is incompatible with the existence of real conflict is quite weak.

The key to understanding how Kant, steeped as he was in the Leibnizian tradition, could estimate this argument so highly is the suppressed premise that causal grounds are real non-logical grounds.[36] Real conflict per se does not pose a special problem for logicism; real conflict, by definition, involves (efficient) *causation* and, according to the argument reconstructed in the earlier part of this section, the relation between a cause and its effect cannot be modeled on logical relations of entailment or conceptual containment. Logicism cannot account for real conflict because it cannot account for real *causal* connections. Conflict itself does not pose any *additional* problems for the logicist.

[35] What I have in mind is that a causal judgment of the form *G causes E* is true just in case the proposition *G+C occurs* entails *E occurs* where C is a condition specified by the context of utterance of the original judgment.

[36] He gives essentially the same argument in the Amphiboly section of the *CPR* (A264–5/B320–1) and in the *Progress* essay (Ak. 20: 282–3).

3.5. The Argument for Real Incompatibility

The discussion in this chapter so far has been at best obliquely related to modal issues. Kant's argument that causal grounds are real non-logical grounds spells trouble for the logicist theory of causation specifically, and grounding more generally, but does not *by itself* conflict with the logicist view that possibility and logical possibility are co-extensive. However, I will show that this is the crucial premise in Kant's argument that there are metaphysically incompatible predicates that are nonetheless logically compatible, which I have been calling *really incompatible* predicates.[37]

It is unclear how Kant argues that there is real incompatibility between some pairs of predicates. As we saw in the introduction to this chapter, he claims there is a real incompatibility between thought and extension, but provides no direct argument for this. However, I think a substantive Kantian argument for real incompatibility can be teased out of his discussion of the rationalist idea of God as the *ens realissimum*. In *Beweisgrund* he writes:

> The data of all possibility must be found in the necessary being either as determinations of it, or as consequences which are given through the necessary being as the ultimate real ground. It is thus apparent that all reality is, in one way or another, embraced by the ultimate real ground. [...] But this is not to be understood to mean that all possible reality is included among its determinations. (*OPG*, Ak. 2: 85)

This is a reference to the argument, which we will analyze extensively in the next two chapters, that God is the ground of real possibility. For now, this passage is important for its negative claim: that God, the most real of all possible beings, is not the rationalist *ens realissimum* that possesses all unlimited realities. Kant goes on to explain why:

> This is a conceptual confusion which has been unusually prevalent until now. All realities are attributed indiscriminately as predicates to God, or the necessary being. That all these predicates can by no means co-exist together as determinations in a single subject is not noticed. The impenetrability of bodies, extension and such like, cannot be attributes of that which has understanding and will. (*OPG*, Ak. 2: 85)

First of all, the *ens realissimum* is not possible because thought and extension, which Kant is here assuming are realities, are not metaphysically compatible. So far, though, he has given no argument that not all logical possibilities are metaphysically possible. He goes on:

> Nor does it help if one seeks to evade the issue by maintaining that the quality in question is not regarded as true reality. The thrust of a body or the force of cohesion are, without doubt, something truly positive. Similarly, in the sensations of the mind, pain is never merely a deprivation. (*OPG*, Ak. 2: 85–6)

[37] This is equivalent to what Chignell (2009*a*) calls 'subject-canceling real repugnance.'

This is a reference to the logicist view that all realities, all positive determinations, are logical complexes of limitations of the unlimited realities possessed by God. This view requires the logicists to hold that predicates like <*pain*> are deprivations, that is, being in pain is a matter of lacking some further reality possessed by God. In context, Kant's point is that there are pairs of really incompatible realities (e.g., thought and extension) and the logicists are wrong in assuming that, in such cases, one of the pairs is not a true reality (e.g., extension) but a complex of limitations of God's unlimited realities. But this does not constitute an *argument* against logicism.

Kant then introduces real conflict:

> A confusion has seemingly justified such an idea. It is said: reality and reality never contradict each other, for both of them are true affirmations; as a consequence, they do not conflict with each other in the subject either. Now, although I concede that there is no logical contradiction here, the real repugnancy is not thereby canceled. (*OPG*, Ak. 2: 86)

He goes on to define what real conflict is, and illustrates it through his preferred example of a body acted on by two equal but opposite forces. These are familiar points by now, but the interesting issue in this context is the implications of real conflict for the idea of God as the *ens realissimum*:

> Now, in the most real being of all there cannot be any real opposition or positive conflict among its own determinations, for the consequence would be a deprivation or a lack, and that would contradict its supreme reality. Since a conflict such as this would be bound to occur if all realities existed in the most real being as determinations, it follows that they cannot all exist in it as determinations. Consequently, since they are all given through it, they will either belong to its determinations or to its consequences. (*OPG*, Ak. 2: 86)

Now we see the beginning of an argument from the phenomenon of real conflict to the claim that not all logically compatible predicates are metaphysically compatible:

(1) For some realities A and B, there is real conflict between A and B.

This is something Kant takes himself to have already established; later we will see reasons to question it.

(2) If there were an *ens realissimum*, a being possessed of every unlimited reality, some of those realities would stand in real conflict. The effects they would normally ground will therefore be canceled. The effects they would normally ground are realities, so the cancellation of them involves a lack of reality.

Kant does not give specific examples of which realities would stand in real conflict, but based on the theory of real conflict in *Negative Magnitudes* it is not hard to determine which ones he has in mind. Although I focused on the real conflict of physical forces, Kant's other main example of real conflict is moral: the will to do evil and the will to do good are distinct positive determinations that stand in real conflict. If there were an *ens realissimum* it would possess both of these realities to the highest possible degree and would thus be in a state of permanent equilibrium, unable to act. This contradicts its nature as the most real possible being.[38] It follows that:

(3) The concept <*ens realissimum*> is not possibly instantiated.

But recall:

(4) The co-instantiation of A and B is a *real ground* of their effect, a lack of overall positive determination in the subject that instantiates them.
(5) The relation between a real ground and its effect is not a logical relation of entailment or concept-inclusion.

This is a consequence of Kant's theory of 'real grounds'—the relation between a cause and its effect cannot be modeled on the logical relations of inclusion (or entailment) between concepts (or propositions). But notice that:

(6) ∴ If the concept <*ens realissimum possessing both A and B*> were logically inconsistent then the real conflict between A and B and their effect, a lack of overall positive determination, would be a logical relation.

If the concept <*ens realissimum possessing both A and B*> were logically inconsistent then the logicist would have an explanation of the 'canceling' that occurs between A, B and the nature of the *ens realissimum*; if one and the same object possesses A and B, then as a matter of logic, this being is not the most real possible being (because <*ens realissimum possessing both A and B*> is logically inconsistent), so it lacks some positive determination. If this concept is logically inconsistent, then the relation between the co-instantiation of A and B and their effect, not being the *ens realissimum*, is a logical relation, contra (4). But since, by hypothesis, A and B are among the unlimited realities:

(7) <*ens realissimum*> = <*ens realissimum* with A and B>.
(8) ∴ The concept <*ens realissimum*> is logically consistent but not possibly instantiated.
(9) ∴ There are really incompatible predicates, those contained in the concept <*ens realissimum*>.

[38] Strictly speaking, this contradicts the nature of the *ens realissimum* (ER) only if we assume that (i) the ER is the most real being in the sense of having the most overall reality of any possible being, and (ii) that some possible being would have more overall reality in the envisaged situation in which two of the unlimited realities of the ER stood in real conflict. Premise (i) is plausible enough, though it is an additional assumption, because possessing every unlimited reality does not strictly *entail* possession of the highest overall degree of reality (as Kant notes in *Danz.RT*, Ak. 28: 1251). Premise (ii) is the more substantive assumption here; Kant's reasoning, I take it, is that a being that had A *and not* B (or B *and not* A) and all of the other unlimited realities—call it ER+—would have more overall reality than ER because, in the example given in the main text, it would have the unlimited will to do good and *that will would have its ordinary effects*. The ER, by contrast, would be in a perpetual state of equilibrium, unable to act. So, while I think Kant's argument is ultimately successful on this point, it does appear to rely on some premises he fails to make explicit.

The basic idea of this argument is that the fact the relation between the grounds and their effects (cancellation of realities) cannot be understood logically—the fact that this is *real* conflict—is used to show that there is a logically consistent concept (that of a being possessing every possible reality) that is not possibly instantiated.

However, this argument relies on Kant's assumption that there are primitive realities (the realities possessed by the *ens realissimum*) that stand in real conflict. Kant's own examples of such pairs of conflicting realities include: pleasure and pain, the will to do good and the will to do evil, etc. But the logicist will simply deny that one half of each of these pairs is a reality at all; the logicist will claim, for instance, that the will to do evil is a mere privation of the will to do good, and pain is a mere privation of pleasure. By relying on his own views about which properties are realities rather than mere privations (e.g., pain), Kant begs the question against the logicist.

However, a structurally identical argument can be made using less controversial examples of realities, like Kant's favorite example of real conflict, the real conflict between two equal but opposite forces acting upon the same body. While the logicist need not admit that forces are logically primitive realities (see section 4), it is not plausible for the logicist to claim that either of these forces is a mere *privation* of reality. Kant's argument, as I reconstruct it, requires only that they are realities, not that they are primitive:

(1) There is real conflict between A and B.

(2) The concept <*body acted upon by forces A and B and no other forces and moved in direction* a> is not possibly instantiated.

I am assuming for the sake of this argument that, in cases of real conflict, it is necessary that *if no other factors are present* the deprivation (lack of positive determination due to real conflict) results. This does not mean that God could not intervene to move the body; in that case, there would be an additional force acting, God's will, so it would not falsify the conditional.

(3) The co-presence of A and B is a *causal ground* of the lack of motion in direction *a* (and of lack of motion in direction *b*).

By definition, real conflict involves causation; premise (3) is just the application of Kant's general principle that causal grounds are real grounds.

(4) A causal ground is not a logical ground of its effects; a causal ground does not logically entail its effect.

This is just the definition of *real ground.*

(5) If the concept <*body acted upon by forces A and B and no other forces and moved in direction* a> is logically inconsistent, then the co-presence of A and B is a logical ground of their effect, lack of motion in direction *a*.

If this concept is inconsistent, then the proposition x *is acted on by forces A and B and no other forces* logically entails the proposition x *is not moved in direction* a; by parity of reasoning, it also entails x *is not moved in direction* b. If this concept were logically inconsistent (due, perhaps, to the highly complex logical structure of the concepts of A and B), then there would be a purely logical explanation of why the body does not move in direction *a*, contra (3) and (4).

(6) ∴ This concept is logically consistent.

This follows from (3)–(5) by *modus tollens.*

(7) ∴ There is at least one concept that is logically consistent but not possibly instantiated. [From (2) and (6).]

This directly contradicts the logicist view of possibility. While Kant never explicitly makes this argument, it is a very natural extension of the argument in *Beweisgrund* that the logicist *ens realissimum*, the being possessed of every fundamental reality, is impossible. In fact, that argument can be naturally extended to *any* example of real conflict to show that the concept of the being possessed of the conflicting realities *and* one of the consequences that is 'canceled' by the real conflict, is a case of a logically consistent but metaphysically impossibly instantiated concept.[39]

3.6. Real Possibility

In the previous chapter we saw that Kant argued for an important anti-logicist claim: there is at least one necessary proposition, that God exists, which is not logically necessary. However, Kant's argument that there are really incompatible predicates shows that there is a much wider class of counter-examples to the logicist equation of possibility and logical consistency. For any pair of really incompatible predicates P and Q, the proposition $\exists x(Px \& Qx)$ will be logically consistent but not possibly true; the existence of really incompatible predicates generates a wide class of counter-examples to the logicist thesis. This means that possibility diverges quite widely from logical possibility, contra the logicists.

A note about terminology. Although Kant does not use the Critical term 'real possibility' in the pre-Critical works to distinguish possibility from logical possibility, it is appropriate to use that Critical term.[40] In the works of the 1760s Kant argues for various non-logical *real* conditions on the possibility of a thing: a lack of *real*

[39] Both this argument, and the argument concerning the *ens realissimum*, are spellings-out of an argument also found in Abaci (2014), 9–10 (cf. the response by Chignell, who basically accepts the argument: 2014, 58). However, Abaci informs me (private correspondence) that the argument derives from my anonymous report on the original version of his paper. So the argument Abaci and Chignell discuss is originally my argument, spelled out in full detail for the first time here.

[40] As Abaci (2014) correctly notes. However, Kant does use the term '*reale Möglichkeit*' in *Refl.* 4196 (Ak. 17: 452), which Erich Adickes dates to 1769–70.

incompatibility among logically compatible predicates and, as I will explain in Chapter 4, an existing *real* ground of the possibility of its predicates. Consequently, I will borrow Kant's Critical term to describe the notion of possibility that, in the pre-Critical works, he distinguishes from logical possibility in terms of its real elements (real compatibility, real grounds).

4

Grounding Possibility

4.1. Introduction

In the previous chapters we critically examined Kant's reasons for rejecting several central logicist doctrines about modality. So far, though, our account of Kant's modal metaphysics has been largely negative: God exists necessarily, but *not* because his existence is grounded in his essence; *not* all logically compatible predicates are really compatible; and hence *not* all logical possibilities are really possible. In this chapter and the next I explore Kant's positive theory of modality in the pre-Critical period.

In this chapter I explore Kant's idea of a *real ground of possibility* and his conception of God as the unique real ground of *all* real possibility. In the next chapter I critically reconstruct Kant's argument that *there is* such a unique real ground of all real possibility. This order of investigation—trying to understand what the conclusion of the argument means before examining the argument itself—might strike some readers as odd. However, it is justified by the unclarity that surrounds Kant's notion of a real ground of real possibility, a notion which is never adequately explained in *Beweisgrund* itself. In fact, I will argue, not only are *all* the extant interpretations of this notion in the secondary literature explicitly rejected by Kant, he even asserts that we *cannot understand* how God grounds all real possibilities. So I have chosen to devote an initial chapter to discussing a notion that is both one of the most important, and most confusing, in Kant's pre-Critical modal metaphysics.

In section 2 I explore two conditions Kant imposes on real possibility: the formal condition that the predicates of a possible thing must be logically consistent, and the 'material' or 'real' requirement that the predicates of a possible thing must themselves be really possible. Predicates are really possible only if they have an existing ground. Consequently, if nothing were to exist, nothing would be (really) possible. I point out that Kant does not explicitly state what relation must obtain between a ground and the possibility it grounds, although he does set down certain criteria, which I explain. In this section I limit myself to what I take to be interpretively uncontroversial aspects of Kant's view in *Beweisgrund*, most importantly: real possibilities require *real* grounds; God is the *real* ground of all *real* possibilities; and there is a distinction between 'fundamental' predicates (instantiated by God himself) and 'derivative' predicates that are possible in virtue of being 'consequences' of God. In section 3 I explore some proposals in the secondary literature for how to understand the Kantian notion of a ground of possibility and I argue that each of them is explicitly rejected in *Beweisgrund*. In section 4 I point out that the mystery surrounding this notion in *Beweisgrund* is no accident: Kant himself claims that God's essence grounds all real possibilities but that we cannot understand *why* God's essence grounds real possibilities. I interpret this to mean: we cannot understand what the relation between God's fundamental predicates and the derivative predicates is, and, hence, we cannot understand why the obtaining of this relation makes those derivative predicates really possible. But this is not the end of the story; in the next chapter I will argue that Kant's argument that there is a unique ground of all real possibility implicitly relies on thinking of the grounding relation in *causal* terms and, consequently, thinking of God as grounding real possibilities through his *power*. We are bound to think of possibility as grounded in God's power even though, according to Kant, we know that this is not so; our cognitive constitution requires us to think of God in a certain way, even though this does not accurately reflect his reality *an sich*. The full resolution of this tension, I will argue in the final chapter, only comes in 1790 in the *Critique of Judgment* and Kant's theory of the intuitive intellect. In the text of the *Beweisgrund* Kant focuses mainly on the possibility of *predicates*. As I argued in Chapter 3, predicates in the pre-Critical period are 'amphibious' entities: they have both a worldly role (they include forces, and can be causally efficacious) and a semantic/representational role (they have an internal logical structure, they can be combined in judgments, and they have a representational content). In these next two chapters I will be mainly focused on their 'worldly' role, but in section 5 I will also consider their role as thought-contents. *Beweisgrund* consists of three Divisions (*Abteilungen*), which are themselves divided into a series of 'Reflections,' which consist in one or more numbered sections. For ease of reference I will refer to parts of *Beweisgrund* by the Division number, followed by the Reflection number, followed by the section number; e.g., II.3.ii will refer to the second Division, third Reflection, second section.

4.2. Two Requirements on Possibility

In the first section of I.2, "On a necessary distinction in the concept of possibility," Kant writes:

> Likewise, in every possibility that which is thought must be distinguished from the agreement of what is thought with the principle of contradiction. A triangle with a right angle is in itself possible. The triangle, as well as the right angle, are the *data* or the material [element] of this possibility, while the agreement of the one with the other according to the principle of contradiction is the formal [element] of possibility. I will also call the latter the logical [element] of possibility, for the comparison of the predicates with their subject according to

the rule of truth is nothing other than a logical relation; the something, or what stands in this agreement, will sometimes be called the real [element] of possibility. (*OPG*, Ak. 2: 77–8)[1]

A predicate (like <*triangular*>) has both a logical form (the logical relation holding among its constituents) and a matter (the more basic predicates standing in that logical relation). For a predicate to be possibly instantiated it must have a logically non-contradictory form. This establishes its logical possibility. For real possibility, it is required that any constituent predicate "is itself something and can be thought" (Ak. 2: 77). This is what Kant calls the 'real' or 'material' requirement on possibility. I take it to be the requirement that the predicates that compose a more complex predicate (the 'data' or 'matter' of possibility) must themselves be possibly instantiated.

It is important not to confuse this requirement with the principle, discussed in the previous chapter, that some logically compatible predicates are really incompatible. The material requirement on possibility is the requirement that the predicates that compose complex predicates must themselves be really possible; it is not a principle about the compatibility of these predicates. Kant is not *rejecting* the 'real compatibility' principle in the opening paragraphs of II.1; he is simply not committing himself to it. He mentions Crusius's view that there are non-logical relations of real incompatibility between predicates ("...Herr Crusius, who does not locate this conflict [of predicates] merely in internal contradiction...") without either explicitly accepting *or* rejecting it.[2]

Later he writes:

You know that a fiery body, a cunning man, and similar things, are possible. If I require nothing more than inner possibility you will not think it necessary that a body or a fire, etc. must exist as the data of their possibility, for they are at least thinkable, and that is enough. The agreement of the predicate *fiery* with the subject *body* according to the principle of contradiction lies in these concepts themselves, regardless of whether they are actual or merely possible things. I also agree that neither a body nor fire may be actual things, and yet a fiery body is nonetheless internally possible. But I venture to ask: in this case, is a body in itself possible? Because you are not basing your answer on experience, you will list for me the *data* of its possibility, namely extension, impenetrability, force, and who knows what more, and point out that there is no contradiction among them. I concede all of this, but you must justify to me why

[1] I follow the translation in Kant (1992*a*) by introducing the dummy noun 'element' to translate '*das Formale/Reale/Materiale/Logische*' (*der Möglichkeit*). I put it in brackets, though, to indicate that it does not correspond to any word in Kant's text.

[2] Ak. 2: 77; cf. *MH*, Ak. 28: 10. This is a reference to Crusius's doctrine of the 'highest material principle of thought,' which states that logically consistent predicates that, given the laws of our understanding, cannot be thought together are not possibly co-instantiated. For instance, <*exists*> and <*does not exist at any time or in any location*> are logically compatible but not possibly co-instantiated. See *Ent.* §§15, 237, 287; *De Usu* §§27, 29; and *Weg* §§250, 261. For critical discussion of Crusius's influence on Kant, see Wundt (1924), 52–81, Heimsoeth (1956), 125–88, and Schönfeld (2000), 225; more recently, this influence has been emphasized by Hogan (2009*a*) and (2005).

you have the right to assume the concept of extension as a *datum*; if it meant nothing, the possibility of bodies, which you have assumed, would be an illusion. (*OPG*, Ak. 2: 80)[3]

The possibility of (the instantiation of) some complex predicate like <*fiery body*> is not settled by the logical compatibility of its constituents. If <*fiery body*> is possibly instantiated then its constituents—the predicates of extension, impenetrability, and force, among others—are themselves possibly instantiated. And what does the possibility of these constituents amount to? He goes on:

Since you must eventually arrive at something whose possibility cannot be further analyzed, suppose that it is the concept of extension that you cannot further analyze into simpler *data* in order to show that there is nothing contradictory in it. The question then arises, whether 'space' or 'extension' are empty words or whether they signify [*bezeichnen*] something.[4] The lack of a contradiction does not settle it; an empty word never designates [*bedeutet*] something contradictory. [...] As long as you prove possibility through the principle of contradiction, you are depending on the thinkable that is given in things and you are considering only connections among it according to logical rules. But when you finally consider how this is given to you, you can base it on nothing other than an existing thing [*ein Dasein*]. (*OPG*, Ak. 2: 80–1)

This shows that the material requirement of possibility is not the requirement of real compatibility among the constituents of a predicate, because the material requirement applies also for atomic predicates, i.e. ones that have no constituents. Kant is not here concerned with compatibility among predicates,[5] but with the possibility of (the instantiation of) the atomic constituents of complex predicates themselves. Kant is not committed to assuming that the predicates he here discusses, e.g., extension, are actually logically atomic ("...suppose that you can now no longer break up the concept of extension into simpler data..."),[6] elsewhere, he expresses skepticism that we could ever discover atomic predicates.[7] He is treating them as logically atomic in order to abstract away from the issue of logical compatibility (and real compatibility, which he has not invoked in this passage) and isolate the issue of the possibility of the atomic predicates themselves. Henceforth, unless stated otherwise, 'predicates' refer exclusively to atomic predicates.

In the same paragraph Kant writes: "if space did not exist, or if space were not at least given as a consequence through something existent, the word 'space' would

[3] The italics on the predicates *fiery* and *body* are mine, not Kant's; he does italicize *data*, though.

[4] I have changed the structure of this sentence slightly to bring out Kant's point.

[5] See Ch. 3.

[6] Chignell (2012), 657 is therefore wrong in claiming that Kant "suggests" that extension is unanalyzable; Chignell attributes this suggestion to Kant, only to then accuse him of "lurching... into apparent self-contradiction" (658) a few pages later. As I argue below, though, the text Chignell cites, Ak. 2: 86, does not even carry the suggestion that extension is unanalyzable.

[7] *Prize*, Ak. 2: 280.

signify nothing at all [i.e. be impossible]." He is referring to his account of what makes predicates possible stated two pages earlier:

> Either the possible is only thinkable insofar as it itself is actual, and then possibility is given in what is actual as its determination, or it is possible because something else is actual, i.e. its internal possibility is given as a consequence [*Folge*] through another existing thing [*ein anderes Dasein*]. (Ak. 2: 79)

By "given in what is actual as its determination" I take Kant to mean: the predicate is instantiated by an existing substance. So, for any possible predicate, there is a substance that either instantiates that predicate *or* that predicate is possible in virtue of being a *consequence* of that substance. Kant does not tell us what kind of 'consequence' it must be, though; since 'consequence' [*Folge*] is Kant's generic term for a thing insofar as it has a ground,[8] I will take this to be equivalent to: the substance is the *ground* of the possibility of the predicate. He does make clear that the possibility of predicates for finite substances like <*extension*> does not depend upon their being instantiated:

> You know that a fiery body, a cunning man, and similar things, are possible. If I require nothing more than inner possibility you will not think it necessary that a body or a fire, etc. must exist as the data of their possibility, for they are at least thinkable, and that is enough. (*OPG*, Ak. 2: 80)

If the possibility of <*fiery*> or <*body*> depended upon the existence of actual fiery substances or bodies (substances instantiating these predicates), then if those substances and those bodies did not exist, those predicates would be impossible. Kant is assuming that the possibility of predicates cannot depend upon the existence of finite substances like these. So predicates of finite substances like these must be possible in virtue of being 'consequences' of some existing substance.

To make things slightly more precise, we can formulate Kant's material requirement on possibility as follows:

> (*Material*) If F is a possible atomic predicate then there is a substance x such that either (i) x instantiates F or (ii) the possibility of F is a consequence of x (x grounds the possibility of F). In either case the possibility of F is 'given' in x.[9]

Kant expresses the material requirement as a requirement on possibility *simpliciter*; as I observed in the previous chapter, Kant never uses the term 'real possibility' [*reale Möglichkeit*] in *Beweisgrund* (or anywhere else in the published pre-Critical writings).[10] The material requirement entails that if there is no existing substance that grounds or instantiates an atomic predicate it is logically possible this predicate is

[8] See *MH* (Ak. 28: 12), *MV* (Ak. 28: 355, 401), *MvS* (Ak. 28: 485), ML_2 (Ak. 28: 549, 625), and *MM* (Ak. 28: 806 f., 820). Connection (*nexus*) is the generic term for the relation of ground and consequence.

[9] *OPG*, Ak. 2: 79.

[10] It does, though, occur in *Refl.* 4196 (Ak. 17: 452), which Adickes dates to 1769–70.

instantiated (because it is atomic) but not possible *simpliciter*. This means there is at least a conceptual distinction between possibility *simpliciter* and logical possibility, and I have introduced the term (one that Kant will use quite frequently in later writings) 'real' possibility to record the fact that possibility *simpliciter* has 'real' or 'material' as well as logical/formal conditions. Henceforth, unless explicitly stated otherwise, 'possibility' in this and the next chapter means *real* possibility. Kant does not explicitly state the requirement that an existing ground of possibility must be a substance (rather than, e.g., a mode of a substance),[11] but this is a plausible assumption. If we follow Kant in defining a substance as something that (to use my terminology) *instantiates* predicates, and which is not instantiated by anything else, then it is plausible to assume that predicates do not ground possibilities 'on their own'; the substances that instantiate them ground those possibilities (perhaps by instantiating the predicates in question).[12]

In II.3 Kant argues from the material requirement to the conclusion that there is a unique substance in which all possible predicates are given, God. I will reconstruct Kant's complex argument for this claim in the next chapter; in the meantime, we need to understand what the conclusion of this argument means.

Kant argues that all possibilities are *given* in God: this means that all possible predicates are either instantiated by God *or* they are possible in virtue of being consequences of God.[13] Some care is needed in unpacking these notions. They invoke a distinction between two kinds of predicates, which I will call the 'fundamental' and the 'derivative' predicates, respectively. It is crucial to understand that God's fundamental predicates are not *made possible* by being instantiated by God. That these predicates are instantiated by God logically entails that they are possible (actuality entails possibility), but, in the terminology of the previous chapter, neither their instantiation by God, nor God himself, is an explanatory ground (what Kant calls an *antecedently determining ground*) of their possibility. There is no explanatory ground of God (see Chapter 1), and God's instantiation of these predicates does not *explain* why they are possible, though it does *entail* it. So when Kant says that God *grounds* all possibilities, he should be interpreted to mean: all possible predicates are *given* in God, but only some predicates have antecedently determining grounds of their possibility. Since Kant uses the 'grounding' terminology much more often than he uses the 'givenness' terminology, I cannot simply abandon it; sometimes I will use 'grounding' in the broader sense to mean 'giving' possibility (either by instantiating a predicate or by being an antecedently determining ground of its possibility) and sometimes in the narrower sense to refer specifically to being an antecedently determining ground of possibility.

[11] At Ak 2: 84 he argues that the ground of all possibility is not an *aggregate* of substances, but he does not argue that it is a substance at all. He seems to think this is obvious.

[12] Cf. *MH* (Ak. 28: 24), ML_2 (Ak. 28:562), *MD* (Ak. 28: 638–9), MK_2 (Ak. 28: 779), *Pöl.RT* (Ak. 28: 1041, 1104 f.), and *Danz.RT* (Ak. 28: 1308).

[13] *OPG*, Ak. 2: 89; see also 86 ("it follows from this ..."), 89, and 125. Cf. *MH* (Ak. 28: 134).

The fundamental predicates are instantiated by God. The derivative predicates are made possible by being 'consequences' of God (in some unspecified sense of 'consequence'). I will take this to mean: they are 'consequences' of God in virtue of his fundamental predicates and represent the 'consequence' relation as a relation between fundamental predicates and derivative predicates. Note that derivative predicates *can* in principle include predicates instantiated by God; what makes these predicates derivative is that they are made possible by being 'consequences' of his fundamental predicates, rather than fundamental predicates whose possibility has no antecedently determining ground.[14]

Furthermore, we need a generalized version of the relation of a substance to a possible predicate it gives (grounds in the broad sense) in order to reconstruct Kant's argument. I propose that for now we just call this 'derivation relation' R and define the 'givenness' of possibility as follows:

> *(Defn.)* Possible predicate F is given in substance *x* (substance *x* grounds the possibility of F in the broad sense) if and only if either (i) *x* instantiates F and there is no further antecedently determining ground of the possibility of F (F is a fundamental predicate of *x*) or (ii) F is possible in virtue of bearing relation R to the fundamental predicates of *x* (*x* grounds the possibility of F in the narrower sense).

A completely worked-out theory of the grounding of possibilities would tell us: (i) what the fundamental predicates are, (ii) what the derivative predicates are, (iii) what the derivation relation R is (allowing us to generate the complete space of derivative predicates from the fundamental ones), and (iv) an explanation of why the obtaining of the derivation relation between a fundamental predicate and a derivative predicate makes that derivative predicate possible. Our investigation into Kant's modal metaphysics will be guided by the attempt to fill in these gaps. Kant argues from the principle that for any possible predicate, there must be a ground of that predicate (a substance in which it is given), to the conclusion that there is a unique substance that grounds *all* possible predicates (in which they are given). In order to reconstruct this argument, then, we need to differentiate between being a ground of *some* possibility (henceforth, a *GSRP*) and being a ground of *all* possibility (henceforth, a *GARP*), even though the ultimate conclusion of Kant's argument will be that anything that grounds some possibility grounds all possibility (every *GSRP* is a *GARP*).[15]

[14] See *OPG*, Ak. 2: 125–6, where, as I read him, Kant claims that God's intellect, will, and power are derivative predicates of the fundamental predicates that make up his essence. Likewise, it is in principle possible for something other than God to possess a fundamental predicate, but that predicate will not be fundamental *to* that other substance, for that predicate is possible in virtue of being instantiated by God, not in virtue of being instantiated by that substance.

[15] *(Defn.)* A substance *x* is a ground of *some* real possibility (*GSRP*) if and only if there is a possible predicate F that is given in *x*.

(Defn.) A substance *x* is a ground of *all* real possibility (*GARP*) if and only if, for any possible predicate F, F is given in *x*.

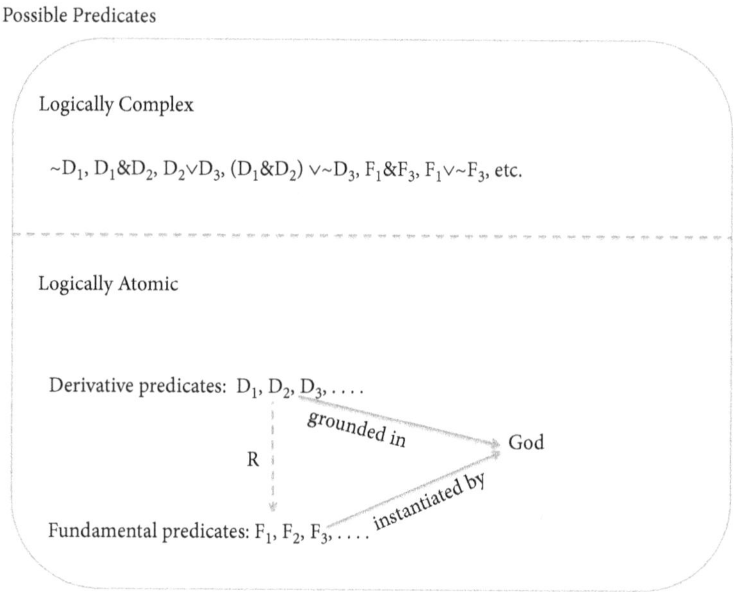

Fig. 4.1. The space of all possible predicates

In the previous chapter we examined Kant's pre-Critical distinction between logical grounds (whose relation or 'nexus' with their consequence is logically intelligible) and real grounds (whose relation to their consequence is not logically intelligible). In *Beweisgrund* Kant claims not only that real possibilities have grounds, but that they have *real* (i.e. non-logical) grounds.[16] He does not explain, though, precisely what kind of real grounds are required. As we will see in section 3, the mere fact that possibilities have real grounds rather than logical grounds will pose problems for the 'intellectualist' model of the grounding of possibility defended by some commentators. Figure 4.1 represents Kant's theory of the space of possible predicates.

The outer box represents all possible predicates; I defer discussion of impossible predicates until section 5. The upper half of the box contains the logically complex predicates; the lower half, logically atomic ones. For now I will only assume that atomic predicates have grounds, an assumption I will revisit in the next section. The fundamental atomic predicates are those that are actually instantiated by God *and* which lack antecedently determining grounds of their possibility. The derivative atomic predicates are those made possible by God (some of which may be instantiated by God). The dashed arrow represents what I have called the *derivation* relation between fundamental predicates and derivative predicates in virtue of which derivative predicates are possible. Since God grounds derivative predicates in virtue of possessing his fundamental predicates, we can think of the 'grounding' relation

[16] *OPG*, Ak. 2: 79, 83, 85, and 87.

between God and the derivative predicates as the 'logical product' of the derivation of derivative predicates from fundamental predicates and the instantiation of those fundamental predicates by God. However, we do not yet know exactly what relation the derivation relation is; determining that will be one of our main tasks.

4.3. Harmony, Power, and Intellect

Perhaps the best place to start in understanding Kant's model of how God grounds all real possibilities is with Andrew Chignell's influential 'real harmony' interpretation.[17] 'Real harmony' is Chignell's term for what I have called 'real compatibility': predicates are in 'real harmony' just in case they are not only logically compatible but really possibly co-instantiated. Chignell reads the 'material requirement' on real possibility in *Beweisgrund* as requiring not only that there must be existing grounds of really possible predicates, but there must be existing grounds of the real harmony among really compatible predicates.[18] Since the co-instantiation of two predicates F and G is just the instantiation of the logically complex conjunctive predicate F&G, the real harmony of predicates is equivalent to the real possibility of conjunctive predicates like F&G. So Chignell's interpretation (see Fig. 4.2) of the 'material requirement' on possibility as requiring grounds of real harmony between compatible predicates is equivalent to extending the range of predicates which need real grounds to arbitrary *conjunctions* of predicates.

It is not clear whether, on Chignell's view, other logically complex predicates, like disjunctive predicates or conditional predicates, also require real grounds, so I will leave them out of consideration.[19] Chignell then reasons that, just as the fundamental predicates must be *instantiated* to be really possible, the real harmony of these fundamental predicates can only be grounded in their co-instantiation: in other words, all of the fundamental predicates must be co-instantiated by God (equivalently, arbitrary conjunctions are instantiated by God).[20] In the next chapter I will criticize Chignell's reconstruction of Kant's *argument* that all real possibilities are grounded in a single being, God, but for now I merely want to investigate his interpretation of what this means.

As an interpretation of Kant, Chignell's model suffers from a lack of textual support, as argued extensively by Abaci (2014) and Yong (2014).[21] Chignell's harmony

[17] Chignell (2009*a*) and (2012); for critical reactions see Stang (2012), 290–1, Abaci (2014), Yong (2014), and Chignell's response, Chignell (2014*a*). Chignell uses the term 'exemplification' for what I am calling 'instantiation.'

[18] Chignell (2009*a*), 176.

[19] Though his invocation of the "modal PSR" Chignell (2009*a*), 176 would seem to require this.

[20] Chignell (2009*a*), 187; (2012), 648.

[21] Yong (2014), 2–33; Abaci (2014), 5–11. As both Yong and Abaci point out, the paragraphs leading up to Kant's assertion of the material requirement on possibility contain no endorsement of a 'real harmony' requirement (Ak. 2: 77–9); Kant does mention Crusius's view that there are logically but not really compatible predicates, but does not explicitly *endorse* it (Ak. 2: 77).

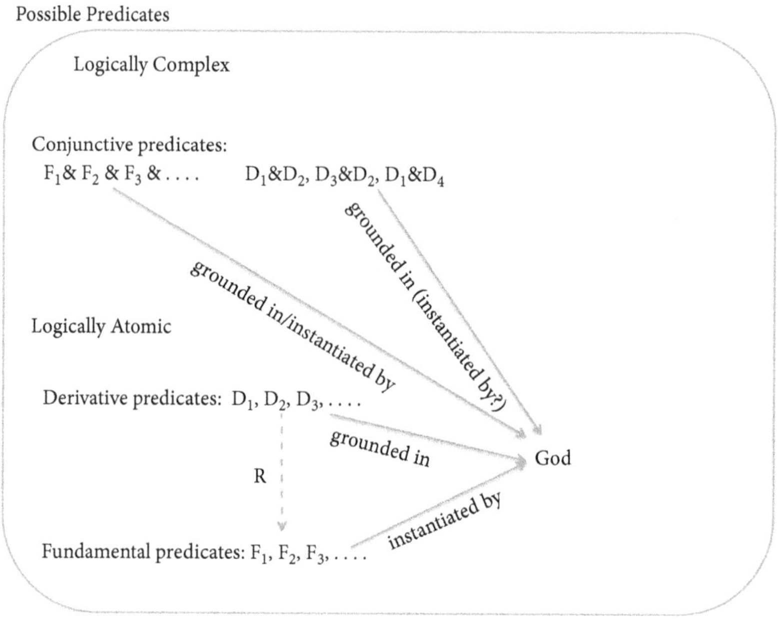

Fig. 4.2. Chignell's space of all possible predicates

requirement—that predicates must not only be logically consistent but really compatible, or lack 'subject-canceling real repugnance'—is obliquely referred to at two points in *Beweisgrund*: Kant's claim that thought and extension are logically, but not really, compatible, and (I argued in Chapter 3) his argument that the *ens realissimum* cannot instantiate every logically atomic predicate.[22] It is not mentioned in the passages in

[22] Both occur at Ak. 2: 85–6. Chignell also detects a reference to 'subject-canceling real repugnance' (real incompatibility) at 2: 89 and 190; Abaci and Yong convincingly argue that these passages concern 'predicate-canceling' real repugnance (Yong 2014, 31; Abaci 2014, 8–9). Chignell responds in (2014*a*), 55–60 by pointing to two texts: 2: 89 and 2: 190. At 2: 89 Kant argues that the being that grounds the world must have intelligence and will because beings in the world have those perfections; if it did not, it would be less perfect than its effects, and this, Kant claims, is impossible. Chignell is right, then, that <*produces a universe of order, beauty, and perfection*> and <*is a blindly necessary ground of other things*> are incompatible; but he has not given us a reason to think these predicates are logically compatible, so he has not shown that this is a case of *real* incompatibility (subject-canceling real repugnance). Chignell is on even weaker ground with 2: 190, where Kant claims that, "for something positive which exists to be cancelled, it is just as necessary that there should be a true real ground" and "in other words, only in so far as an equal but opposed real ground is combined with the ground of *a* is it possible for *a* to be cancelled." Kant regards the ground of the cancellation of *a* as "– *a*" and he makes clear earlier in *Negative Magnitudes* that "– *a*" refers *not* to the logical negation of *a* (~*a*) but to a reality that is equal in magnitude but 'opposite in direction' to *a* (e.g., a force in the opposite direction, pain as opposed to pleasure, etc.). Chignell's attempt to assimilate the 'conflict' between *a* and – *a* to the metaphysical conflict between predicates like *being water* and *being XYZ* (2009*a*, 173) founders on the fact that predicates like *being water* do not have 'opposites' (other than their logical contradictories, which is not relevant here); there is no relevant sense in which *being XYZ* is the 'opposite' of *being water* (any more than *being cyanide* is), so they cannot be values of *a* and – a in Kant's schematic description of conflict. As for the first quote ("for something positive which exists to be cancelled, it is just as necessary that there should be a true real ground"), he needs to read

which Kant states and argues for the material requirement on possibility, and is only evoked after Kant has argued that there is a unique ground of all possibility (Ak. 2: 84). Furthermore, Chignell provides no textual support for the claim that the compatibility among predicates (their harmony, in Chignell's sense) requires a *ground* (or that this ground is God, rather than the natures of the predicates in question).

Aside from its lack of textual support,[23] there is a deep philosophical problem with Chignell's model as well. As Yong (2014) points out, a problem arises when we apply the co-instantiation principle to the derivative predicates themselves: what grounds their 'real harmony'? For obvious reasons, logical consistency will not do. What about the real harmony of the fundamental predicates of which they are derivative?[24] This leaves us without an answer to the question of why the derivation relation between fundamental and derivative predicates cannot introduce sources of real incompatibility not present in the fundamental predicates. If D1 and D2 are derivative predicates of, respectively, fundamental predicates F1 and F2, and if F1 and F2 are really compatible, why does it follow that D1 and D2 are really compatible?[25]

To satisfy the 'real harmony' requirement via the co-instantiation principle, then, Chignell is forced to admit that derivative predicates are really harmonious (conjunctive derivative predicates are really possible) only if they are co-instantiated (if the conjunctive predicate is instantiated).[26] Since, as we have seen, the instantiation of a derivative predicate by a finite substance does not ground the possibility of that predicate, the substance whose co-instantiation of these really harmonious derivative predicates grounds their real harmony must be God himself. So God must instantiate not only all of the fundamental predicates but all of the co-harmonious pairs of derivative predicates. But this is absurd for two reasons. First, it means that if, for example, being liquid and being extended are really harmonious, God must instantiate both, and thus be extended. Secondly, it leads to inconsistency: there are pairs of harmonious derivative predicates that are not mutually harmonious. For example,

this not just as extending to the existent but to the possible as well: for something positive which is possible to be canceled, it is just as necessary that there should be a true real ground. But this is not Kant's view: the cancellation of the possibility of some positive determination (e.g., *extension*) can result from the mere *absence* of its ground (e.g., God). I agree with Chignell that subject-canceling real repugnance (real incompatibility, in my terms) is a part of Kant's pre-Critical modal theory (as I argued in Ch. 3); I disagree that it plays any crucial role in Kant's theistic argument in *Beweisgrund*.

[23] Chignell also cites Kant's remark in *ND* that "nothing can be conceived as possible unless whatever is real in every possible concept exists and indeed exists necessarily" (Ak. 1: 395; Chignell 2012, 177), but in *ND* Kant explicitly rejects real non-logical incompatibility: "the principle of contradiction... is in fact nothing but the definition of the impossible" (Ak. 1: 391). If the principle of contradiction is the definition of the impossible then *a fortiori*, if no contradiction arises from the co-instantiation of F and G, it is *possible* that F and G be co-instantiated.

[24] Or the real harmony of fundamental with derivative predicates.

[25] Yong (2014), 33–4.

[26] As Chignell himself admits (2012), 665–7.

being liquid and being extended are harmonious, and thinking and being attentive are harmonious, so it would follow that God has the following predicates: *liquid*, *extended*, *thinking*, and *attentive*. But *extended* and *thinking*, we already know, are not really compatible, so there would be what Chignell calls a 'subject-canceling' real conflict in God.[27]

Nor is Chignell correct in assimilating the distinction between fundamental and derivative predicates to the distinction between logically atomic and logically complex predicates.[28] On Chignell's reading, the derivative predicates are logical 'constructions' from the fundamental predicates, using the standard eighteenth-century rationalist operations of negation, conjunction, and limitation. This entails that the predicates of finite beings are negations, conjunctions, and limitations of the atomic (hence unlimited and positive) predicates of God. This means that if there is a non-logical relation of *real* incompatibility among any of the fundamental (Chignell: logically atomic) predicates then there is a 'subject-canceling' real conflict in God, and God is impossible.[29]

In fairness to Chignell, he notes all of these problems and takes them to be Kant's reasons, in the Critical period, for rejecting this modal argument as an objectively valid proof of God's existence. But Kant himself notes these problems, both in *Beweisgrund* itself and in contemporary texts, and offers them as reasons to *reject* the view on which God instantiates all logically atomic predicates *for the very reasons Chignell gives*: there would be a real incompatibility among God's predicates.[30]

In a passage I analyzed in some detail in Chapter 3, Kant explicitly rebukes the logicists for a "conceptual confusion":

[27] Yong (2014), 34. Chignell (2012), 668 suggests that Kant might be able to avoid this conclusion by accepting a Spinozistic view of the divine attributes: mutually incompatible predicates are instantiated by distinct divine attributes. But the very same objection applies as before. If <*liquid*> and <*water*> are possibly co-instantiated and <*liquid*> and <*xyz*> are possibly co-instantiated then the same attribute instantiates <*water*> and <*xyz*>, which by hypothesis is impossible. Yong (2014), 35 points out that if these attributes are instantiated by God, then incompatible predicates instantiated by distinct attributes lead to incompatible predicates instantiated by God. However, if the relation between attributes and God is not instantiation then the co-instantiation thesis, which, according to Chignell, drives the argument, is being jettisoned.

[28] Chignell (2009*a*), 166 n. 19, 182.

[29] Chignell's conflation of atomic (logically fundamental) predicates with (metaphysically) fundamental predicates leads him to claim (2012, 657–8) that Kant's claim at Ak. 2: 86 that God cannot possess the predicate of extension because he has a will, contradicts his "suggestion" (six pages earlier) that extension is unanalyzable. Kant never claims extension is unanalyzable; Kant's consistent pre-Critical view is that to be extended is to be composed of substances in interaction. And his claim that <*extended*> does not characterize God only entails that <*extended*> is analyzable if we are assuming that God instantiates all unanalyzable (atomic) predicates. But that is precisely what Kant is denying in this passage.

[30] Some care is required here. In the passage I quote below Kant asserts a real incompatibility between extension and thought, something the logicists deny. But his argument, as I reconstructed it in Ch. 3, does not beg *this* question: he argues there would be a real conflict (predicate-canceling real conflict) among God's predicates, and, because of this, a real incompatibility (subject-canceling real conflict). However, as I pointed out, even the claim that there is real conflict among divine predicates is potentially question-begging.

> But this should not be understood to mean that all possible realities belong among its [the necessary being's] determinations. This is a confusion of concepts that until now has been unusually prevalent. One attributes all realities without distinction to God as predicates, without realizing that they can never occur in the same subject as determinations. The extension, impenetrability, etc. of bodies cannot be predicates of that which has understanding and will. Nor can one evade this issue by claiming that the predicates in question are not genuine realities. (*OPG*, Ak. 2: 85)[31]

Kant's point in this passage is that God cannot possess every reality (every 'positive' predicate) because some of them are metaphysically incompatible (e.g., extension and understanding). But, as Kant well knew, no one in the rationalist tradition claimed that God possesses every reality *period*, much less that every reality *period* is compatible; Leibniz, Wolff, and Baumgarten claim that God possesses every logically fundamental reality, every unlimited reality, of which the realities of finite beings are complex conjunctions, negations, and limitations. So Kant's point here would be dialectically *irrelevant* unless he means: God does not possess every *logically* fundamental (atomic) reality. Hence his claim, "nor does it help if one seeks to evade the issue by maintaining that the quality in question is not regarded as true reality." He is envisaging a logicist response: your putative examples of incompatible realities, Herr Kant, are beside the point because they are not logically fundamental (atomic) realities; they are complexes of negations and limitations of atomic realities. Kant's point is that there would be a real incompatibility among God's predicates if he possessed every logically fundamental (atomic) reality. If his point is merely that there is a conflict among logically *derivative* realities, then he is diagnosing a "conceptual confusion" that afflicted precisely no one in the German rationalist tradition (as he would have known).

Kant then gives a series of examples of positive realities that stand in real conflict and concludes:

> Since, if all realities were present in [the necessary being] as determinations, a conflict of this sort would arise, they cannot all be in it as predicates. Consequently, because they are all possible through [the necessary being], they belong either among its determinations or among its consequences. (Ak. 2: 86)

Chignell is correct that Kant rejected this view of God as instantiating every logically atomic reality, and he is correct about why: such a God would have a real conflict among his predicates. But he is wrong about *when*: Kant had already rejected the view of God Chignell attributes to him in 1763, and so the developmental story Chignell tells—Kant realizes that we cannot prove a lack of incompatible predicates in God

[31] In the Herder metaphysics lectures from the mid-1760s, the period of the composition of *Beweisgrund*, Kant explicitly and repeatedly denies that God instantiates all realities: Ak. 28: 128, 132–3, and 150 (see also MK_2, Ak. 28: 781–2, and 28: 917).

and thus gives up on the modal theistic argument of *Beweisgrund*—cannot be upheld.[32]

Chignell's basic interpretive misstep was to ignore the fact that God is a *real* ground of possibility and Kant's pre-Critical view of *real* grounds: grounds whose relation to their consequence cannot be understood logically. If the derivative predicates are conjunctions, negations, and limitations of God's fundamental predicates then we can understand logically how the former make the latter possible.[33] Kant's primary example of real non-logical grounds, as we saw in the previous chapter, are *causal* grounds. This suggests a possible interpretation of the distinction between fundamental and derivative predicates: derivative predicates are *causally* derivative of God's fundamental predicates and, in virtue of this causal relation, they are really possible.

On this interpretation (which I will ultimately reject), the fundamental predicates are God's fundamental causal predicates; the derivative predicates are the predicates he can cause to be instantiated, in virtue of possessing these fundamental predicates. This definition actually fits the text of *Beweisgrund* quite naturally; Kant argues that *intellect* and *will* must be among the fundamental predicates rather than the derivative predicates (i.e. each must be possessed by God rather than merely a 'consequence' of God), writing:

> Leaving aside all the causes which are responsible for the generation of plants and trees, everyone knows that regular flowerbeds, avenues, and such like, are only possible as a result of an understanding which conceives the plan and a will which executes it. In the absence of understanding, no power [*Macht*] or generative force [*Hervorbringungskraft*], nor any other data of possibility, are adequate to render the possibility of such order complete. (*OPG*, Ak. 2: 88)

On one very natural reading of this passage, Kant is making a familiar theistic claim: nothing that lacked an intellect and will could create a world with organized living beings. This lends credence to the *causal* interpretation of the fundamental/derivative distinction.

On this reading, God's fundamental predicates ground the possibility of the derivative predicates they *causally* bring about (cause to be instantiated). Recall, though, that Kant denies that the non-instantiation of a derivative predicate renders it impossible; if a derivative predicate like <*extended*> were never instantiated, it would remain possible. So the derivative predicates grounded in God's causally fundamental predicates are not the predicates God *actually* causes to be instantiated; if they were, any non-instantiated derivative predicate would be impossible. The derivative predicates grounded by God's causally fundamental predicates are the predicates God *could* cause to be instantiated. The really possible derivative predicates are the ones God has the *power* to cause to be instantiated.

[32] Chignell (2009*a*), 188–92.

[33] *OPG*, Ak. 2: 79, 83, 85, and 87.

There are passages in *Beweisgrund* that strongly suggest this 'powers' interpretation of the distinction between fundamental and derivative predicates, for instance: "not only the manner of their combination, but the things themselves are only possible through this being, that is, they can exist only as its effects [*Wirkungen*]" (*OPG*, Ak. 2: 125). Furthermore, Crusius, who was an important influence on Kant in this period, distinguishes logical from real possibility, and grounds real possibility in God's power to produce a given state of affairs: "the ideal [logical] possibility of a non-existent thing would mean nothing, if it did not at least contain in itself enough reality that God, at least, would be a sufficient cause of everything in the thing [...] if he were to make use of his omnipotence [*Allmacht*]."[34] The Crusian 'powers' view of possibility thus bears a certain similarity to Kant's view: we can distinguish between the fundamental predicates (God's fundamental powers) and the derivative predicates (the powers and predicates that are made possible by God's powers) they make possible.

It is crucial to distinguish this view from the voluntarist view on which God's actual choices ground what is possible and what is not.[35] That voluntarist view, famously endorsed by Descartes, grounds possibility in God's *will*; the 'powers' view grounds possibility in God's *power*. The distinction can be thought of in the following way: God has the power to actualize various worlds, but he wills only to actualize one of them, this world. God's will chooses among the space of possibilities, it does not generate or ground that space. On the powers interpretation, the space of real possibilities is constituted by what God has the *power* to choose, not by what he *does* choose. Commentators who have argued against the 'powers' view by assimilating it to the voluntarist view have conflated an important distinction.[36]

Kant clearly rejects the Cartesian voluntarist view: "the will makes nothing possible, but rather chooses things whose possibility is already presupposed" (Ak. 2: 100).[37] What is more, in a passage I will analyze in some detail in the next section, he also rejects the *powers* view:

> But granted that the ground, which underlies not only the essence of all things but also the essence of wisdom [*Weisheit*], goodness [*Gute*], and power [*Macht*], is a unity, it follows that all possibility must of necessity harmonize with these predicates. (*OPG*, Ak. 2: 125)

This is an exceptionally rich passage, and fully analyzing it will take some significant work in the next section, but for now I just want to point out that Kant is here

[34] *Ent.* §56.

[35] Descartes' voluntarist views about modality can be found in his correspondence (CSMK III: 22–6, 235, 358–9) and the Sixth Objections and Replies (CSMK II: 281, 291–2); for critical discussion, see Curley (1984), Van Cleve (1994), Kaufman (2002), and Easton (2009).

[36] E.g., Chignell (2009*a*), n. 41. In Stang (2010), 296–7 I explicitly distinguish them. Chignell (2012), 671 n. 52 describes that paper as attributing to Kant a "quasi-voluntarist" view of modality. I no longer hold the 'powers' interpretation, but I do think it should be strictly distinguished from the voluntarist view.

[37] Cf. the rest of that paragraph and the next, as well as Ak. 2: 91 and 28: 134.

claiming that the unitary ground of the essence (real possibility) of all things, which is in God (I take this to refer to the totality of God's fundamental predicates), also grounds his own omniscience (*Weisheit*), his omni-benevolence (*Gute*), and his omnipotence (*Macht*).[38] *A fortiori*, none of them is the ground of possibility; none of them is what makes everything else possible. In particular, God's unlimited *power* is not the ground of all real possibilities. The reason the 'powers' interpretation is wrong is not, as some commentators have claimed, that it is equivalent to voluntarism (which it clearly is not), or that it is incoherent (which is more debatable); it is simply not *Kant's* view.[39]

Finally, some readers of *Beweisgrund* have attributed to Kant an 'intellectualist' model of the grounding of possibility, on which God grounds all real possibilities by thinking of them, and all of the relations among them, as a totality.[40] However, the intellectualist reading is also incorrect, for two reasons. First of all, the intellectualist model falls prey to the same textual problem that the powers model and the voluntarist model do: Kant explicitly rejects it. Consider this passage, which immediately precedes the passage quoted above:

> If one were to ask 'how do these natures depend on this being, and in such a way that I can understand their agreement with the rules of wisdom?' I would answer 'they depend on something in this being, which, in virtue of containing the ground of the possibility of things, is also the ground of the being's own wisdom; for its wisdom presupposes the possibility of things.' (Ak. 2: 125–6)[41]

I will analyze both passages in detail in the next section, but for now a few remarks will suffice. In this passage Kant claims that real possibilities depend upon something in God upon which his intellect also depends; it follows that his intellect is not the ground of all real possibility.[42] Kant reiterates this point in a footnote:

[38] One might think that Kant is claiming that God's unitary essence makes the wisdom, goodness, and power of *other* things possible, but in the previous sentence he makes clear he is talking about God's *own* wisdom: "auch der Grund seiner eigenen Weisheit" (2: 125 f.). By parity of reasoning, I think Kant is also talking about God's *own* goodness and power.

[39] I defended the 'powers' interpretation in Stang (2010), which has received sharp criticism from Newlands (2014), 175 n. 52, and Chignell (2009*a*) and (2012). Note, though, that neither Chignell nor Newlands present what I take to be the 'proof texts' against the 'powers' interpretation: Ak. 2: 125–6 and 152–3 (see section 4). My present attitude towards that interpretation is complex: I think it is not adequate to Kant's text (as I argue in this chapter), but I will argue in the next chapter that Kant's argument is best reconstructed assuming the powers interpretation. On Kant's view, we are simultaneously forbidden and required to think of God as *causally* grounding all possibilities. This is what I refer to as the tension in Kant's pre-Critical modal thought. I discuss it further in the next chapter.

[40] E.g., Watkins and Fisher (1998), Insole (2011), and Yong (2014). There are passages in Leibniz, Wolff, and Baumgarten that support such an intellectualist view, on which God grounds all possibilities by representing them in his intellect; see *Mon.* §§43–44, *Th.* §184, *Dt.Met.* §974, and *Meta.* §864. See Newlands (2013) for critical discussion.

[41] Thus, Chignell is wrong to claim that Kant is "more or less silent about why" the intellectualist model does not work (Chignell (2009*a*), 182).

[42] Cf. Ak. 1: 119: "unity is thus derived from a wise being, but not through his wisdom."

> Wisdom presupposes that agreement and unity in the relations [among possible things] are themselves possible. A being which is of a completely independent nature can only be wise in virtue of containing in itself the ground of the harmony and perfection it itself has the option of realizing. If there were no such relation to order and perfection to be found in the possibility of things, wisdom would be a chimera. If possibility were not grounded in the wise being itself, its wisdom would not be independent in every respect. (*OPG*, Ak. 2: 126 n.)

There is a lot going on in this footnote, but Kant's primary point is that possibility must be grounded in God himself, otherwise his omniscient intellectual awareness of all possibilities would depend upon something external to him, the hypothesized external ground of possibility; this would make God's intellect dependent upon an external source, which Kant assumes is impossible. But the first sentence of this note—"Wisdom presupposes that agreement and unity in the relations are themselves possible"—also entails that God's intellect, his '*Weisheit*', is not the ground of possibility itself.[43] Instead, Kant here claims, the agreement and unity among possibles must be independently grounded in God, and God's intellectual apprehension of that agreement and unity among possibilities is thus a form of self-cognition. God's intellect is not itself the ground of possibility; the unity in his essence is the ground of all possibilities, and this itself makes his intellect possible; it is because God's essence grounds all possibilities that God is capable of having omniscient intellectual awareness of all possibilities, through self-cognition of his essence.

But the problems with the intellectualist model do not end there; very deep commitments of Kant's pre-Critical metaphysics make the intellectualist model unavailable to him. While it is clear that, according to Kant in the *Beweisgrund*, God is a not merely an epistemic ground of real possibility (something through which real possibilities can be known), it is less commonly appreciated that, given Kant's pre-Critical commitments in rational cosmology, this precludes the intellectualist model of how God grounds possibility.

A major controversy of eighteenth-century German metaphysics concerns the relations substances must bear to one another to be members of a common world (or 'world-mates' in contemporary lingo), a unified totality of finite substances.[44] Leibniz (and Leibnizians like Baumgarten) held that pre-established harmony among the representational states of substances, which they conceived of as mind-like simples (monads), makes substances into world-mates: the world is a totality of substances whose representational states harmonize with one another.[45] From his earliest writings on, Kant defends the anti-Leibnizian position that world-mates must really interact with one another; substances x and y are world-mates if and only if the

[43] Noam Hoffer has suggested to me in conversation that '*Weisheit*' here does not refer to God's intellect but to his intellectual representation of the harmony and 'purposive unity' (as Kant would later call it in *CJ*) among possibilities; I do not have space to respond to this intriguing suggestion here, though. Cf. Hoffer (forthcoming).

[44] See Watkins (2005), 21–100 for a detailed discussion of this issue.

[45] AG 33, 42, 46–7, 143–4, *Meta.* §§448–65; for discussion, see Watkins (2005), 24–8, 74–8.

forces in x are causal grounds of accidents in y and vice versa.[46] Kant rejects the pre-established harmony theory (PEH) because he believes that a world is a *real* unity, i.e. the parts of a world are antecedently determining grounds (not merely epistemic grounds) of one another's alterations. The representational or intellectual harmony that obtains among the states of substances (Leibnizian monads) on the PEH theory by itself does not make them antecedently determining grounds of one another's alterations. The complete representational state of such a substance provides an appropriately informed agent (e.g., God) with complete *epistemic* grounds for predicting the subsequent states of all other substances in the same world; however, because they do not interact, the states of these substances are not antecedently determining grounds of one another.[47]

By parity of reasoning, though, Kant is committed to rejecting the intellectualist model of how God grounds possible predicates because it would entail that God is not an antecedently determining (explanatory) ground of their possibility. If God's intellect grounds real possibilities *solely* by representing them, then God's mind could be an *epistemic* ground of beliefs about real possibilities—*per impossibile*, were we able to 'see into' God's mind we would know what is and is not really possible—but his mind does not explain *why* the represented possibilities are really possible. If God's intellectual representation of possibilities is an antecedently determining ground of their possibility, then, by parity of reasoning, Kant would have to admit that the representational unity among Leibnizian monads could make them into antecedently determining grounds of one another's states, and this would undermine one of his core objections to pre-established harmony. Kant does not merely happen not to endorse the intellectualist view of the ground of real possibility; one of his deepest commitments in metaphysics (rejection of PEH as an explanation of the world-mate relation) is incompatible with it.[48]

4.4. "This thought rises far higher than a created being can reach"

Having argued against all extant interpretations of the notion of a real ground of real possibility, including my own previous 'powers' interpretation, I want to pause and reassess my conclusions up to this point. Kant builds an entire metaphysics in

[46] *MH*, Ak. 28: 39, 45, and 51–2.

[47] Although Kant thinks that world-mates are real grounds of one another's accidents, his argument against the PEH theory of the unity of a world does not depend on this assumption.

[48] Some 'intellectualist' readers might object that Kant grounds the possibility of interacting substances in God's intellect in *ND* (Ak. 1: 413, 414). But Kant there says that the divine intellect is the source of the *existence* of substances ("the self-same schema of the divine understanding, which gives existence") and of their dynamic relations, so he is not, at least there, claiming that the divine understanding is what makes substances really possible.

Beweisgrund on the twin principles that (i) God is the real ground of all possible predicates, and (ii) possible predicates are either (a) instantiated by God (the fundamental predicates), in which case their possibility has no antecedently determining ground, or (b) consequences of God (the derivative predicates), in which case they are possible because of the grounding relation they bear to God (and his fundamental predicates). But Kant never tells us what exact relation obtains between the derivative predicates and the fundamental predicates (or God himself) and why the obtaining of this relation makes the derivative predicates really possible. This has spawned an entire secondary literature in which commentators have tried to answer this crucial interpretive question.

What none of *Beweisgrund*'s commentators, myself included, have noticed is that it is no accident that Kant fails to identify what this feature of God is; he explicitly claims that we *cannot* say anything more than that it is essential to God to do so. In a convoluted passage that I began to analyze in the previous section, he rejects several of the models that have been attributed to him. He begins by raising a question: "how do these natures depend on this being, and in such a way that I can understand their agreement with the rules of wisdom?" (Ak. 2: 125). The context is Kant's argument that the unity and harmony that we observe in the essences of things (their real possibility) points to there being a single ground of all possibility. Kant now asks, on behalf of his reader, what relation is there between these essences and that unique ground in virtue of which those essences are possible (essences of possible things)? We might reasonably expect, then, that Kant will tell us (in my terminology) what the derivation relation is and why it makes derivative predicates (and the essence they compose) possible. But he does not do so:

> I would answer, 'they depend on something in this being, which, in virtue of containing the ground of the possibility of things, is also the ground of the being's own wisdom; for its wisdom presupposes the possibility of things.' (*OPG*, Ak. 2: 125)

I interpret this to mean that God has an essence (a set of fundamental predicates) that grounds all other possibilities (all other possible predicates) and that this *essence* is itself the ground of the possibility of his wisdom (his intellect). So God's intellect cannot itself be the ground of possibility. He goes on: "but granted that the ground, which underlies not only the essence of all things but also the essence of wisdom, goodness, and power, is a unity, it follows that all possibility must of necessity harmonize with these predicates" (*OPG*, Ak. 2: 126). So the divine essence, the ground of the possibility of all derivative predicates, is also the ground of God's goodness (his omni-benevolent will) *and* his power. Since these predicates of God are made possible by his essence, they are not themselves the ground of possibility. Kant is here explicitly repudiating the voluntarist, the intellectualist, and the 'powers' interpretation of how God grounds possibility; these divine attributes are themselves grounded in some more fundamental divine essence.

Not only does Kant explicitly repudiate all of the models for how God grounds possibility that have been attributed to him in the secondary literature, he elsewhere claims that we cannot *in principle* understand how God grounds possibility:

> Nor is my wonder at the unity and harmony in things diminished once I have convinced myself that this is only possible because there is a being which is not only the ground of actuality but also of possibility. Though one can form, through an analogy with human action, some concept of how a being can be the cause of everything actual, one can form no concept of how such a being could contain the ground of the inner possibility of other things. It appears that this thought rises far higher than a created being can reach. (*OPG*, Ak. 2: 152–3)

It is important to be clear on what question Kant is here denying we can ever answer. God grounds possible predicates in two ways: (i) either as determinations (fundamental predicates), or (ii) as consequences (derivative predicates). Fundamental predicates have no 'antecedently determining' ground of their possibility; properly speaking, there is no ground *why* they are possible, they just *are* instantiated. Derivative predicates are possible because of the relation they stand in to God's fundamental predicates. I take it that Kant is not merely denying that we can understand why the obtaining of that derivation relation makes those derivative predicates possible. He is also denying that we can ever understand what that derivation relation is, for we can "form no concept of how that being can contain the ground of the internal possibility of other things." The way in which possibilities are grounded in God is literally incomprehensible to us.

This is a dissatisfying result; it is supposed to be. However, it is the beginning of a story that will end, I hope, in a less dissatisfying result. In the next chapter I reconstruct Kant's argument that there is a unique ground of all real possibility. One of the conclusions of my reconstruction will be that Kant's argument succeeds, but only if (i) we grant him certain premises about rational cosmology, and (ii) we implicitly think of the grounding of real possibility in *causal* terms. At the end of the next chapter I will argue that if we look back at the argument we have reconstructed we will see that we—and Kant himself!—have been implicitly thinking of God as grounding possibility through his powers, *precisely* what Kant denies is the case at Ak: 2: 153 (quoted above). So in *Beweisgrund* Kant argues that there is a unique ground of real possibility, by implicitly thinking of this ground as a *causal* ground of real possibility, but then denies that we can understand its role in grounding possibilities in causal terms, or any other terms for that matter. This tension in Kant's thinking about the ground of possibility will not be resolved, I will argue, until 1790, in the *Critique of Judgment*. Explaining how Kant resolves it, though, will require the rest of this book.

4.5. Possibility, Thought, and Content

Before continuing, in the next chapter, to discuss Kant's *argument* that there is a unique ground of all possibility, I want to briefly comment on Kant's apparent equation of the possible with the 'thinkable' in *Beweisgrund*.[49] Although Kant frequently uses these expressions in such a way as to suggest that he thinks of them as *identical*, he does not claim that everything possible is thinkable by our minds, and, as we have seen, even God's intellect does not *ground* possibility: rather, it is because God's essence determines a certain space of possibilities that God's intellect is intentionally directed at those possibilities. Even for God, thinkability presupposes, and is grounded in, possibility.[50]

What is the relation between a predicate being *thinkable* and it being *possible*? A thinkable predicate, I take it, is one that can figure in the content of a contentful thought. It would be highly problematic for Kant to assume that all thinkable predicates are possible, if, by the 'possibility' of a predicate we mean the possibility of its instantiation,[51] for this would entail that every thinkable predicate is possibly instantiated. For the sake of clarity:

(1) Predicate F is thinkable ⊃ predicate F is possible.
(2) Predicate F is possible ⊃ possibly, predicate F is instantiated.
(3) ∴ Predicate F is thinkable ⊃ possibly, predicate F is instantiated.

The conclusion, (3), is implausible, for at least two reasons. First of all, I can have prima facie contentful thoughts, e.g., *there is a largest prime number*, involving predicates, in this case, <*largest prime number*>, that are not possibly instantiated. Secondly, if thoughts involving impossibly instantiated predicates lack content, then any two such thoughts have the same content, namely, nothing, and that is highly implausible. For instance, *there is a highest prime number* and *there is a counter-example to Fermat's Last Theorem* are thoughts involving impossibly instantiated predicates, but clearly they have distinct contents. After all, they can be disproven (it can be proved that their content does not accurately represent their subject matter), and they stand in different inferential relations with uncontroversially contentful thoughts.

We might be tempted to reject (2), but we should be cautious: Kant must hold on to some connection between the possibility of a predicate and its possible instantiation. If the possibility of a predicate has nothing to do with its possible instantiation, then in proving that God is the ground of the possibility of all predicates he would not thereby have shown anything about their possible instantiation; it would be compatible with the conclusion of Kant's argument that *no* such predicates are possibly instantiated, which is surely absurd. Fortunately, there is a far less drastic

[49] See *OPG*, Ak. 2: 77, 78, 80, 82, 83, 85, 87, and 162.

[50] I was prompted to think again about the relation between thinkability and possibility in *Beweisgrund* by the fascinating discussion of these issues in the final chapter of Yong (forthcoming), which connects them to the opening of Hegel's *Science of Logic*.

[51] This should not be read to mean: possibly, there is a predicate F. For, as I will shortly argue, there are *impossible* predicates: predicates that combine atomic predicates in ways that are necessarily uninstantiated.

way to avoid the implausible conclusion, for we can understand the relation between what is thinkable and what is possible by amending (1) as follows:

> (1*) F is a thinkable predicate ⊃ (i) F is atomic and possibly instantiated or (ii) F is complex, the atomic predicates of which F is composed are possibly instantiated, and it is logically possible that F is instantiated.

Note that (1*) does not include the requirement that the atomic constituents of F be *really* compatible. This delivers what I take to be the correct result that we can have contentful thoughts about impossible co-instantiations of (individually possibly instantiated) atomic predicates. Assuming for the moment that <*extended*> and <*thinking*> are atomic predicates, the predicate <*extended & thinking*> is a thinkable predicate, a predicate of which we can have a contentful thought; after all, the early modern materialist (e.g., Hobbes) who thinks that some bodies think may be wrong, indeed necessarily wrong, but his thought is not devoid of content.

What is more, there is good textual ground for attributing this view to Kant; he consistently describes God as providing the *data* for thought.[52] If we take 'data' in its original meaning as 'that which is given,'[53] we can plausibly read him as claiming that God is the ground of the material of thought, the possible atomic predicates; the thinkability of some complex predicate (the contentfulness of thoughts about it) requires that its data be possible, and that they be combined in a logically consistent way, but it does not require that they be really compatible. We can have contentful thoughts about impossibilities as long as the basic contents of our thoughts (atomic predicates) are really possibly instantiated. A contentful atomic predicate, on the other hand, must be possibly instantiated.

To anticipate somewhat the discussion of Chapter 6, this also strengthens the continuity between *Beweisgrund* and the *CPR*, for in both works Kant is interested in the question: what is the ground of thinkable content? Important differences remain, though. First, in the Critical philosophy there are two importantly different ways in which thinkable content can be given to us (spontaneously, in concepts, or receptively, in intuition), and in accounting for thinkable *conceptual* content (thinkable predicates, in *Beweisgrund*) we have not yet accounted for how objects (which instantiate those concepts) are given to the mind. Second, in accounting for the metaphysical basis of thinkable content—atomic predicates are possibly instantiated, in virtue of standing in the appropriate relation to God—Kant, in *Beweisgrund*, has not yet explained how those predicates come to be contents of concepts available to discursive intellects like ours. In particular, he has failed to account for how our concepts come to represent possible (atomic) predicates, and thus acquire content, without depending upon experience (*a priori*).

[52] *OPG*, Ak. 2: 77, 79, 80, 82, 84, 85, 88, and 100.

[53] Cf. *OPG*, Ak. 2: 81.

5

Kant's Modal Argument

5.1. Introduction

In the previous chapter I explored Kant's notion of a real ground of real possibility and the meaning of his claim that there is a unique real ground of all real possibilities: God. In this chapter I reconstruct his argument for that claim. The conclusion of Chapter 4 was largely negative: Kant's own explicit view is that we cannot say what it is about God in virtue of which he grounds all real possibilities. In particular, we cannot identify the fundamental predicates of God, the derivative predicates (the predicates made possible by God's fundamental predicates), or the 'derivation' relation that makes the latter possible. My reconstruction of Kant's argument will attempt to respect these constraints: as much as possible, it will be based on premises about the structure of the derivation relation itself that does not attempt to say what that relation is or why its obtaining makes the derivative predicates possible. Ultimately, though, I will argue that we must violate Kant's own scruples if the argument is to succeed; reconstructing a crucial section of the argument relies on implicitly thinking of grounds of possibility as *causal* grounds.

In section 2 I explore the other central modal notion of *Beweisgrund*: absolutely necessary existence. I argue that this should be interpreted 'hyperintensionally': not all beings that exist necessarily exist with absolute necessity.[1] In particular, it should be interpreted in terms of a counterpossible relation of 'cancellation': if a being exists with absolute necessity, then were it not to exist nothing would be (really) possible. In section 3 I reconstruct several important steps in Kant's modal argument for the existence of a unique ground of real possibility. Given plausible principles, I argue, Kant's argument is successful up to a certain point: he can prove relatively unproblematically that there is a unique ground of all real possibility, and that this being exists with absolute necessity. However, this alone does not prove that this being is a single substance, rather than a plurality of substances. While this may seem a small

[1] The core idea of 'hyperintensionality' is that possible worlds are not 'fine-grained' enough for all philosophical purposes; for instance, 2 and {2} exist in all possible worlds, but the latter exists in virtue of the former. I will mainly be interested in hyperintensional relations between propositions; a relation R between propositions is hyperintensional just in case for some *p* and *q*, *p* bears *R* to *q* but (i) *p* does not bear *R* to *q** for some *q** that is necessarily equivalent to *q*, or (ii) *p** does not bear R to *q*, for some *p** that is necessarily equivalent to *p*. For instance, *grounding* is a hyperintensional relation: if *p* grounds *q* it does not follow that *p** grounds *q*, or *p* grounds *q** (where *p** and *q** are as above).

point, it is potentially devastating, for many metaphysicians would accept that there is a plurality of entities that collectively ground all real possibilities.[2] The 'plurality' objection, as many commentators on *Beweisgrund* have observed, poses a major threat to Kant's argument, and his own stated solution is quite weak. In section 4 I survey the secondary literature for solutions to the plurality objection and argue that none of them succeed. In section 5 I reconstruct a Kantian solution to the plurality objection using a (non-Humean) modal recombination principle and some core principles of Kant's cosmology. In section 6 I observe that, on my reconstruction of Kant's argument, we must implicitly think of God as *causally* grounding all real possibility, something, as we saw in Chapter 4, Kant explicitly denies. I argue that this is a consequence of Kant's claim that we have "no concept" of how God could ground the real possibility of things: because we do not understand the real grounding relation in which God stands to real possibility, we have no choice but to model it on a species of real grounding that we understand better: causal grounding. In section 7 I summarize key points of Kant's pre-Critical modal metaphysics that will be crucial to the second half of this study.

5.2. Absolute Necessity

Kant begins *Beweisgrund* I.3 with a familiar theme, necessary existence:

> The absolutely necessary is that whose non-existence is impossible in itself. This is undoubtedly a correct nominal definition. However, when I ask: why is it that the non-being of a thing is absolutely impossible? what I am looking for is the real definition, which alone can serve our purpose. (*OPG*, Ak. 2:81)

As I argued in Chapter 1, a *real definition* of 'necessary existence' tells us what it is in virtue of which a necessarily existent being necessarily exists. One candidate answer is the logicist one—it necessarily exists because its non-existence is logically impossible—but Kant rejects it:

> All of our concepts of the inner necessity in the predicates of possible things, of whatever kind they may be, derive from the fact that [*laufen darauf hinaus, daß*] the opposite contradicts itself. However, in the case of absolutely necessary existence, one would have little success in understanding it through that characterization. Existence [*Dasein*] is not a predicate, nor is the cancellation of existence the negation of any predicate, through which something in a thing is cancelled and an internal contradiction might arise. (*OPG*, Ak. 2: 81)

Kant is referring us back to the First Reflection, where he argued that, because existence is not a real predicate, nothing exists with logical necessity.

[2] For instance, on the view of Fine (1994*a*), facts about necessity are grounded in the *essences* of things; since there are multiple things whose essences ground the necessity facts, in Kantian terminology, the ground of real possibility is a plurality of beings rather than a simple individual.

Logically necessary existence is a *contradictio in adjecto*, so if anything exists necessarily it exists with *real* necessity. Kant therefore needs a real definition of really necessary existence, and he goes on to provide one in terms of the concept of *absolute necessity*:

> If I now consider for a moment why that which contradicts itself should be absolutely nothing and impossible, I find that through the cancellation of the law of contradiction, the ultimate logical ground of all that can be thought, all possibility vanishes, and there is nothing left to think. The conclusion immediately follows that, when I cancel all existence whatever and the ultimate real ground of all that can be thought therewith disappears, all possibility likewise vanishes, and nothing any longer remains to be thought. Accordingly, something may be absolutely necessary either when the formal element of all that can be thought is cancelled by means of its opposite, that is to say, when it is self-contradictory; or, alternatively, when its non-existence eliminates the material element and all the data of all that can be thought. The former, as has been said, never occurs in the case of existence. It follows that, since there is no third possibility, either the concept of absolutely necessary existence is a deceptive and false concept, or it must rest on the fact that the non-being of a thing is at the same time the negation of all the data that can be thought. (*OPG*, Ak. 2: 82)

This passage contains a number of important claims and themes and, before continuing, we need to separate and analyze them.

For Kant, as for the logicists, the PNC is the highest principle of logical possibility. The principle of non-contradiction grounds the whole space of logical possibility. Therefore, Kant reasons, if you eliminated or 'canceled' the PNC, nothing would be logically possible. This is what it means to say that the principle of non-contradiction is *absolutely* logically necessary. By contrast, a proposition is conditionally logically necessary just in case it follows logically from the principle of non-contradiction. All logically necessary propositions, i.e. all propositions whose negations entail contradictions, are consequences of the principle of non-contradiction, and hence conditionally logically necessary. To say that they are conditionally logically necessary is not to impugn their necessity: it is only to draw a distinction between these logically necessary truths, whose truth follows from the principle of non-contradiction, and the principle of non-contradiction itself.[3]

Kant understands absolute *real* necessity on the model of absolute *logical* necessity: it is absolutely really necessary that p just in case the 'cancellation' of p would cancel all real possibility. An absolutely necessary proposition is a proposition that grounds the whole space of real possibility. Likewise, an absolutely really necessary being is one whose existence governs and gives rise to the entire domain of real possibility. If there is such a being its non-existence would cancel all real possibility; without it, nothing would be really possible. Conditionally really necessary truths are truths that

[3] This is a hyperintensional relation of grounding *within* the space of logical possibility. See n. 1.

follow by real necessity from absolutely really necessary truths. There may very well be really necessary consequences of absolutely really necessary truths; likewise, there may be beings whose existence follows with real necessity from the existence of the absolutely necessarily existing being (if there is one).[4] This in no way impugns the necessity of these truths, or the necessary existence of these beings; the propositions could not have been false, and the beings could not have failed to exist.[5] It merely draws a distinction between those propositions and those beings whose truth and whose existence follow necessarily from an absolutely necessary being, on the one hand, and absolutely necessary truths and the absolutely necessary being, without which nothing would be really possible, on the other.

In light of these points, we can state Kant's definition of absolute necessity:

(*Defn.*) It is *absolutely necessary* that p if and only if $\neg p$ cancels all possibility,

and distinguish two different kinds of absolute necessity:

(*Defn.*) It is *absolutely logically necessary* that p if and only if $\neg p$ cancels all logical possibility.

(*Defn.*) It is *absolutely really necessary* that p if and only if $\neg p$ cancels all real possibility.

As we know from the previous chapter, logical possibility is a necessary but not a sufficient condition for real possibility. But this appears to entail that anything that cancels all logical possibility cancels all real possibility, so anything that is absolutely logically necessary is absolutely really necessary. This cannot be correct, however, because, while the negation of the principle of contradiction cancels all *real* possibility, it does so by canceling the logical element in real possibility. Kant is interested in whether there exists a being whose non-existence would cancel the real non-logical

[4] Whether there are necessary consequences of God's existence is related to the complex issue of whether God necessarily creates the actual world, or whether there is some contingency in God's choice of this world. This, of course, is tied up with difficult questions about divine freedom and whether freedom requires the possibility of doing otherwise. In *Beweisgrund* itself Kant does not come down definitely on either side of this issue; at Ak. 2: 153–4 he appears to argue that necessarily, if God creates a world, he creates the actual world (because it is the most perfect possible world), but this is far from a full-throated endorsement. The issue of whether the actual world is the most perfect possible world is one on which Kant worked extensively in this period. *Refl.* 3703–5 contain Kant's notes for a planned essay on Pope's optimism, the topic of the 1755 prize essay set by the Berlin Academy; the editors' Introduction to Kant (2011) contains a detailed discussion of Kant's work on optimism in this period (pp. xxxii–xli). Kant continues to uphold the view that the actual world is the most perfect possible in the mid-1760s; see *Refl.* 3854, 3880, and 4226. Matters are complicated by the fact that *Beweisgrund* is a transitional text between the compatibilist, indeed, even necessitarian, theory of freedom Kant defends in *ND* (Ak. 1: 398–405) and the more incompatibilist conception of freedom he will defend in the *CPR* (see Ch. 7). In a number of *Reflexionen* from the 1760s Kant seems to endorse a 'mixed' view, which combines an incompatibilist theory of human freedom and a compatibilist theory of divine freedom; see *Refl.* 3855–9 (Ak. 17: 313–15, from the years 1764–8), as well as *Refl.* 4033, 4034, 4156, 4157, 4180, 4184, 4218, 4226, 4228, 4334, and 4338, which Adickes dates to 1769. If this is correct then the actual world exists necessarily, although it does not exist *absolutely* necessarily. For critical discussion, see Kain (forthcoming) and Hogan (2009*a*).

[5] This is a hyperintensional relation within the space of real possibility.

element in real possibility, so we should amend the definition of absolute real necessity slightly to read:

(*Defn.**) It is *absolutely really necessary* that p if and only if $\neg p$ cancels the real element in all real possibility.

What does it mean for possibility to be 'canceled'? The most straightforward reading of the relation of 'canceling' in this definition is that it refers to a necessary conditional, i.e. $\neg p$ cancels q if and only if $\Box(\neg p \supset \neg q)$. However, interpreting cancellation in terms of a necessary conditional reduces Kant's definition of absolute necessity to necessity *simpliciter*: for any necessarily true p, $\Box(\neg p \supset \neg q)$ is true for any q. If 'cancellation' is interpreted this way, then any necessary falsehood cancels *any* proposition whatsoever (including necessary falsehoods). The 'necessary conditional' reading of cancellation eliminates the specific connection between absolute necessity and the cancellation of all *possibility*.

I think we can do better if we adopt a counterfactual interpretation of 'canceling'; but we have to be careful about how we understand counterfactuals. For example, if we interpret 'cancellation' in counterfactual terms and assume that $\Box(\neg p \supset \neg q)$ entails $\neg p \Rightarrow \neg q$ we still run into problems.[6] This assumption is equivalent to assuming that all counterfactuals with impossible antecedents are (vacuously) true. If we make these two assumptions, then for any necessarily false p, p cancels all real possibility, which once again reduces absolute necessity to necessity *simpliciter*: any necessarily true proposition is *absolutely* necessary. Consequently, we should not *both* accept a counterfactual interpretation of 'cancellation' *and* assume that all counterfactuals with impossible antecedents are true.

The situation improves if we reject the assumption that all counterfactuals with necessarily false antecedents are vacuously true and interpret Kant's notion of a 'cancellation of possibility' as involving the notion of a non-vacuous counterpossible conditional. In other words, on Kant's conception, some things would be 'canceled' if a necessary truth were false, while others would not be. For instance, there may be a necessary truth p ($\neq$ God exists) such that if p were false, some, but not all, real possibilities would cease to be possible. This means that the relation of cancellation between propositions is 'hyperintensional': necessarily equivalent propositions cannot be substituted into the sentence 'p cancels q' *salva veritate*. In the idiom of possible worlds, cancellation is more 'fine-grained' than possible worlds.

While the Lewisian orthodoxy that once reigned in analytic metaphysics rejected hyperintensional relations and non-trivial counterpossible conditionals, they are finding increasing acceptance.[7] One relevant motivation for accepting

[6] Where '$p \Rightarrow q$' stands for the 'would' counterfactual conditional: if p were the case, it would be the case that q.

[7] I call this orthodoxy 'Lewisian' because Lewis (1986) identified propositions with sets of possible worlds. However, other philosophers made room for hyperintensionality by introducing 'impossible worlds,' e.g., D. Nolan (1997), Vander Laan (2004), Kment (2006), and Brogaard and Salerno (2013).

hyperintensionality is dependence relations between necessary existents. Intuitively, both the number 2 and the singleton set {2} necessarily exist, but the latter exists *in virtue of* the former. This is a non-causal relation of metaphysical dependence; the number 2, a causally inert object, cannot cause {2}, another causally inert object, to exist, and the existence of 2 does not logically entail the existence of {2}.[8] Translating this into the Kantian language of cancellation, we could say: *per impossibile*, the non-existence of 2 would cancel the existence of {2}. This counterpossible relation of cancellation is not trivial, though, because it does not hold between the non-existence of 2 and just any impossible state of affairs: the non-existence of 2 would not cancel the fact that nothing is simultaneously red all over and green all over. The precise nature of such hyperintensional relations is a difficult and controversial matter in contemporary metaphysics.[9]

This newfound interest in counterpossible conditionals is, in a certain sense, quite Kantian, because Kant himself, I have been arguing, accepts non-trivial counterpossible conditionals, and what is more, thinks these counterpossible conditionals are explained by *grounding* relations between necessary truths. To jump ahead slightly, there is a unique absolutely necessary being, God, and the reason God's non-existence cancels all real possibility is that God's existence grounds all real possibility: a grounding relation between necessary truths (God exists, something is really possible) explains a counterpossible cancellation relation between them (God's non-existence, nothing being really possible).

I have gone on at some length about this because some readers will be tempted to dismiss the notion of 'absolute necessity' as I have defined it and reduce it to necessity *simpliciter*.[10] This definitely has the merit of making Kant interpretation easier. However, it does so at the expense of conflating two notions that Kant seems to want to keep distinct. At various points in *Beweisgrund* Kant distinguishes between things whose non-existence cancels *some* real possibilities and things whose non-existence cancels *all* real possibilities. For instance, "that existence, by means of which the material element of all that can be thought is not given, and in the absence

Lange (2009) develops a non-Lewisian (i.e. not world-based) theory that also allows for non-trivial counterpossible conditionals. However, there is also a growing interest in hyperintensional metaphysics, centering around the notion of 'grounding': see Fine (1994*a*), Schaffer (2009), Rosen (2010), Audi (2012*a*) and (2012*b*); skeptics about the grounding relation include Wilson (2014) and Koslicki (forthcoming). For a survey of these issues, see Bliss and Trogdon (2014) and Raven (forthcoming).

[8] That inference requires the axioms of set theory, which are taken by most philosophers to be extra-logical.

[9] For discussion see D. Nolan (2014) and Jago (2014).

[10] Cf. Chignell (2009*a*), 167. By necessity *simpliciter* I mean necessity independent of source or ground; in eighteenth-century terms, it is necessary *simpliciter* that *p* if and only if either *p* is *in se* necessary or is a necessary consequence of some *in se* necessary truth. Kant's theory of absolutely necessary existence is a theory of *in se* necessary existence, not of necessary existence *simpliciter* (which would apply also to the necessary consequences of an *in se* necessary being), much less of merely hypothetically necessary existence.

of which, therefore, there is still something left to be thought, that is to say, still something possible—the opposite of such an existence is possible in the real sense, and in the same sense it is also contingent" (*OPG*, Ak. 2: 83).[11] Since Kant is also committed to the principle that *possibly p* entails *necessarily possibly p*,[12] anything whose non-existence cancels *some* real possibility exists necessarily. If we collapse absolute necessity and necessity *simpliciter*, then by definition, all such beings exist absolutely necessarily; there would then be no *modal* distinction between things whose non-being cancels some possibility and the unique being whose non-existence cancels *all* possibilities. In my reconstruction of Kant's argument in the next section, I bear in mind the possibility of such a 'reductive' reading and argue that it would not help Kant significantly. Interpretive charity, and fidelity to the text, therefore support distinguishing absolute necessity and necessity *simpliciter*.

We are now in a position to understand Kant's definition of absolute necessity:

(*Defn.**) It is *absolutely really necessary* that p if and only if $\neg p$ cancels the real element in all real possibility,

which, I have argued, should be understood as:

(1_C) It is *absolutely really necessary* that p if and only if, were it the case that $\neg p$, there would be no real element in any real possibility.

An absolutely necessary proposition is a proposition whose negation cancels all possibility. This entails that:

(1_E) For any x, x exists absolutely necessarily if and only if, were x not to exist, there would be no real element in any real possibility.

Absolutely necessary existence is always absolutely *really* necessary existence, since nothing exists with logical necessity. Since I have been focusing, as Kant does in the first Division of *Beweisgrund*, on the possibility of predicates, and, specifically, on the possibility of *atomic* predicates, the specific version of this claim that we will need to reconstruct Kant's argument is:

(1_{AP}) For any x, x exists absolutely necessarily if and only if, were x not to exist, there would be no really possible atomic predicates.

Principle (1E) is stated at the end of I.3.1, which begins, as we have seen, with Kant inquiring into the *real definition* of necessary existence. What is the relation between (1_E) and the project of giving a real definition of necessary existence? (1_E) is a principle about *absolutely* necessary existence, which is related to necessary existence *simpliciter* as follows:

[11] Cf. *OPG*, Ak. 2: 82, 99–100, and 162 f.

[12] *OPG*, Ak. 2: 84. This principle, the characteristic axiom of S5, is discussed in more detail below in §3.

(2) *x* necessarily exists if and only if (i) *x* exists absolutely necessarily or (ii) $\exists y$ such that *y* exists absolutely necessarily and *y* is necessarily the ground of the existence of *x* (*x* is conditionally necessary).

In other words, a being exists necessarily just in case it is absolutely necessary or is a consequence of an absolutely necessary being. Clause (ii) has the paradigm form of a real definition: if *y* exists necessarily and is necessarily the ground of the existence of *x* then we have an explanation of why *x* exists necessarily. We might then ask 'why does *y* exist necessarily?' and for this we are referred back to (i). But is (i) (or for that matter (1_{AP})) a real definition? No, for it does not tell us, of a being that exists with absolute necessity, *why* that being exists with absolute necessity.

The real definition of *absolutely necessary existence* is provided by Kant's concept of a ground of all real possibility (*GARP*): an absolutely necessary being is absolutely necessary (its non-existence cancels all real possibility) because it is a ground of all real possibility. The concept of a ground of all real possibility can provide a real definition of absolutely necessary existence *only if* the grounding relation in which God stands to all real possibility *explains* (is the ground of) the counterpossible 'cancellation' relation in which God stands to all real possibility. This means that grounding entails cancellation:

(3) For any *x*, if *x* is a ground of all real possibility (*GARP*) then the non-existence of *x* would cancel all really possible atomic predicates.

But if being a *GARP* is going to be the real definition of absolutely necessary existence it must be the case that all absolutely necessary beings are *GARP*s, for otherwise there could be a being whose absolute necessity is not explained by its being a *GARP*. In other words:

(4) For any *x*, if the non-existence of *x* would cancel all really possible atomic predicates, then *x* is a ground of all real possibility (*GARP*).

With these two pieces in place, we can state the *real definition* of absolutely necessary existence:

(*Real. Def.*) For any *x*, if *x* exists absolutely necessarily, then this is so in virtue of the fact that *x* is a ground of all real possibility (*GARP*).

This, combined with the definition of necessary existence *simpliciter* from above, constitutes Kant's complete alternative to the logicist real definition of necessary existence (the containment of existence within a being's essence).

5.3. The Only Possible Ground of Proof

In this section I reconstruct Kant's modal argument for the existence of God. This is what Kant calls a 'proof-ground,' a set of premises that entail the existence of God, but which have not yet been turned into an explicit demonstration (a logically valid argument) for that conclusion.[13] Each sub-section corresponds to a sub-section of Kant's argument in *Beweisgrund* I.2–3.

(i) It is absolutely necessary that something exists (I.2.3)

The first thing Kant does is argue that it is absolutely necessary that something exists, which he does in I.2.3, after having introduced the 'material' requirement on possibility. His argument is relatively straightforward:

(1) It is absolutely necessary that *p* if and only if, were it the case that $\neg p$, no atomic predicates would be really possible. [By definition – see section 2.]
(2) If F is a really possible atomic predicate, then there exists a substance that grounds its real possibility.
(3) If F is a really possible atomic predicate and a substance grounds its real possibility, then the non-existence of that substance cancels the real possibility of F.
(4) ∴ If nothing were to exist, no atomic predicates would be really possible. [From (2) and (3).]
(5) ∴ It is absolutely necessary that something exists. [From (1) and (4).]

The most important claim here is (1); since Kant is effectively introducing the concept of absolute necessity through this premise, I think we should understand it as a stipulative definition, and grant it to him. The other main premise in this argument is (2) the principle that possibility requires an existing ground in a substance. While not philosophically uncontroversial, it is the central assumption of his argument, so I think we should grant it to Kant and see whether the argument he builds upon it succeeds. The justification for (3) is that, if a predicate is possible, then it is necessarily possible (Ak. 2: 84) and it is plausible to assume that possibility facts have their grounds essentially; nothing but the actual ground of F could ground its real possibility.[14] At this point in the argument, no specific view about what the existing grounds of real possibility are needs to be appealed to, other than that they are substances (rather than predicates of substances); even someone who holds the 'intellectualist' model (see Chapter 4.3) should accept (5).[15]

[13] As Kant explains at Ak. 2: 66, this work does not contain a demonstration of the existence of God but only the 'proof-ground' (*Beweisgrund*) for a possible demonstration (cf. Ak 2: 65, 67). As Kant explains in his logic lectures, this refers to a technical distinction between the proposition to be proved, the proof-ground (argument), and the logical form by which the conclusion follows from the proof-ground (*WL*, Ak. 24: 892; cf. *JL*, Ak. 9: 71, and *LPö*, Ak. 24: 561). Chignell (2009*a*), 161 n. 9 points out that this is a discussion of §191 of Meier's *Auszug*. In this book, Kant is only giving a *Beweisgrund*, a set of premises from which a conclusion (the existence of God) follows, without giving it the 'form' of a demonstration, i.e. without turning into a formally valid proof (cf. Ak. 2: 88–9).

[14] Thanks to Catharine Diehl and Tobias Rosefeldt for pressing me on this point.

[15] Cf. Schönfeld (2000), 201–6.

(ii) There is an absolutely necessary being (I.3.2)

Having proven the *de dicto* claim that (absolutely) *necessarily something exists*, Kant's next argumentative task is to prove the *de re* claim that something necessarily exists.[16] But, given Kantian premises, this is relatively straightforward. Note the short space of text in which Kant moves from the *de dicto* to the *de re* claim:

> all possibility presupposes something actual, in and through which the thinkable is given. Therefore [*demnach*], there is a certain actuality whose cancellation would itself cancel all possibility. However, that whose cancellation or negation eliminates all possibility, is absolutely necessary. Therefore, something exists absolutely necessarily [*existiert etwas absolut notwendigerweise*] (*OPG*, Ak. 2: 83)

The first sentence should be read as making the *de dicto* claim that (absolutely) necessarily something exists (if nothing existed, nothing would be really possible), the conclusion of the immediately preceding section; but by the end of this passage, without much argument, Kant is making the *de re* claim that some being exists (absolutely) necessarily. This may give the impression that Kant, disappointingly, is just conflating these claims, but this is unlikely for two reasons. First, he made the *de dicto* claim several pages earlier in I.2.3, a section whose title is "It is absolutely impossible that nothing exists," which is naturally read *de dicto*. The quoted passage occurs under the heading "there is an absolutely necessary being," which is naturally read *de re*. Secondly, as I will now argue, given his premises, the *de re* claim is an almost immediate consequence of the *de dicto* claim.

If we take all of the beings that are grounds of some real possibility (*GSRPs* in the terminology of the previous chapter) and treat them collectively as one being, then *that* being is a ground of all real possibility (*GARP*) and exists with absolute necessity. All really possible atomic predicates are grounded by some part of this collection and if this collection were to be completely annihilated (if all of its parts were annihilated) it would cease to exist. Some care is required here. We need this collection to be such that annihilating any proper part of it does not annihilate the collection itself; we need it to be such that the only way of annihilating it is to annihilate *every one* of its proper parts. In contemporary metaphysics there are various ways of doing this; perhaps the easiest is to take what Fine (1994*b*) calls the *aggregate* of the *GSRPs*.[17] The aggregate of the *X*s exists if and only if *at least one* of the *X*s exists; unlike classical mereological fusions, aggregates are as mereologically *non-rigid* as can be: they can exist if even one of their parts exists. Call the aggregate of all *GSRPs* Ω. The

[16] As Abaci (2014), 14–15, points out.

[17] See Fine (1994*b*) and (1999). As I am using the term, the aggregate of the *X*s exists if and only if at least one of the *X*s exists. Fine originally introduced aggregation in terms of tense: the aggregate of the *X*s exists at time *t* if and only if at least one of the *X*s exist at *t*. My notion of aggregate is a generalized non-temporal version of the Finean notion.

absolute necessity of Ω is secured by the following principle, which was premise (3) in the argument of I.2.3:

(3) If F is a really possible atomic predicate and substance *x* grounds its real possibility, the non-existence of *x* cancels the real possibility of F.

By definition, Ω exists if and only if at least one *GSRP* exists, so the cancellation of Ω is equivalent to the annihilation of all *GSRPs*, which would cancel all really possible atomic predicates. So it follows that Ω exists with absolute necessity. Since Ω contains all *GSRPs*, it grounds *all* real possibilities (it is a *GARP*).

(iii) It is unique (I.3.3)

Kant's next step is to prove that there is a unique absolutely necessary being:

> Let A be a necessary being and B another. By definition [*vermöge der Erklärung*] B is only possible, insofar as it is given as the consequence of another that grounds it, A. By assumption, B itself is necessary, so its possibility is in it as a predicate and not as a consequence of another, but yet, according to what was just shown, it is a consequence, and this is a contradiction. (Ak. 2: 84)

Kant formulates this argument in terms of one substance grounding another, but his modal metaphysics up to this point has been formulated in terms of a substance grounding the possibility of predicates. Let us attempt, then, to reconstruct this argument in terms of substances grounding one another's predicates:

(1) Assume A and B are absolutely necessary beings and A≠B.
(2) Anything that exists absolutely necessarily is a ground of all real possibility (*GARP*). [See section 2.]
(3) If *x* is a *GARP* then for any possible predicate F either (i) F is instantiated by *x* and the possibility of F has no ground (F is fundamental) or (ii) the possibility of F is grounded in *x* (F is derivative). [Defn. of *GARP*—see Ch. 4.2.]
(4) ∴ The fundamental predicates of A have no ground and the fundamental predicates of *A* are grounded in B. [From (1)–(3).] This is a contradiction.
(5) ∴ If A and B are absolutely necessary beings then A=*B*.

In section 2 I argued that Kant is entitled to (2). However, even without (2) he can prove directly that there cannot be two distinct *GARPs*, by using (3)–(5) to reduce to absurdity the assumption that:

(1*) Let A and B be *GARPs* and A≠B.

Note, though, that the reductive reading, on which absolute necessity is equivalent to necessity *simpliciter*, should reject (2) as a mistake on Kant's part. For (2), on the reductive reading, entails:

(2*) Any necessary being is a *GARP*.

But this has the problematic consequence that if *x* is a *GARP* and *x* necessarily causes *y* to exist, then *y* is a *GARP* as well. This is not a consequence that Kant, or any metaphysician, should embrace. For it would entail that, if God necessarily creates the actual world, as, on some theories of the divine will, he does, then the world itself is a *GARP*. While Kant is not committed to such a necessitarian theory of creation in *Beweisgrund*,[18] he should not be forced to the conclusion that if God necessarily creates the actual world then the actual world is a *GARP*. A charitable reading of Kant should eschew commitment to (2*); consequently, we should avoid *if possible* a reading of Kant on which he is committed to identifying absolute necessity with necessity *simpliciter*.

(iv) It is simple[19]

In particular, Kant must show that the *GARP* is a single substance, rather than an aggregate of substances. Unless he can show this, Kant will not have shown anything that many atheist metaphysicians would reject. For instance, some contemporary metaphysicians (e.g., Fine) hold that possibilities are grounded in *essences*; the aggregate of these essences is, in Kant's terminology, absolutely necessary: if you cancel all of them, you cancel all possibility. But few such metaphysicians would think there is a single such essence that grounds all real possibility.[20] This is the final stage of the argument that I will reconstruct; I will leave out of consideration Kant's further argument that the *GARP* has the other traditional divine attributes: intellect, will, etc.

Kant must argue against an alternative picture on which possibility is parceled out among a plurality of substances, each of which grounds some possibilities (*GSRP*) but not all possibilities (*GARP*). I will call this modal *pluralism*. Kant's principal argument against the modal pluralist is the following:

> If one were to appeal to the definition of the necessary being and say that in each part the data of some inner possibility is given, but in all of the parts together all possibility is given, one would be imagining something wholly, albeit covertly, incoherent. For if one then thought that some inner possibilities could be canceled, while others, given through other parts, remain, one would have to suppose that it is in itself possible for inner possibility to be negated or canceled. But it is absolutely unthinkable and contradictory that something be nothing, and this means that canceling any inner possibility eliminates all that is thinkable. It is apparent from this that the data for all that is thinkable must be given in the thing whose cancellation is the negation of

[18] See Ak. 2: 153–4; cf. *Refl.* 3830.

[19] Kant represents the conclusion of this argument as the *simplicity* [*Einfachheit*] of the unique *GARP* but he never argues that the unique *GARP* lacks parts entirely; he argues that the unique *GARP* is not composed of parts, each of which grounds some subset of the space of possible predicates. It is compatible with everything for which Kant explicitly argues that the unique *GARP* has parts, as long as they play no *separable* role in its grounding of possibility. Thanks to Manish Oza for pressing me on this point.

[20] Whether they would agree that essences are substances in Kant's sense—bearers of predicates/properties that are not themselves predicates/properties—is another matter (see Fine 1995, 66–7). But such metaphysicians might be able to accept all of Kant's arguments up to this point provided we delete the assumption that the grounds of possibility are substances (rather than bundles of predicates/properties).

all possibility; therefore, that which contains the ultimate ground of any inner possibility contains the ground of all possibility whatsoever, and this ground cannot be divided into distinct substances. (*OPG*, Ak. 2: 84)

Kant argues against the plurality view, in this passage, by claiming that if something's non-existence cancels some possibilities then it cancels all possibilities, and concluding from this that if something grounds some possibilities then it grounds all possibilities (it is the *GARP*). More formally, his reasoning could be represented as follows:

(1) Let x be a ground of *some* real possibilities (*GSRP*).
(2) If x grounds *some* real possibilities, its non-existence cancels some real possibilities (those it grounds). [By premise (3) in section (i) above.]
(3) If x's non-existence cancels some real possibilities, its non-existence cancels all real possibilities ("...this means that canceling any inner possibility eliminates all that is thinkable").
(4) ∴ If x grounds some real possibilities, it exists absolutely necessarily. [From (2) and (3) and the definition of absolutely necessary existence from §2.]
(5) If x exists absolutely necessarily, then it is a ground of *all* real possibility. ("It is apparent from this that the data for all that is thinkable must be given the thing whose cancellation is the negation of all possibility.")
(6) ∴ For all x, if x grounds *some* real possibilities, it grounds *all* real possibilities ("that which contains the ultimate ground of any inner possibility contains the ground of all possibility whatsoever"). [From (4) and (5).][21]

From this conclusion, Kant can easily prove that Ω, from above, has no parts. From (6) it follows that any *GSRP* is a *GARP* and Kant has already proven (see section (iii)) that there cannot be two *GARP*s. So there is only one *GSRP* and Ω is identical to it (it is the aggregate of only one object). In fact, (1)–(6), if successful, would allow Kant to bypass entirely the proof in (ii) that required the introduction of Ω in the first place, for it would allow Kant to argue as follows:

(7) Assume: some atomic predicate F is really possible.
(8) If F is really possible there exists an x such that x is the ground of the possibility of F (*GSRP*).
(9) Any *GSRP* is a *GARP*. [Conclusion of previous argument.]
(10) There cannot be more than one ground of all real possibility. [Proved in section (iii).]
(11) ∴ There is a unique *GARP*.

However, the original argument (1)–(6) above is not successful, so the point is moot. Premise (3), in particular, begs the question against the pluralist. According to

[21] Cf. the corresponding argument in *ND*, Ak. 1: 395.

pluralism, the complete space of possibilities is 'parceled out' among a plurality of grounds, so that the non-existence of any one of them would cancel only some possibilities; it would beg the question to assume that anything that cancels some possibilities cancels *all* of them. Kant does not succeed in establishing his key claim, (6), at Ak. 2: 85.[22]

A Kantian rejoinder to this objection might go as follows. In the passage quoted above, Kant writes: "it is entirely inconceivable and self-contradictory that something should be nothing." Since 'something,' in the context of *Beweisgrund*, means 'something (really) possible,' and 'nothing' means 'not (really) possible,' we could interpret this as a variant of the S5 axiom in modal logic: all possibilities are necessarily possible.[23] If there is a ground of some real possibilities (a *GSRP*) then its non-existence cancels *some* possibility, and since possibilities are necessarily possible, the non-existence of this ground is impossible. So any *GSRP* exists necessarily. A being exists necessarily just in case its non-existence cancels all possibility. If a being is such that its non-existence cancels all possibility, this is because it is a ground of all possibility; so any being that grounds some possibility grounds all possibility.

There are several places, however, where this argument is vulnerable. Using the S5 axiom we can prove straightforwardly that any *GSRP* exists necessarily; that much of the argument is valid. But to prove that it exists with *absolute* necessity we must either explicitly collapse absolute necessity and necessity *simpliciter* or make assumptions that entail this collapse. For instance, if we assume that, if a *GSRP* exists necessarily then the non-existence of a given *GSRP* stands in the counterpossible cancellation relation to *any* proposition whatsoever (including the proposition that something is really possible),[24] then we can prove that any *GSRP* exists with absolute necessity. It is true that in the context of this argument Kant does seem to collapse this distinction ("this is tantamount to saying that canceling an internal possibility is the same thing as eliminating all that can be thought"). But even if we grant Kant this modal collapse the argument *still* begs the question against the pluralist. For the modal pluralist can grant that any *GSRP* exists with absolute necessity but very plausibly deny that it *grounds* all real possibility.[25] The idea would be this: there are multiple grounds of possibility, the cancellation of any one of which would cancel some (real) possibilities; since possibilities are necessarily possible (S5 axiom), these grounds of possibility each exist necessarily. The non-existence of any of them is impossible, and any counterfactual with an impossible antecedent is trivially true (thus collapsing the necessary/absolutely necessary distinction), so each of these grounds of possibility satisfies Kant's definition of absolute necessity. *However*, although each of these grounds stands in a counterfactual relation to *all* of possibility,

22 This problem is pointed out by Adams (2000), 433.

23 $\Diamond p \supset \Box\Diamond p$. Cf. Ak. 17: 252.

24 Equivalently, assuming that $\Box(\neg p \supset \neg q)$ entails $\neg p \Rightarrow \neg q$.

25 The modal pluralist does not need to accept (4) from §2. Adams (2000) makes an equivalent point.

none of them stands in a grounding relation to *all* of possibility. The grounding relation between these substances and possibilities is more fine-grained than the counterfactual relation; each substance bears the former relation only to a subset of the possibilities. So reducing absolute necessity to necessity *simpliciter* and adopting the S5 axiom does not actually help Kant's argument here. I think, therefore, that the most charitable interpretation is that, although Kant collapses that distinction in this passage ("this is tantamount to saying that canceling an internal possibility is the same thing as eliminating all that can be thought") he does so because he (mistakenly) thinks it will help his argument, and, in light of the many considerations that support this distinction (section 2), his considered view is that absolute necessity is not equivalent to necessity *simpliciter*.[26]

I am not the first commentator to notice that the 'plurality' objection poses significant problems for Kant's argument. In the next section I explore some solutions that have been proposed in the secondary literature.

5.4. Prior Replies to the Plurality Objection

The secondary literature on *Beweisgrund* has spawned some sophisticated proposals for answering the plurality objection on Kant's behalf.[27] Perhaps the best place to start is Chignell's solution, which rests on his 'real harmony' principle:

> (*Real Harmony*) If F and G are fundamental predicates and F and G are really compatible then the real compatibility of F and G must be grounded in their co-instantiation by a single substance.[28]

In the previous chapter I criticized Chignell's interpretation on textual and philosophical grounds. In this context it suffices to point out that Real Harmony will only defeat the plurality objection if we assume additionally:

> (*Compatibility*) If F and G are fundamental predicates (predicates whose possibility is given in a substance that instantiates them) then F and G are really compatible (it is really possible that they are co-instantiated).

[26] Nor is the slightly different argument Kant offers earlier in the same paragraph any more successful against modal pluralism (Ak. 2: 84); it is a variant of the idea, deriving from the fifth of Aquinas' 'five ways' (*Summa Theologica* I.q2.a3), that if all beings are contingent then it is possible that nothing exists. For reasons of brevity I forgo further discussion of it here.

[27] To my knowledge, the first to point out that it is a problem were Watkins and Fisher (1998), who write: "while a contingent being can ground some possibilities, that is, those that arise from the material elements given through it, it cannot ground its own possibility or the absolute possibility of anything else" (Watkins and Fisher 1998, 375). However, this appeals to a different conception of contingency. Kant needs to show that an absolutely contingent being—a being whose non-existence does not cancel all possibility—cannot ground some possibilities. Adams (2000) also contains a highly influential discussion of this problem.

[28] Chignell (2009*a*), 187; (2012), 648.

But Chignell gives no reason why the pluralist must admit this. Note that Compatibility does not say merely that some fundamental predicates are really compatible; it says that all fundamental predicates are collectively compatible, that is, they can *all* be instantiated by the same substance. So Compatibility cannot be established by pointing to a few examples, such as the compatibility of understanding and will, which Kant asserts at Ak. 2: 87. Far from being a weak or uncontroversial principle, Compatibility is in fact extremely strong, and Kant would need a strong argument to justify it. In particular, the pluralist might hold that distinct fundamental predicates are instantiated by distinct substances (distinct *GSRP*s) and are *not* compatibile.

Perhaps Chignell would respond to the pluralist by asking what grounds the real harmony of derivative predicates on this picture. As we saw in the previous chapter, this is precisely the point on which the coherence of Chignell's own view is threatened.[29] But notice the pluralist does not need to admit that derivative predicates are compatible if they are grounded in different *GSRP*s. The pluralist could hold a modified Cartesian view on which each of the attributes is a separate substance in its own right that instantiates various fundamental predicates and grounds various derivative predicates. So, for instance, the attribute of *Thought* can ground the possibility of various modes of thought, and the attribute of *Extension* can ground the possibility of various modes of extension, but no thought mode is compatible with an extension mode. This is not an especially attractive picture of how possibilities are grounded, but Kant claims not merely that it is rationally unappealing but that it is demonstrably false.

Yong (2014) steps into the breach with an innovative proposal: Kant is searching for the ground not merely of individual possibilities (really possible predicates) but of the space of possibilities as a totality and, thus, the modal relations among possibilities (such as compatibility, incompatibility, entailment, etc.). Yong makes some intriguing remarks about the idea that possibility as such constitutes a unified totality, but it is unclear how to make this idea dialectically effective against the pluralist, who, after all, might *deny* that the space of possibilities constitutes a unified totality.[30]

Yong's more concrete proposal appeals to the doctrines of Kant's pre-Critical rational cosmology to defeat the plurality objection.[31] Let us assume that A and B are distinct grounds of real possibility, so A grounds some really possible predicates (call them the a-predicates) and B grounds a distinct set of predicates (call them the b-predicates). Yong points out that A and B both *exist* (by definition of ground of

[29] Cf. Yong (2014), 33–4, and Chignell (2012), 665–7.

[30] Yong (2014), 38–40. Yong's reconstruction also threatens to collapse the distinction between Kant's proof from the mere fact of possibility (contained in *Beweisgrund* I) and the proof from the unity in the essences of things (contained in *Beweisgrund* II). See Ak. 2: 155 for more on the distinction. The harmony among possibilities is presented as a premise only of the latter, while Yong claims it as a premise of the former as well.

[31] Yong (2014), 41–4.

possibility) and therefore stand in *relation* to one another. He then appeals to Kant's cosmological doctrine that substances' relations to one another are not *brute* facts but must be grounded in some causal relation between A and B. In particular, for A and B to be *related* to one another, their relation must be grounded in a third substance C. Yong goes on to argue that, on pain of infinite regress (what grounds the relations among A, B, and C?), we should identify A and B and accept that there is only one ground of possibility.

While I think Yong is right to invoke Kant's pre-Critical cosmology to respond to the plurality objection, the details of his reconstruction are problematic. First of all, Kant's doctrine that substances do not 'relate' to one another solely in virtue of existing, but require an external ground of their 'relation,' applies only to substances that are members of the same *world*. In other words, Kant's notion of *relation* is more substantive than the contemporary post-Russellian logical notion of *relational predicate*. In fact, it is one of Kant's consistent pre-Critical contentions that merely in creating a plurality of existing substances God does not thereby make them into members of a single *world*.[32] But Yong has given no argument that the pluralist must admit that distinct grounds of possibility are members of the same *world* in Kant's specific technical sense: a totality of mutually interacting substances.

It is crucial to realize that Kant's notion of a *world* (k-world) is very different from the notion of a 'possible world' in modal logic (m-world). The modal logical concept of a possible world is ultimately the concept of a 'modal index,' an element in a model theory for modal logic. By contrast, a Kantian *world* is a concrete object: a collection of substances unified by mutual causal interaction. Perhaps the clearest way to see the difference is to note that the necessity of God's existence means that God exists 'in' every m-world (every counterfactual situation) but God is not a part of any possible k-world (because he does not reciprocally interact with other substances—their causal dependence on him is one-way). Similarly, Kant holds that God can create two separate non-interacting k-worlds; in contemporary terms, this means there is an m-world in which there are two k-worlds (and God).[33] K-worlds are not modal indices (the original function of m-worlds in modal logic) or even truth-makers of modal claims (a service into which they were pressed by post-Kripkean discussions of modal metaphysics).[34] To return to Yong and the plurality objection, if two grounds of possibility exist then they are members of the same m-world (trivially), but it does not follow that they are members of the same k-world. In the next section I try to improve on Yong's proposal and explain how Kant's pre-Critical cosmology can be used to answer the plurality objection. In what follows, 'world' will always mean

[32] *LF*, Ak. 1: 21–2; *ND*, Ak. 1:410, 415; *ID*, Ak. 2:407, 410. [33] Ak. 1: 414.

[34] This does not mean concrete worlds cannot play the role of modal truth-makers; in David Lewis's modal theory, possible worlds are conceived in a way very similar to k-worlds (maximal mereological sums of spatiotemporally connected concrete objects) and modal operators are quantifiers over these worlds. The differences between these two non-Kantian notions of 'possible worlds' are brought out very clearly in Burgess (2012), 143–56.

k-world (neither possible worlds as modal indices, nor possible worlds as the truth-makers of modal sentences, will play any role in my argument).

5.5. *E Pluribus Unum*

In the previous section I criticized Yong for assuming that if there are multiple grounds of possibility (*GSRPs*) then they exist in the same world. I will argue that a Kantian response to the plurality objection should appeal instead to a weaker but similar principle:

> (*Recombination*) It is really possible for there to be a world in which all the fundamental atomic predicates are instantiated (though not necessarily co-instantiated).

Intuitively, the fundamentality of predicates entails that they do not stand in modal exclusion relations with one another: if fundamental predicate F is instantiated in a k-world, this does not necessitate that fundamental predicate G is not instantiated in that world (though it may necessitate that F and G are not instantiated by the same object). The pluralist view is that there is a plurality of distinct substances, each of which possesses some of the fundamental predicates, and no single substance that possesses all of them (no *GARP*). I do not think it begs the question against that view to assume Recombination.

In fact, we need to reformulate Recombination slightly because, as stated, it begs the question against Kant himself, for Kant thinks that God cannot be part of any world, so any fundamental predicate that *only* God can possess will be a counter-example to Recombination as stated. Fortunately the repair is quite simple:

> (*Recombination**) It is really possible for there to be a world in which all of the fundamental predicates that can be instantiated by a being other than God (a *GARP* that lacks *GSRP*-parts) are instantiated (though not necessarily co-instantiated).[35]

Since the only *GARP* the pluralist accepts is an aggregate of all the *GSRPs*, this claim is equivalent, for the pluralist, to Recombination above. It is this recombination principle that I will use to reconstruct my response to pluralism. Furthermore, I take the following constraint on fundamental predicates to be prima facie plausible and not question-begging against the pluralist:

> (*Fundamentality*) If F is a fundamental predicate and x is the ground of the existence of something that instantiates F then x instantiates F.

[35] Recombination should not be confused with various stronger Humean Recombination principles that Kant would reject, e.g., that "patching together parts of different possible worlds yields another possible world" (Lewis 1986, 87–8). Neither Recombination nor Recombination* entail Humean Recombination.

This is merely the idea that nothing that lacks a fundamental predicate can 'introduce' that fundamental predicate into the space of possible predicates (e.g., by grounding the existence of an instance). The pluralist should embrace this principle, for it excludes a view on which the *GSRPs* that instantiate various fundamental predicates themselves have grounds that *lack* these predicates.

Before I explain my own reply to the plurality objection, I want to explore in a little more detail two principles of Kant's pre-Critical cosmology that will be crucial to my argument. The first, which we have already touched upon, is Kant's definition of *world*: a world is a totality of mutually interacting substances. That a world is a totality means that it is maximal: anything that interacts with a part of a world is a part of that world.[36] That substances in worlds mutually interact means that they are reciprocal causes of one another's accidental predicates.[37] The second major piece of Kant's pre-Critical cosmology to which I will appeal is a principle about the possibility of mutually interacting substances, which he calls the 'principle of co-existence':

> Finite substances do not, in virtue of their existence alone, stand in a relationship with each other, nor are they linked together by any interaction at all, except insofar as the common principle of their existence, namely the divine understanding, maintains them in a state of harmony in their reciprocal relations. (Ak. 1: 412–13)[38]

This might not seem like an auspicious premise for Kant's *Beweisgrund* argument, because it explicitly mentions God, the very being whose existence he is trying to establish. But Kant's *reasons* for holding this principle do not depend upon the conclusion of that argument. He holds the principle of co-existence in order to solve a standard early modern problem in the metaphysics of causation: how to reconcile the modal independence of finite substances with the possibility of causation as a *necessary* connection between them.[39] His solution is that causation between finite substances is not a necessary connection *tout court*; finite substances can exist without interacting with one another. For a set of finite substances to interact, there must be a substance that causes them to exist and 'places' them in interaction. Kant, of course, thinks this third substance that grounds the interaction

[36] *MH*, Ak. 28: 40. God is saved from world-membership by the fact that his causal influence is one-way: he causally influences finite substances, but they do not influence him.

[37] *MH*, Ak. 28: 25–6.

[38] The principle of succession (Ak. 1: 410), which states that a finite substance can undergo internal change only if it is in interaction with other finite substances, will not play a role in my reconstruction of Kant's argument, so I do not discuss it here.

[39] See *ND*, Ak. 1: 413, where Kant argues for the principle of Co-existence, which I reconstruct in what follows; cf. the parallel discussion of the principle of co-existence in *MH* (Ak. 28: 51–2, 54). I agree with Watkins (2005), 140–9 that the crucial issue is the modal independence of substances (rather than the reducibility of relations); Langton (1998), 107–9 focuses on the reducibility of relations and interprets Kant's primary target as Leibniz (rather than Crusius's doctrine of existential grounds—see Watkins (2005), 145. For further critical discussion see Laywine (1993), 37–49, Schönfeld (2000), 149–54, and Insole (2011).

among finite substances is God, but for our purposes we do not need to assume that, nor do we need to understand *how* God grounds the interaction of substances.[40] All we need is the principle that if substances interact then there is a substance that is the ground of their existence. In fact, we need a modal corollary of that principle:

> (*Modal corollary*) If there exists a plurality of substances that really possibly interact then there exists a substance that is the ground of their existence.

This is a natural corollary of the Co-existence principle, given Kant's background commitment to the principle that real possibilities require existing grounds: if A and B possibly interact, some substance *makes* this possible; by the Interaction principle it is not A or B themselves, so it must be some being that is the ground of their existence. This reasoning can be easily extended to an indefinite plurality of interacting substances. This is not a principle that the modal pluralist *as such* is committed to rejecting, so, by appealing to it, Kant would not be begging any crucial questions against the pluralist.

With the Interaction principle and the Recombination principle in place, it is not difficult to formulate a Kantian argument against modal pluralism:[41]

(1) Let S be the set of all fundamental predicates instantiated by all the *GSRPs*.
(2) ∴ It is really possible that there is a world in which all of the predicates in S are instantiated. [By Recombination*.]
(3) Necessarily, if the predicates in S are instantiated in a world, then their grounds (the *GSRPs*) exist in that world and are world-mates. [Assumption.]
(4) Necessarily, any two world-mates interact. [By the definition of 'world'.]
(5) ∴ Possibly, the *GSRPs* interact. [From (2)–(4).]
(6) ∴ There is a substance that grounds the existence of the *GSRPs*. [From (5) and the Modal Corollary.]
(7) Every fundamental predicate is instantiated by some *GSRP*. [By definition.]
(8) ∴ There is a substance that instantiates every fundamental predicate. [From (6) and (7), by Fundamentality.][42]

Call this substance θ. To show further that θ is a *GARP* Kant would have to show not only that it instantiates every fundamental predicate but that every fundamental predicate is fundamental *to it*, that is to say, that there is no distinct being that makes those fundamental predicates possible by instantiating them and grounding the existence of θ. But Kant does not need to show that θ itself is a *GARP*; he only needs to show that *there is* a *GARP*. What Kant needs to exclude is the possibility that there is an infinite ascending chain of beings, each of which instantiates every

[40] Kant attributes this to a "schema of the divine understanding" at Ak. 1: 414. For discussion see Watkins (2005), 149–55 and Insole (2011), 413–19.

[41] This argument is offered as a refinement of Yong (2014)'s proposal.

[42] Compare my reconstruction of this argument to Kant's refutation of Manicheanism in *ND*, Ak. 1: 414–15.

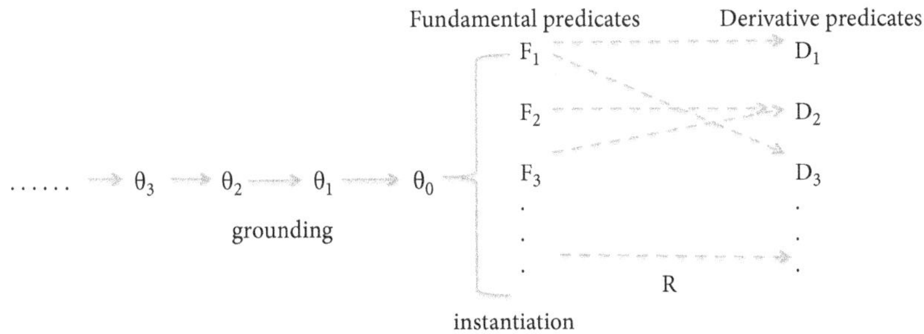

Fig. 5.1. An infinite ascending chain of maximal beings

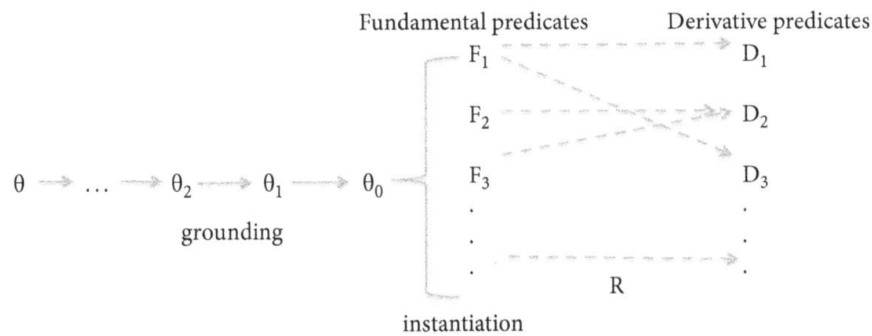

Fig. 5.2. A series of maximal beings terminating in a GARP

fundamental predicate; for the sake of simplicity, I will refer to such beings as 'maximal' beings. Kant needs to exclude, therefore, the view shown in Figure 5.1.

From the Kantian perspective, none of the beings on the left is *absolutely necessary*;[43] if you canceled any one of them, the beings further to the left would still exist, and every really possible predicate would still be possible (fundamental predicates would still be instantiated, and the derivation relations would still hold). If Figure 5.1 represents the metaphysical structure of the grounds of real possibility, there is no *GARP*.

To exclude such a view, Kant would have to appeal to a strong form of the principle of sufficient reason, for note that in Figure 5.1 every derivative predicate is grounded in some fundamental predicate, every fundamental predicate is instantiated, and every maximal being is grounded in another maximal being. So Kant would have to appeal not only to the principle that everything has a ground but one to the effect that there are no infinite ascending chains of grounds, or, in the terminology he will deploy in the *CPR*, that every series of conditions (grounds) terminates in an unconditioned condition: a ground that has no ground.[44] This would license Kant in concluding that the structure of grounding relations is as shown in Figure 5.2.

[43] Though the *aggregate* of them is; see section 3.

[44] A416–8/B444–6.

The being that terminates the chain of grounds on this picture is a *GARP*: it instantiates every fundamental predicate and it has no ground. If it were not to exist, nothing would be really possible (because none of the maximal beings would exist, and thus none of the fundamental predicates would be instantiated).

Before continuing, I want to consider an objection to my reconstructed Kantian argument against modal pluralism. The most questionable premise in this argument, from the pluralist point of view, is (3). Why must the pluralist admit that if substances A and B ground the instantiation of F and G in the same world (by substances that are world-mates) then A and B are members of that same world?[45] The pluralist could claim instead that the finite substances whose instantiations of F and G are grounded in A and B are world-mates but A and B are not parts of any world (a pluralist version of the Kantian doctrine that God grounds the existence of substances in worlds but is not a part of any world). A and B, after all, do not interact; their grounding role is one-way: they ground instantiations of F and G, and derivative predicates (see Fig. 5.3).

Although on such a view A and B do not technically *interact* with any finite substances (or with another) because interaction is *mutual* and finite substances are not the grounds of any accidents in A and B, Kant could point out that they bear the following relation to one another: they are the grounds of substances that interact, and he could reasonably ask what is the ground of the fact that they stand in this relation to one another. After all, the fact that A and B are able to ground substances that causally interact cannot itself be *groundless*. There must be some ground of this channel of influence and it cannot be A or B (because the pluralist is arguing that A and B do not interact). Kant will reply that it must be some third substance that grounds the existence of A and B.[46]

Perhaps a better pluralist response would be to reject the Recombination principle altogether and claim that fundamental predicates are never instantiated by

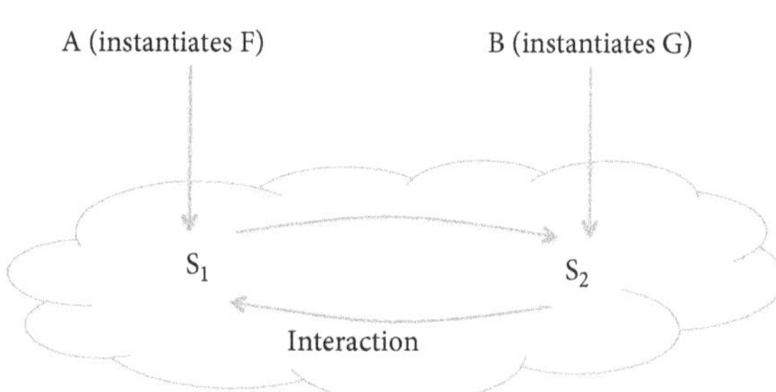

Fig. 5.3. Extramundane grounds of possibility

[45] This is a version of the problem I pointed out for Yong (2014)'s reconstruction.

[46] Cf. "suppose that [the substances that compose a world] are caused by a number of necessary beings; the effects, the causes of which are free from reciprocal relation, would not be in interaction" (*ID*, Ak. 2: 408).

substances that compose worlds but are instead instantiated by extramundane substances that stand outside of worlds but ground substances in worlds. In Figure 5.3, this means thinking of substances S_1 and S_2 as instantiating only derivative predicates. But the Kantian response to this move will be the same: although on this view substances A and B do not mutually interact (they do not cause alterations in one another), they are grounds of substances that interact. What grounds the fact that the very substances whose predicates (e.g., modes of thought) are grounded by the fundamental predicates of substance A (e.g., the attribute of thought) can interact with beings whose derivative predicates (e.g., modes of extension) are grounded by the fundamental predicates of substance B (modes of thought)? The pluralist might reject the assumption that these modes interact at all, but the pre-Critical Kant takes himself to have proven, contra Leibniz and other rationalists, that members of the same world must mutually interact. So there must be a ground of the fact that A and B can ground members of the same world, and, Kant will argue, the only being capable of grounding this fact is a being that grounds both of them. As he says in the Herder metaphysics lectures: "if several causes [*Urheber*] posit several things, then these causes must be connected among themselves [*verknüpft unter sich*], because their consequences are connected, and thus they must not be absolutely necessary, because they have a common cause [*alle zusammen eine Urheber haben*]" (Ak. 28: 129). Since A and B, by hypothesis, instantiate fundamental predicates, their common ground must possess both predicates (see Fig. 5.4).

Obviously, this argument rests on Kant's own pre-Critical cosmology, so any pluralist who rejects that metaphysical theory can escape this argument. However, as I argued above, that cosmology is not prima facie question-begging against the modal pluralist. The point of this reconstruction, though, is not to show that Kant decisively demonstrated that there is a unique ground of all real possibilities. It is to show that he could have taken himself, plausibly, to have good reasons, from within

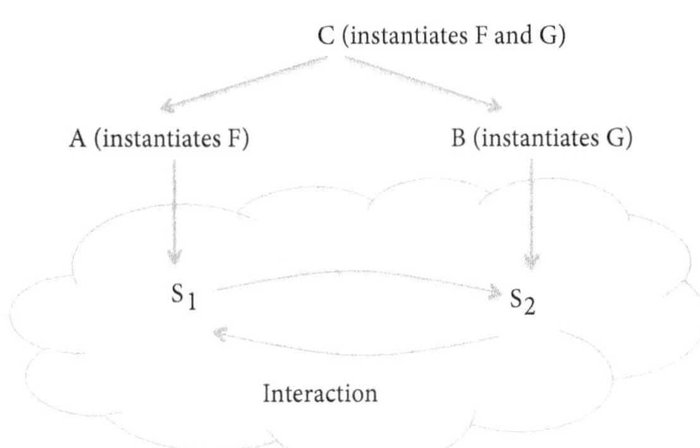

Fig. 5.4. Extramundane grounds with a common ground

his own metaphysical system, for asserting that possibility could not be 'parceled out' among a plurality of grounds.[47]

5.6. How (Not) to Represent God

My Kantian rejoinder to the plurality objection appealed to a modal corollary of the Co-existence principle and a constraint on fundamentality:

> (*Modal corollary*) If there exists a plurality of substances that possibly interact then there exists a substance that is the ground of their existence.
> (*Fundamentality*) If F is a fundamental predicate and *x* is the ground of the existence of something that instantiates F, then *x* instantiates F.

But note that the modal corollary is a principle about *causal* grounding: there must be a common *cause* of the existence of a plurality of (possibly) interacting substances.[48] If the argument from section 5 is to be valid, then Fundamentality must also apply to causal grounding (though it may also apply to grounding in general). But this has the effect of making God *causally* fundamental with respect to the *GSRPs* and making God's fundamental predicates his *causally* fundamental predicates. The reason, according to this argument, why the fundamental predicates instantiated by GSRPs are not fundamental *to them* (the instantiation by GSRPS of those predicates is not what makes them possible) is that they are not *causally* fundamental to them; the existence of those substances, and their instantiation of those properties, is causally grounded in God and his instantiation of those fundamental predicates. This has the effect of treating God's fundamental predicates as his causally fundamental predicates, the predicates without which he would not be able to cause the instantiation of these predicates (*Fundamentality*), and thus be unable to causally ground the possibility of the interaction of *GSRPs* that instantiate them (*Modal corollary*). So this argument works by implicitly treating God's modally fundamental predicates as his causally fundamental predicates and thus treating God as the *causal* ground of all really possible predicates.

If God's modally fundamental predicates are his causally fundamental predicates, then derivative predicates (predicates that are possible in virtue of being 'consequences' of God's fundamental predicates) are causally derivative of God. But this leaves us with two options: either the (really possible) derivative predicates are *actually caused to be instantiated* by God or they are possible in virtue of being *such that God has the power to cause them to be instantiated*. If we take the former

[47] The argument of the previous three paragraphs can be seen as a supplement to the argument of Yong (2014).

[48] In *ND* Kant attributes their common ground to the "schema of the divine understanding" which he describes as the "origin of existence" and as "preservation" (Ak. 1: 414). I take this to mean that the common ground of their existence is the divine intellectual apprehension that *causes* them to exist and causes them to *continue* to exist (preservation).

route, then there will be no non-instantiated but possibly instantiated predicates. But Kant himself claims that a derivative predicate like <*extended*> is still possible even if it is not instantiated; in this context this means that <*extended*> is possible even if God does not actually cause any instantiations of it. So the grounding of derivative predicates in God through *actual* causation is incompatible with Kant's own views about the possibility of non-instantiated predicates.

The alternative is to hold that derivative predicates are grounded, not in God's actually causing them to be instantiated, but in God's *power* to cause them to be instantiated and that God has the power to cause instantiations of predicates that are not actually instantiated. The conclusion of this line of reasoning is that *if* the argument in section 5 is indeed Kant's best possible reply to modal pluralism *and* it implicitly treats God's fundamental predicates as *causally* fundamental then, given that Kant does not want to be committed to the view that (necessarily) there are no contingently uninstantiated derivative predicates, we must implicitly think of God's fundamental predicates as his *powers* and think of derivative predicates as predicates God has the *power* to cause to be actualized.

But this, of course, is precisely how, according to Kant, we should *not* represent God as grounding all possible predicates. As I discussed in the previous chapter, Kant explicitly says that the essence of God, in virtue of which he grounds all really possible predicates, itself makes possible his power [*Allmacht*]; I take this to mean that God's power is itself a derivative predicate, something that is grounded in whatever fundamental predicates of God ground all possibilities. Intuitively, God has the power to actualize various possibilities *because* they are possibilities; the fact that God has the power to cause predicate D to be instantiated is itself partly grounded in the fact that D is possibly instantiated, not constitutive of that fact.[49] Elsewhere in *Beweisgrund* Kant makes clear that we cannot conceive of God grounding possibility through his *powers*:

> Though one can form, through an analogy with human action, some concept of how a being can be the cause of everything actual, one can form no concept of how such a being could contain the ground of the inner possibility of other things. It appears that this thought rises far higher than a created being can reach. (Ak. 2: 152–3)

Kant's point is that we can coherently represent God as the ground of the existence of everything actual through an analogy with our will (God's will stands to the actual world in a relation analogous to that between our wills and the products of our actions), but he denies that this analogy can be extended to represent God as the ground of the possibility of things. The modal analogy Kant has in mind, I take it, is that, just as various actions are possible for us because we have the power to perform them, so various things are possible because God has the power to create or actualize

[49] As Newlands (2013) and Chignell (2012), quite rightly, objected to the 'powers' interpretation in Stang (2010).

them. But this is precisely the model of God's grounding of possibility that, I have argued, the reply to modal pluralism in section 5 relies on: conceiving of God's fundamental predicates as causally fundamental *powers*. Kant does not explicitly state *why* we cannot, using an analogy with human action, form a concept of how God contains the inner ground of the possibility of all things, but his reasoning is not difficult to reconstruct: plausibly, it is just the idea, discussed above, that God's own power to create something is itself partly explained by the fact that this thing is possible and hence cannot be part of the explanation of that fact (on pain of circularity).

This is the basic tension in Kant's pre-Critical modal metaphysics: in order to understand his modal argument we need, implicitly, to think of God as grounding real possibility through his causal power, but we have independent reason to think this cannot be the case. Some readers might wonder whether the requirement to think of God grounding possibility through his powers arises only on my idiosyncratic reconstruction of Kant's argument against modal pluralism, that is, whether this is a tension of my own construction. However, recall that Kant maintains in *Beweisgrund* that God is the real ground of the real element in real possibility, but fails to specify what kind of real grounding is involved here (causal or otherwise). Nor do Kant's other writings (published or unpublished) contain an answer to this question. What is more, it is not clear what category of real grounding other than causal grounding is available to Kant, for he does not develop a theory of real non-causal grounds in any of these writings.[50] It is no wonder, then, that in reconstructing Kant's arguments we need to rely implicitly on the notion of causal grounding. Kant says that God is the real ground of possibility, but explicitly forbids us from representing God's relation to real possibility using the only species of real grounding about which he has anything substantive to say. But if we take seriously Kant's claim that we "can form no concept of how [God] could contain the ground of the inner possibility of other things," then we are left with the conclusion that, through implicitly assimilating it to causal grounding, we obtain no adequate way of representing God's relation to all possibility. *Beweisgrund*, therefore, raises a question about our capacity to represent possibilities that it does not answer: how is it possible for us to represent the relation between the space of real possibility and its unique ground in such a way

[50] Kant's examples of real grounds in *NG* (Ak. 2: 190, 202) and *MH* (Ak. 28: 12, 24–6, 37, 43, 49, 103, 844–5, 888) are all examples of causal grounds (primarily, forces—at Ak. 28: 102 he even appears to equate these concepts). Nor is the other main source for Kant's notion of real ground, Crusius, particularly helpful on this score. In *Ent.* §36 Crusius divides real grounds into efficacious causes and "inefficacious causes or existential grounds." Clearly, the former category will be no help in understanding non-causal real grounds; his main examples of existential grounds are a wedge that has the power to overcome a resistance, whether or not it ever exercises this power, and "the three sides of a triangle and their relations to each other [which] constitute a real ground of the size of the angle" (*Ent.* §36, translation from Watkins 2005, 84–5; see also *Ent.* §79). The wedge example is clearly useless in understanding non-causal real grounding, and, while I think the mathematical example is a clear precursor to Kant's notion of a non-causal real ground of possibility (*ratio essendi*) in the Critical period, it is not particularly helpful to Kant in the 1760s, for he has not yet developed his Critical theory of space as the form of intuition that grounds the possibility of mathematical objects. See Ch. 7 for more discussion.

that would allow us to formulate and understand Kant's own argument for the existence of such a ground? In the second half of this study I will argue that Kant does not fully answer this question until 1790, in the *Critique of Judgment.*

5.7. Kant's Pre-Critical Modal Metaphysics

As I pointed out in section 1.3, Kant sometimes uses 'ontological proof' in a wider sense to refer to any *ontotheological* argument for the existence of God, one that attempts to prove his existence using modal concepts alone. In this broader sense, Kant's modal argument in *Beweisgrund* is an ontological argument because it is an argument in ontotheology.[51] But the *Beweisgrund* argument is an argument from a consequence—that something is possible—to an antecedently determining ground of that consequence, a real ground that explains why it obtains. Kant's argument provides *a posteriori*$_G$ knowledge of God's existence: it demonstrates his existence on the basis of a consequence of that existence, the fact that there are really possible predicates. This is because Kant's view is that there is no such *a priori*$_G$ (antecedently determining) ground of God's existence. As he says in the Herder metaphysics lectures: "that which absolutely necessarily exists does not exist because of a cause [*Ursache*], but rather because its non-existence cannot be thought. But this is merely a ground of knowledge [*Erkenntnisgrund*] and not an antecedently determining ground [*ratio antecedenter determinans*]. In short: it exists" (*MH*, Ak. 28: 55).

Nor is there an antecedently determining ground of God's possibility:

> Now the necessary being is the sufficient real ground of everything else which is possible, apart from itself. Consequently, the necessary being will possess that predicate, through which everything else, apart from itself, can become actual, in agreement with these relations [order, beauty, perfection]. (*OPG*, Ak. 2: 88)
>
> Its own possibility is originally given in its existence. It is of other possibilities that the necessary being contains the real ground. It follows from this in accordance with the law of contradiction that it cannot be the real ground of the possibility of the most real being itself [...] (*OPG*, Ak. 2: 86)[52]

In the first passage, Kant draws attention to the fact that God is not the ground of his *own* possibility; God is "the sufficient real ground of everything else which is possible, apart from itself." In the second passage Kant writes that "its own possibility is originally given in its existence." I take this to mean: God exists, therefore God is possible. However, if God's existence is not an antecedently determining ground of

[51] See *OPG* (Ak. 2: 155), *MV* (Ak. 28: 454), and ML_2 (Ak. 28: 599).

[52] The next paragraph begins "accordingly, the possibility of all other things, in respect of that which is real in them, rests on the necessary being" (Ak. 2: 87) Cf. Ak. 2: 88 and *ND*, Ak. 1: 396, where Kant writes, "of all beings, God is the only one in which existence is prior to, or, if you prefer, identical with possibility." Later on the same page he draws the conclusion that God is exempt from the law that everything has a determining ground.

his possibility, in what sense is God's possibility 'given' in his existence? The relation between God's existence and his possibility, for Kant, is epistemic and logical: God's existence entails that he is possible and is thus is a ground of *knowing* that he is possible (a 'consequently determining' ground).[53] There is no ground that explains *why* God is possible, because such a ground would have to be either God himself or a being distinct from God. No being distinct from God can ground the possibility of God, since God is himself the ground of the possibility of all other beings. Nor can God be the ground of his own possibility, since Kant is adamant that nothing can ground itself.[54] Therefore, just as there is no *a priori* ground of God's existence, there is no *a priori* ground of his possibility.[55]

The logicist view was that God exists necessarily in virtue of his essence containing existence. Kant rejects this as a ground of his necessary existence and offers instead a new concept, that of *absolutely necessary existence*, and a real definition of it: God exists absolutely necessarily (= his non-existence cancels all possibility) *in virtue of* the fact that he is the real ground of all possibility. On this view, God has a hyperintensional counterfactual relation to possibility in virtue of the fact that he has a grounding relation to possibility. It is because God instantiates the fundamental predicates that, were he not to exist, nothing would be really possible. There is no ground of God's existence, but there is a ground of God's possession of the fundamental predicates; it follows from God's essence that he possesses them (although that is not how Kant derives them in his argument; God's essence appears not to be an epistemic ground *for us* of his fundamental predicates). Given that God exists and has these predicates, this grounds his absolute necessity.

So far, we have explored in some detail Kant's rejection of crucial aspects of the logicist account of modality. These include:

(*Anti-Logicism* 1) The idea of a being whose existence is logically grounded in its essence is incoherent. Consequently, no existential proposition is logically necessary. If there are necessarily existing beings, they do not exist necessarily in virtue of a logical relationship between their essence and existence. This is because existence is not a determination; the idea of merely possible but non-existent objects is incoherent.

[53] Cf. Kant's claim that God's possibility "is in him as a predicate" (Ak. 2: 84).

[54] Cf. *ND*, Ak. 1: 394 and *MH*, Ak. 28: 13–14.

[55] There is at least one passage where Kant suggests that we can have *a priori* knowledge of God's existence: "the argument [*Beweisgrund*] that we are giving is based merely on the fact that something is possible. Consequently, it is a proof which can be conducted entirely *a priori*" (Ak. 2: 91). However, I think Kant's point here is simply the point he makes in the passage quoted above, where he writes: "all grounds of proof [*Beweisgrund*] for the existence of God must derive either from the concepts of the understanding of the merely *possible* or from the empirical concept of the *existent*" (Ak. 2: 155–6). When Kant claims that his argument for the existence of God is *a priori* he means that it "derives from concepts of the understanding." I take this to mean that his argument for the existence of a first real ground of possibility is *a priori*. It does not depend upon experience of the determinate predicates of existing things, like their harmonious causal powers, as his alternate argument in Division Two of *Beweisgrund* does.

(*Anti-Logicism* 2) There is real, non-logical incompatibility between predicates. Consequently, some states of affairs are logically consistent but not possible.

But in this and the previous chapter we have also seen Kant's positive modal metaphysics in the pre-Critical period. We can represent this modal metaphysics, first, as the conjunction of three very general claims about real possibility:

(*Non-Logical*) *Real possibility is distinct from logical possibility.* That some proposition p is logically consistent ($\neg p$ does not entail a contradiction) does not entail that p is really possible. So logical possibility and real possibility are at least *conceptually distinct*; it is not a conceptual truth that $\Diamond_L p$ iff $\Diamond_R p$. In fact, some logical possibilities are not really possible; for some propositions p, $\Diamond_L p$ & $\neg\Diamond_R p$. Equivalently, for some p, $\neg\Box_L p$ & $\Box_R p$.
(*Ground*) *Real possibilities have real grounds in actuality.* If it is really possible that p, then there is actually an existing object that grounds the fact that it is really possible that p. The relation between *grounds of real possibility* and the *real possibilities they ground* is not a logical relation; if x grounds the real possibility that p, this is not because of the conceptual containment of p within the concept of x.
(*Worldly*) *Real possibility is a form of metaphysical, or world-based, possibility.* Real possibilities in general are grounded in facts about the world that do not depend upon how our minds are constituted or how we experience or conceptualize the world.
(*GARP*) There is a unique real ground of all real possibility (God). Because it is the real ground of all real possibility, this being exists absolutely necessarily.[56]

I will argue that, in the Critical period, Kant retains Non-Logical and Ground as requirements on real possibility in *general* and retains Worldliness as a requirement on some, but not all, kinds of real possibility. Furthermore, I will argue in Chapter 9 that Kant retains *GARP* not as an article of knowledge, but as a belief grounded in the nature of our faculty of theoretical reason. I have also argued that Kant is committed to, but does not explicitly endorse:

(*Tension*) Given our cognitive constitution, we cannot understand how the *GARP* (God) grounds all real possibilities. However, given those cognitive limitations, we must think of God as grounding possibility through his powers.

In Chapter 10 I will argue that within the Critical philosophy Kant not only remains committed to this, but embeds it within a sophisticated theory of the discursive nature of our intellect, as opposed to the intuitive nature of God's, and why it obliges us to *think* of God as grounding possibility in this way.

[56] I have formulated these as principles about the possibility of *propositions* to facilitate comparison with Kant's Critical views, for, I will argue, Kant gives up his pre-Critical fixation on the possibility, specifically, of predicates to consider the predicative structure of possibility: the possibility of an object instantiating a concept.

PART II

Kant's Critical Modal Metaphysics

6

Real Possibility and the Critical Turn

6.1. Introduction

As we have seen in detail in the previous chapters, Kant rejects the logicist metaphysics of modality. By doing so, he must also reject the logicist account of our knowledge of modality. If not all logically consistent concepts are really possibly instantiated, by analyzing our concepts and ascertaining their logically consistency we do not thereby come to know that they are really possibly instantiated. What then is the source of our knowledge of real possibility? This problem in modal epistemology poses a significant challenge to metaphysics, as Kant and his rationalist predecessors had practiced it. Ontology (*metaphysica generalis*) was conceived by Wolff and Baumgarten and, in some texts, by Leibniz himself, as the science of all possible beings *qua* possible beings. Consequently, a challenge to our modal epistemology, for both Kant and his predecessors, is also a challenge to the very idea that metaphysics is a *science* [*Wissenschaft*], a body of knowledge [*Wissen*], rather than merely rationally grounded conjectures or hypotheses.

This chapter traces Kant's struggle with this and related problems, from the 1760s through the publication of the *Critique of Pure Reason* in 1781 (and the B edition in 1787). I begin, in section 2, with Kant's modal epistemology in the 'Prize Essay' of 1764, and argue that the epistemology and methodology for metaphysics Kant defends in that essay cannot be squared with the positive metaphysical theory he defends in *Beweisgrund*. In section 3 I examine Kant's 1770 Inaugural Dissertation in light of his continuing engagement with problems about modality. While the distinction between sensible and intellectual cognition Kant makes in that work will permanently transform how he conceives of these problems, the Inaugural Dissertation does not contain a satisfactory answer to the modal problems Kant had been engaged with since at least the mid-1760s. In section 4 I argue that the problem of the 'relation' of *a priori* concepts to objects that Kant described to Marcus Herz in his famous letter of 1772 is in fact a question about the relation of *a priori* concepts (the categories) to *really possible* objects. That problem (what grounds the 'relation' of a representation to a really possible object?) becomes the problem of how the categories, the fundamental concepts of metaphysics, can be used to 'cognize' [*erkennen*] sensibly given objects in the *CPR*. In section 5 I argue that, because <*possible*> and <*exists*> are not real predicates of objects, these modal questions should be formulated at the level of concepts, e.g., how do we know *a priori* that our concepts are really possibly instantiated? In section 6 I return to where this study began in the Introduction: Kant's remarks about the centrality of modal concepts to 'transcendental philosophy' in the *CPR*. I argue that questions about our capacity to represent real possibilities we can *prove* to be really possible are at the center of Kantian transcendental philosophy. In section 7 I reconstruct Kant's positive account of how *a priori* knowledge of really possible phenomena (objects that can be given in sensible intuition) is possible. In section 8 I reconstruct his negative argument that we cannot know what is really possible for 'noumena,' objects that cannot be given in sensible intuition. It follows that we cannot know all (really) possible beings *qua* possible (the objects of general metaphysics or ontology), and, in particular, that we cannot know the real possibility of any of the objects of 'special' metaphysics: the soul, the cosmos, or God.[1]

6.2. Modal Epistemology in the Prize Essay

In the 'Prize Essay,'[2] written at the same time as *Negative Magnitudes* and *Beweisgrund* but published in 1764, Kant offers a methodology for metaphysics. However, that methodology fails to explain how, given his anti-logicist modal metaphysics, we can come to know modal facts.

Kant distinguishes the method of mathematics from the method of philosophy in general, and metaphysics more specifically, which he describes as "nothing other than the philosophy of the fundamental principles of our cognition" (Prize, Ak. 2: 283). Mathematics proceeds synthetically; it starts with a small stock of undefined concepts and undemonstrated principles (axioms), defines new concepts by arbitrarily combining (synthesizing) the undefined concepts, and proves theorems involving those concepts by using their definitions and the axioms. In philosophy, we analyze concepts given by the nature of our intellect (e.g., <*substance*>, <*cause*>, <*space*>), rendering them more distinct, that is, making apparent their logical structure.[3] Consequently, definition plays a different role in the two disciplines. Mathematics begins by defining its concepts, while a definition of a philosophical concept, if attainable at all, is attained at the end of a process of philosophical analysis.[4]

[1] My argument in this chapter, that Kant's doctrine of noumenal ignorance is motivated by concerns in modal epistemology, agrees in broad outlines with Chignell (2010), (2011), and (2014*b*). However, there are many differences of detail in our interpretations, which I do not have space here to discuss fully.

[2] This essay is Kant's submission for the Berlin Academy's 1763 essay competition. Kant received second place; first place went to Moses Mendelssohn's essay "On Evidence in Metaphysical Sciences" (Mendelssohn 1997, 251–306). See the editor's introduction to Kant (1992*a*), lxii–lxiv for a detailed account of the Prize competition.

[3] Prize, Ak. 2: 276.

[4] Prize, Ak. 2: 284.

We cannot begin our philosophical inquiry by defining our concepts because defining them is precisely the task of philosophy. But neither can we expect that the outcome of philosophy will be a definition of all of our concepts; we will eventually come upon concepts we do not know how to analyze any further. Having reached the limit of analysis, we temporarily suspend the project of definition and consider indemonstrable propositions involving these unanalyzable concepts.[5] These propositions will either be indemonstrable *merely by us*, because we cannot further analyze their constituent concepts, or because there is no logical analysis of their constituent concepts that would show them to be true. In the context of Kant's rejection of logicism, it is important to note that he allows that there are indemonstrable necessary truths: necessarily true propositions which are not made true by the internal logical structure of their concepts. These are the precursors of synthetic *a priori* judgments in the *CPR*, but Kant has not yet developed his sophisticated Critical theory of how such judgments are possible. In the Prize Essay he claims only that we "immediately perceive" them.[6]

Kant's proposed methodology for metaphysics in the Prize Essay fits some of the remarks he makes about his methodology in *Beweisgrund*. In that work, Kant eschewed the task of presenting a demonstration of God's existence in strict logical form in favor of presenting the *matter* of such a demonstration, the premises that entail the conclusion (a *Beweisgrund* or 'proof-ground').[7] He spares himself the task of defining various concepts in *Beweisgrund*, claiming that, even without a complete analysis of them (which may be impossible for us), certain fundamental propositions involving them are immediately certain to us, albeit indemonstrable; in the terminology of the Prize Essay they are the indemonstrable fundamental judgments of reason. One such instance is the principle, discussed in Chapter 4, that understanding and will are compatible; Kant gives no argument for this, but claims merely that "an immediate judgment of the understanding forces one to admit the truth of this contention, even though it cannot properly speaking be given the distinction required of a logically complete proof" (*OPG*, Ak. 2: 87). His analysis of existence in Division One also conforms to his epistemology of conceptual analysis in the Prize Essay.[8] He analyzes the concept <*existence*> into the slightly simpler concept of absolute positing, not because he thinks such analyses are in general required in philosophy, but because "the present case is one in which [omitting such an analysis] could occasion confusion and lead to serious errors" (Ak. 2: 70).[9]

Nonetheless, Kant's Prize Essay epistemology has severe limits when the anti-logicist metaphysics of *Beweisgrund* is taken into account. Kant credits us with an

[5] Prize, Ak. 2: 280. [6] Prize, Ak. 2: 280. [7] *OPG*, Ak. 2: 65–7.

[8] See esp. I.1.1, which reproduces the methodology of the Prize Essay exactly.

[9] Cf. *OPG*, Ak. 2: 70. However, given Kant's theory of absolute positing as the positing of an object for a concept rather than the positing of a conceptual mark, it is not clear whether the idea of a further conceptual analysis of absolute positing even makes sense.

ability to immediately know real non-logical modal relations between predicates, for instance, that <*extension*> and <*thinking*> are really incompatible.[10] But Kant does not explain how we have access to these non-logical real modal relations among predicates. As we saw in Chapters 4 and 5, Kant thinks that real possibilities have *real* grounds. A real ground is one whose effects, or range of possible effects, is not contained in or logically derivable from its concept. We cannot come to know the real grounds of possibility, or the way in which they ground real possibilities, merely by analyzing our concepts, or performing any other broadly logical operation. In *Negative Magnitudes* Kant's answer to how we know non-logical grounding relations is more of a restatement of the question than anything: "all of our cognitions of [the relation of a real ground to its consequence] reduce to simple, unanalyzable concepts of real grounds, the relation of which to their consequences cannot be rendered distinct at all" (Ak. 2: 204). We simply do know that some grounds posit their consequences, even though they do not logically entail their consequences, and *a fortiori* we do not know this positing relation through logical analysis of the ground.[11]

6.3. Modal Epistemology in the Inaugural Dissertation

For our purposes, the most important text Kant published between the Prize Essay (1764) and the 'silent' decade that preceded the publication of the *CPR* in 1781 is *De mundi sensibilis atque intelligibilis forma et principiis*, the Latin dissertation he composed on the occasion of his appointment as Professor Ordinarius of Logic and Metaphysics at the University of Königsberg in 1770.[12] The 'Inaugural Dissertation' (ID), as it is commonly known, is a crucial text in Kant's development, for it is the first publication which contains the distinction between concepts and intuitions, the doctrine that space and time are *a priori* forms of intuition, and the doctrine that they present objects as they appear to our sensible faculty, not as they are in themselves. It thus anticipates many of the distinctively Critical themes that will occupy us in the rest of this study. However, it makes few advances on the problems of *Beweisgrund* itself: the nature of real possibility and how it is given to our minds.

The concept/intuition distinction, and the doctrine that space and time are pure forms of intuition, will be discussed in further detail in sections 5–7, but for now I want to discuss them briefly in order to note a modal theme that is sounded in the

[10] *OPG*, Ak. 2: 85 f.

[11] Strictly speaking, Kant's theory of real grounds is compatible with the concept of a real ground *G* containing the concept of its effect *E*, although this conceptual containment relation does not explain the grounding relation between *G* and *E*. Similarly, it might be contained in the concept of (the possibility of) *E* that it is grounded by *G*, but this conceptual relation would not explain why *G* grounds *E*, assuming that *G* is a real ground of (the possibility of) *E*. See Chapter 3.2.

[12] A more complete study would also include discussion of Kant's 1766 work, *Dreams of a Spirit-seer Elucidated Through Dreams of Metaphysics*; for reasons of space, though, I forgo discussion of *Dreams*.

ID and will resurface later in the *CPR*. Kant distinguishes *concepts*, which represent indefinite pluralities of objects in virtue of their sharing common marks, from *intuitions*, which immediately present particular objects to our minds.[13] He points out that there are concepts whose instances cannot be intuitively presented to us, but that we cannot infer from this that these concepts are meaningless or that they are not possibly instantiated. For instance, we have the concept of an infinite composite whole but we cannot be intuitively presented with an instance; because the successive intuition of parts of a whole takes place in time, we can only ever intuit a finite series of the parts of the whole.[14] We have to distinguish, for any given concept, two kinds of modal questions: 'is the object of this concept possible?' and 'can this concept be given an object in sensible intuition?' This distinction in modal orders—which we might think of as a distinction between 'possibility in its own nature' and 'presentability-to-the-mind'—is made especially clearly in the ID, and will become an important theme in the *CPR*.

Consonant with *Beweisgrund*, Kant further distinguishes the first question—whether a given concept is possibly instantiated—into two distinct questions: is it logically possible that it is instantiated? And, is it possibly instantiated? Although Kant does not explicitly introduce the language of 'real' possibility or the 'real' element of possibility in the ID, it is clear this is what he has in mind when he argues that we cannot infer possibility from lack of internal contradiction (logical possibility).[15] He makes an associated distinction between the *logical* use of the understanding, which orders concepts in logical relations according to the principle of contradiction,[16] and the *real* use of the understanding, of which he writes: "the fundamental concepts of things and of [their] relations, and the axioms themselves, are given in a fundamental fashion by the [real use of] pure understanding itself" (Ak. 2: 411). Although Kant does not make this connection explicitly, I think it is clear that the *real* use of the understanding provides us cognitive access to *real* possibility; if what is really possible for things is determined by their fundamental concepts (essences) and principles (their grounds of possibility) then in cognizing those concepts and principles we cognize what is really possible for them.

However, Kant does little more than drop a few suggestive hints about what the real use of the understanding is and how it is possible. These hints foreshadow important Critical doctrines that will concern us in later chapters, so I want to mention them briefly. Throughout the ID Kant contrasts our receptive form of intuition—objects are given to us in intuition by causally affecting our sensible faculty—with the purely spontaneous manner in which God intuits objects.[17] Kant

[13] ID, Ak. 2: 397, 399, 402, and 405.

[14] ID, Ak. 2: 388–9. It is clear from Kant's discussion that the infinite whole in question is an infinite *compositum*, an infinite whole that depends upon its parts; an infinite *totum*, a whole that is prior to its parts, is given to us in our *a priori* intuition of space. See A438/B466.

[15] ID, Ak. 2: 416. [16] ID, Ak. 2: 411. [17] E.g., ID, Ak. 2: 389 n., 396.

notes two main differences between our intuition and God's. First of all, because God is a completely independent being, his intuition of the world does not depend upon anything external to him; God spontaneously creates and intuits the world and, in some passages, Kant seems to identify God's creation of the world with his intuition of it. Secondly, God immediately intuits the whole world *as a whole*, unlike us, who must successively intuit parts of the world and synthesize them into (necessarily limited) intuitions of (successively larger parts of) the world. The theistic cognitive-semantic problem posed by *Beweisgrund* (how we represent God's grounding of possibility) is not addressed here, but Kant does make an interesting new suggestion.[18] He concludes the fourth section by writing that "[Malebranche's view], namely, that we intuit all things in God, is very close indeed to the one which is expounded here" (Ak. 2: 410), but does not further explain *in what respects* Malebranche's view is similar to his own. The Malebranchean view that we perceive objects through ideas in God's mind, translated into this context, means that we cognize real possibilities through the real understanding by participating, in some limited fashion, in divine intellectual intuition. This is significant, because Kant's Critical account of our cognition of real possibility will rest on rejecting such a Malebranchean view; I return to it below in section 8.

6.4. Relation to an Object

Sometime after 1770 Kant discovered a problem in the foundations of metaphysics, which he first poses to Marcus Herz in his famous letter of 1772:

> I noticed that I still lacked something essential, something which, in my long metaphysical studies, I, as well as others, had failed to consider, and which in fact constitutes the key to the whole secret of metaphysics, hitherto hidden even from itself. I asked myself this question: what is the ground of the relation [*Beziehung*][19] of that in us which we call 'representation' to the object? (Corr., Ak. 10: 129)

Kant here raises a problem about 'the relation' of representations to objects and goes on to consider two possible models of this relation: either the object is the ground of the representation, or the representation is the ground of the object. He concludes that in the case of the 'intellectual' concepts involved in the real use of the understanding, concepts given by the nature of our mind itself rather than by abstraction from experienced instances, no explanation has yet been given of their relation to objects.

[18] ID, Ak. 2: 126 n.

[19] Translation (with minor modifications) from Kant (1999), 133. Ellington (Kant (1977), 117) translates *Beziehung* as 'reference.' However, I think this mistakenly prejudges which *Beziehung* Kant has in mind, because 'reference' has a more specific philosophical meaning in English than *Beziehung* does in German.

This passage, and the dichotomy Kant poses (either the object grounds the representation, or vice versa), parallel the opening of the "Transition to the transcendental deduction of the categories" in the *CPR*:

> There are only two possible cases in which synthetic representation and its object can come together, relate to one another necessarily, and, as it were, meet each other: either if the object alone makes the representation possible, or if the representation alone makes the object possible. (A92/B124 f.).[20]

Kant goes on to apply this dichotomy to *a priori* concepts (concepts given by the nature of the mind itself) and their relation to objects. The problem of the relation of *a priori* concepts to their objects is what Kant will call, in the *CPR*, the problem of the 'objective validity' of the categories. The Transcendental Deduction will argue that categories have 'relation' to (objective validity for) sensibly given objects and, what is more, that they have 'relation' *only* to (objective validity for) sensibly given objects.

This close textual parallel has led many commentators to read the problem of the 'objective validity' of the categories in the *CPR* as the successor of the problem of the 'relation' of intellectual representations to objects in the Herz letter.[21] On this reading, the problem that Kant describes to Herz in the 1772 letter is the problem that motivates the Transcendental Deduction. In the remainder of this chapter I am going to follow this interpretive tradition and argue that the original problem in the Herz letter *and* the problem of the objective validity of the categories is a *modal* problem: how is it possible for us to know *a priori* that it is really possible for objects to instantiate the categories?

In both texts Kant raises a question about representation in general, but, at least since ID, he has distinguished two different kinds of representation: concepts and intuitions. He regards the 'relation' of *a priori* intuitions to their objects as less problematic,[22] so I will follow Kant in focusing on the more problematic case of *concepts* and their relation to their objects. It is quite unclear, however, both in the Herz letter and in A92/B124, what the 'relation' in question is.[23] One natural thought would be that the relation in question, of representation to object, is simply the relation *x represents y*. There are good reasons to reject this natural thought, though. First, in the *CPR* Kant will never deny that categories *represent* non-sensible objects

[20] Cf. the parallel passage in *MM*, Ak. 29: 796.

[21] E.g., Cassirer (1981), 126; Kemp Smith (1984), 186–7; Guyer (1987), 25; Longuenesse (1998), 17; and Förster (2012), 2.

[22] Since he had already explained this relation, in ID, through the Critical doctrine that space and time are forms of intuition. See section 3.

[23] Nor is it uncontroversial what *objects* are in question here. Carl (1989*b*) argues that the problem Kant poses in the letter to Herz is the problem of how intellectual concepts relate to empirical objects; Beck (1989) argues, on the contrary, that the problem is how intellectual concepts relate to intelligible objects (noumena). However, Ciovacki (1991) shows that Carl and Beck are both mistaken: in the letter to Herz, Kant is raising a perfectly general problem about the relation of intellectual concepts to objects *überhaupt*. Both restrictions of the problem are misguided.

in the minimal sense that they can be used to think *about* such objects and thus have some (however minimal) intentional content for those objects. But he will deny that the categories *relate to* (in the relevant sense) objects that cannot be given in sensible intuition. Secondly, consider the list of concepts Kant explicitly considers in the Herz letter: mathematical concepts of magnitude, purely intellectual concepts (like *<substance>*), and empirical concepts. One thing all of these concepts have in common, for the pre-Critical Kant, is that they are concepts of really possible objects. If Kant were interested in representation *in general* he would presumably be just as interested in how logically consistent concepts *überhaupt* represent their objects, but he is not. The 'relation' in question seems to arise only for representations of really possible objects. Another natural thought would be that the 'relation' in question here is the relation between an object and a concept it instantiates. However, as I will argue more fully below, Kant holds that mathematical concepts have 'relation' to objects in virtue of being possibly instantiated by objects given in intuition, whether or not they are actually instantiated by any objects. So the 'relation' in question cannot be actual instantiation of a concept by an object.

I take this to be sufficient reason to entertain the hypothesis that Kant is concerned, in both passages, with the question of how a representation 'relates' to *really possible* objects: what is the ground of possibility of our representation of really possible objects? But this suggestion needs to be refined further. If the 'relation' between representation and object at issue in the Herz letter is merely *being the representation of a really possible object* then it does not have an epistemic dimension; a concept, for instance, can be the concept of a really possible object without our knowing it, or even being in principle capable of knowing it. The relation could obtain 'behind our backs,' so to speak. Kant considers one such model, and rejects it on the basis that it is circular.[24] A few paragraphs later in the Herz letter, Kant describes Malebranche and Crusius (as well as Plato) as explaining the 'relation' between representation and objects by appeal to a harmony implanted by God. If the relation in question is simply possible instantiation, this means: our concepts are possibly instantiated because God has created us with concepts of really possible objects (Crusius) or allowed us to see ideas in his mind of really possible objects (Malebranche). But Kant's objection—that it is circular—then becomes unintelligible: Malebranche and Crusius do not offer a circular explanation of why our concepts are concepts of really possible objects.

Kant's circularity objection comes into clearer focus, though, if we recognize that the 'relation' in question is not purely modal but has an epistemic dimension as well: what is the ground of the possibility of our representing really possible objects that we can know *a priori* to be really possible? Malebranche and Crusius (according to Kant) attempt to prove the real possibility of the objects of our concepts by assuming that

[24] Corr., Ak. 10: 131.

some of our other concepts, e.g., <*God*> and <*cause*>, are actually instantiated. I will examine the nature of this circularity objection in detail in section 8 below, but for now I just want to point out that on this modal-epistemic interpretation of the 'relation' of representation to object, Kant's 'circularity' objection becomes at least intelligible (though not necessarily decisive).

If this is correct, then the question whose discovery motivated the 'Critical' turn during the 'silent decade' is implicitly a modal question and, what is more, the modal-epistemic question that I argued is elicited by, but not explicitly posed in, *Beweisgrund*.[25] This also clarifies Kant's claim in the Herz letter that "I, and others, in my long metaphysical labors had neglected" the question of the relation of representations to objects. It is implausible that Kant would accuse his predecessors of neglecting to consider how representations *represent* objects, because Kant's predecessors in the German rationalist tradition developed extensive theories of representation.[26] This remark becomes clearer, however, if we interpret the relation as I have done, for we can read Kant as casting an eye back on his logicist predecessors and his own pre-Critical writings and saying: "they assumed that by representing something as logically possible one thereby represents it as really possible, and although I carefully distinguished these things, I did not have an account of how we represent something as really possible or how we can prove of really possible objects that they are really possible objects." If this is the 'relation' between representation and object that Kant is here concerned with, he is right that he, as well as many others (though not perhaps all), neglected it.

One term Kant uses in the *CPR* for representations of knowably really possible objects is *cognition*:

> To cognize an object it is required that I be able to prove its possibility (whether by the testimony of experience from its actuality or *a priori* through reason). But I can think whatever I like, as long as I do not contradict myself, i.e. as long as my concept is a possible thought, even if I cannot give any assurance whether or not there is a corresponding object somewhere within the sum total of possibilities. But in order to ascribe objective validity to such a concept (real possibility, for the first sort was merely logical) something more is required. (Bxxvi n.)[27]

A concept is a cognition, according to this characterization, only if I can prove that it is really possible for it to have an object (i.e. to be instantiated). Kant characterizes cognition in different terms in other contexts (e.g., as thinking an intuited object

[25] This is one possible route from ID to the *CPR* during the 'silent decade.' In this chapter I cannot hope to do justice either to all of the other factors motivating Kant's thought in this period (e.g., his rereading of Hume, or the discovery of the Antinomies), nor can I engage in detail with the vast scholarly literature on the subject. For other narratives of Kant's development in this period see de Vleeschauwer (1962), Cassirer (1981), Werkmeister (1981), Carl (1989*a*), Kreimendahl (1990), Guyer (1987), and Kuehn (2001).

[26] Most notably, Leibniz's theory of representation in terms of expression; see L 339.

[27] Kant goes on to claim that there can be practical as well as theoretical sources of cognition, but in this book I am focusing almost exclusively on theoretical cognition. See Chignell (2010) and Schafer (unpublished) for more on how practical reason can fill the 'real possibility' requirement on cognition.

under a concept), but for now I want to focus on this modal concept of cognition (m-cognition, for short). The problem of the Herz letter can now be stated in Critical terms as follows: how is it possible for the categories, concepts given by the nature of the mind rather than abstracted from experienced instances, to be *a priori* cognitions? This problem will be our guiding thread for the rest of this chapter.

6.5. Intuition, Existence, and Possibility

Before continuing, I want to make an observation about the transformative role that the intuition/concept distinction will play in how Kant so much as *formulates* questions about possibility. To do so, I need to explain a little more about what intuitions are, and what distinguishes them from concepts.

Kant defines intuitions as *singular* and *immediate* representations of objects, whereas concepts are *general* and *mediate* representation of objects.[28] A concept is a *general* representation because it represents *all* of the objects that share certain common *marks*; these objects make up the extension (in the contemporary sense)[29] of the concept, while the marks of the concept make up its *intension*.[30] We can *use* a concept to pick out a single object by combining the concept with a demonstrative or indexical expression (e.g., '*that* chair') but the concept itself remains a general representation; 'singular concept,' for Kant, is a contradiction in terms.[31] By contrast, intuitions are singular in their very content. I interpret this to mean that an intuition is 'of' a single object. To any intuition we can ask 'what is the object of that intuition?' and the answer must be a single object, rather than a plurality of objects (those two ducks) or a conceptually specified set of objects (all the ducks in the pond).

I interpret the immediacy of intuitions to mean that an intuition does not represent its object via some mediating representation; the immediate object of an intuition is an object itself, not a further representation of that object (e.g., a mental image).[32] Concepts, by contrast, represent their objects by means of some mediating representation, either a concept or another intuition.[33] The clearest case of mediating representations are the conceptual 'marks' that compose a concept. For instance, *<body>* represents bodies by means of the marks of which it is composed—*<extended>*, *<impenetrable>*, and *<lifeless>*—which themselves represent objects

[28] A19/B33, A50/B74, A320/B77, and A713/B741. On the singularity and immediacy of intuitions see also *WL* (Ak. 24: 904–5), *MM* (Ak. 29: 880), and *MK*$_3$ (Ak. 29: 971). There is a vast literature on the Kantian notion of intuition, which I cannot hope to engage with in detail here: Hintikka (1969), Parsons (1969) and (1992), Thompson (1973), Howell (1973), Falkenstein (1995), Smit (2000), Hanna (2008) and (2011), Allais (2009) and (2015), McLear (2011), (2015), and (forthcoming), Tolley (2013) and (forthcoming), and Stephenson (forthcoming).

[29] Kant sometimes identifies the extension of a concept with the set of concepts in which it is contained as a mark, rather than the set of objects that it applies to; see *JL* (Ak. 9: 95–100), *LB* (Ak. 24: 240), and *LP* (Ak. 24: 569), as well as Longuenesse (1998), 86–7 and Anderson (2004), 512–13 for discussion.

[30] *JL*, Ak. 9: 95–6. [31] *JL*, Ak. 9: 91.

[32] Cf. Smit (2000), 263 and Allais (2015). [33] A19/B33 and A68/B93.

(perhaps through the mediation of further marks). Kant does not explain how intuitions can play the mediating function that conceptual marks do, but the mediating role of intuitions will not be crucial for my account so I forgo further discussion of it here.[34]

The first thing I want to explore is the crucial connection between intuition and existence as 'absolute positing.' The immediacy of intuitions means that intuitions can never be 'uninstantiated' the way that *mediate* concepts can be; an 'uninstantiated' intuition would not be about anything, so it would not be a representation of a singular object, contrary to its definition. This entails that for every intuition *there is* an object of that intuition, which entails that *the* object of that intuition (a definite description, due to the *singularity* of intuition) exists in the 'absolute positing' sense.[35] Furthermore, intuition is a necessary epistemic basis for making judgments of 'absolute positing.' This is made especially clear in "On the impossibility of an ontological proof of God's existence" in the *CPR*, in which Kant restates the 'absolute positing' theory of existence from *Beweisgrund*:

> Now if I take the subject (God) together with all his predicates (among which omnipotence belongs), and say **God is**, or there is a God, then I add no new predicate to the concept of God, but only posit the subject with all its predicates, and indeed posit the **object** in relation to my **concept**. (A599/B627)

By contrast, when I judge that *God is omnipotent* I only posit the 'relation' between two concepts: anything that falls under <*God*> falls under <*omnipotent*> (because the latter is a mark of the former). For us, the only way to make an existential judgment with epistemic warrant is through intuition, which for us is necessarily sensible (receptive):

> Thus whatever and however much our concept of an object may contain, we have to go out beyond it in order to attribute existence to it [*diesem die Existenz zu erteilen*]. With objects of

[34] Atomic concepts, concepts that are not logically composed of further concepts, by definition do not have marks, hence their representation of their objects is not mediated by marks. Since atomic concepts' representation of their objects is not mediated by conceptual marks, it stands to reason that it would be mediated by intuition. On this interpretation, atomic concepts represent objects by being abstracted from intuitions that present instances of them. For instance, the concept <*red*>, assuming it is atomic, represents objects via an intuitive representation of a red object: it represents all objects that resemble the intuited instance in respect of color, but lacks any marks. While Kant is skeptical of our ability to discover any atomic concepts (concepts composed of no marks), his account of conceptual representation should not be *inconsistent* with their existence. Cf. Smit (2000).

[35] The idea that the object of an intuition always *exists* is liable to provoke the objection that in various passages Kant seems to allow that the object of an intuition might not exist (e.g., when we hallucinate or dream); B278, *Prol.* (Ak.4: 281–2), B151, and *Anthr.* (Ak. 9: 153, 167–8). But notice that the notion of existence involved in those passages is not mere absolute positing: it is the positing of an object that exists in space outside the subject and is causally efficacious. But that notion of existence is a real predicate, for it "adds to the concept of an object." I am not claiming every intuition has an object that exists in that sense; I am making the more minimal claim that for every intuition *there is* (absolute positing) an object of that intuition. In hallucination, we mischaracterize the nature of the object; we mistake, for instance, a mere visual image for an object that exists in space outside us. See Stephenson (forthcoming), McLear (forthcoming), and Allais (2015) for discussion.

sense this happens through the connection [*Zusammenhang*] with some perception [*Wahrnehmung*] of mine in accordance with empirical laws; but for objects [*Objecte*] of pure thinking there is no means whatever for cognizing their existence, because it would have to be cognized entirely *a priori*, but our consciousness of all existence (whether immediately through perception or through inferences connecting something with perception) belongs entirely and without exception to the unity of experience, and though an existence outside this field cannot be declared impossible, it is a presupposition that we cannot justify through anything. (A601/B629)

Epistemic warrant for an act of absolute positing (i.e. the judgment *there is an F*, or, equivalently, *an F exists*) must come from a consciously apprehended intuition (perception)[36] of the object absolutely posited, or from a perception of a distinct object, from whose existence we can infer the existence of another object, through known empirical laws ("connection [*Zusammenhang*] with some perception of mine in accordance with empirical laws"). Since, by definition, the former is lacking for objects that cannot come before our senses (noumena—see section 7), and we cannot satisfy the latter for such objects because we cannot cognize the concept <*cause-effect*> when applied to them (the argument I reconstruct in section 8), we lack epistemic warrant for acts of absolute positing with respect to objects that cannot be sensibly presented to us (like God). But we should not, therefore, assume *there are no such* objects, or, equivalently, that they do not exist (in the absolute positing sense).

Note that these passages occur in the section of the *CPR* where Kant raises his famous objection against the ontological argument (reconstructed in Chapters 1 and 2): "being is obviously not a real predicate" (A598/B626). This is because *the epistemic indispensability of intuition is the downstream consequence of the denial of ontological proofs*: if ontological proofs were possible, it would be possible to justifiedly make *at least one* existential judgment (there is a God) without intuiting any objects. Conceptual analysis of internally logically consistent objects would be enough for at least some existential judgments.

Having sharply separated existence (the absolute positing of an object) from possibility, and sharply separated intuitions of objects from concepts, Kant now has to interpret questions about *possibility* differently. The question of whether, for instance, spatiotemporal substances are possible is not a question about *objects* (of absolute positing) but a question about the concept <*spatiotemporal substance*>: is it possibly instantiated *by an object*? We cannot intelligibly interpret this question as a question about objects of absolute positing (the objects there are/the objects that exist) because that would be tantamount to interpreting it as the question: are the spatiotemporal substances *there are* possible or impossible? That question would either receive a trivial 'yes' (every object *there is* is possible) or it would involve

[36] See Tolley (forthcoming) on Kant's notion of *Wahrnehmung*.

entertaining the *absolute positing* of impossible objects, e.g., *there are* spatiotemporal substances, but are they possible?

This means that <*possibility*> is not a real predicate of objects any more than <*existence*> is: "the modal categories have this special feature: they do not augment the concept to which they are ascribed, as determinations of the object, but rather express only the relation to the faculty of cognition" (A219/B266).[37] It will take the rest of this book to fully explicate what it means that the modal categories "express the relation to the faculty of cognition," but for now I want to focus on the first part of this sentence. That the modal categories do not "augment the concept to which they are ascribed, as determinations of the object" means that the modal categories are not determinations or 'real predicates,' or, in the terms of Kant's definition later in the *CPR*, they are not "predicates that go beyond the concept of the subject and enlarge it" (A598/B626). Recall my interpretation of what a 'real predicate' is from Chapter 1: a real predicate is one such that, possibly *there is* an object (absolute positing) that does not instantiate it. Assuming my interpretation there was right, this means: it is not possible for there to be (absolute positing) an object that does not instantiate the concept <*possible*>. This means that questions about possibility are not first-order questions about the objects there are (objects of absolute positing) but second-order questions about concepts: is a given concept possibly instantiated, i.e. is it possible for there to be objects of absolute positing that instantiate it? In other words, the distinction between the possible and the impossible is not an extensional distinction between two domains of *objects* (of absolute positing) but a distinction between *concepts*: a distinction between concepts possibly instantiated by objects of absolute positing, and concepts not possibly instantiated by such objects. Likewise, the distinction between the existent and the non-existent is not a distinction between two domains of objects but one between concepts: those that are instantiated and those that are not.[38]

Our epistemic access to absolute positing is via sensible intuition, or so Kant will argue. We must be careful not to assume *at the outset* that all objects *there are* are sensibly intuited or intuitable by us; there may be objects that could be absolutely posited by a being with an intellect different from ours (e.g., a non-sensible or 'intuitive' intellect). And we must not assume *at the outset* that a concept is possibly instantiated if and only if it is possibly instantiated by objects of the kind we have epistemic warrant to posit absolutely: sensible objects. And we should not assume *at the outset* that if there are objects of a certain kind then there is an intellect to posit them absolutely. That there are such objects means that if some intellect were to posit them absolutely then that act of positing would be correct or true; it does not (without additional argument) entail that any intellect *does* posit them or even *could* posit them. Consequently, when I refer to objects as 'objects of absolute

[37] Cf. A74/B100. [38] Cf. Abaci (2013), 18.

positing' I am reminding the reader that 'object' here means: appropriate target of absolute positing (q-object, cf. Chapter 1.3). I am not *necessarily* claiming that any intellect engages in an act of absolutely positing these objects (e.g., judging that there are such objects).[39]

6.6. Transcendental Philosophy and the Concept of an Object

In thinking about the continuity, and differences, between Kant's conception of modality in the *CPR*, his pre-Critical works of the 1760s, and the older logicist metaphysics, a particularly productive place to start is the 'Table of Nothings,' a short section at the end of the Transcendental Analytic, which has not received the attention I think it deserves from scholars.[40] That section begins:

> The highest concept with which one is accustomed to begin a transcendental philosophy is generally the division into the possible and the impossible. Since all division, however, presupposes a concept that is divided, a yet higher concept must be given, and this is the concept of an object in general (taken problematically, and unspecified whether it is something or nothing). (A290/B346)[41]

Kant's claim that the highest distinction in transcendental philosophy is "generally" [*gemeiniglich*] taken to be that between the possible and the impossible is a reference *back* to Wolff and Baumgarten; as we saw in Chapter 1, the highest concept of ontology (*metaphysica generalis*) is the concept *ens/Ding*, possible thing or possible being, and Wolff and Baumgarten begin by dividing possible beings from impossible ones (*nihil, Nichts*).[42] But 'transcendental philosophy' is Kant's term for the philosophical discipline whose preparatory critique is being given in the *CPR*,[43] so Kant is here announcing that the *CPR* is both a *successor* to ontology and an *improvement* in this respect: it begins with the more general concept <*object*>, which it then further divides into <*possible*> and <*impossible*>.[44]

It is important from the outset to realize that these are not *extensional* distinctions between different sets of objects; Kant is not claiming that the extension of <*object*> includes both possible and impossible objects, just as Wolff and Baumgarten are not

[39] There may be forms of absolute positing that are not judgmental; e.g., the intuitive intellect may absolutely posit its objects without making judgments (because judgment is a form of discursive unity). See Ch. 10 for more.

[40] It merits only one paragraph in Willaschek (1998) and is not mentioned in Guyer (2010); Longuenesse (1998), 303–5 is a welcome exception.

[41] The 'Table of Nothings' is mirrored by several similar passages in Kant's lectures on metaphysics in the 1780s, corresponding to Baumgarten's discussion of possibility and impossibility in *Meta.* §§7–18; see *MV* (Ak. 28: 414), ML_2 (Ak. 28: 543–4), *MD* (Ak. 28: 628), *MM* (Ak. 29: 811–12), and MK_3 (Ak. 29: 960–2).

[42] *Meta.* §7, *Ont.* §57, *Dt.Met.* §28.

[43] A12/B25–6.

[44] Cf. Kant's claim at A246–7/B303 that the "proud name of ontology... must give way to the modest one of a mere analytic of the pure understanding." Cf. Ak. 29: 752.

claiming that *impossible beings* are a kind of *being*. He is making an intensional distinction among concepts in virtue of their representational content: those that are possibly instantiated (concepts of something) and those that are not (concepts of nothing). This also means that Kantian transcendental philosophy operates at one level of abstraction higher than Wolffian ontology; rather than concerning itself in the first instance with objects (beings) it concerns itself with our representations of objects. Its highest genus is *<object of representation>*, which Kant then divides into the concepts *<possible object (of a representation)/something>* and *<impossible object (of a representation)/nothing>*.

Kant then divides the concept of nothing—the concept *<impossible object (of a representation)>*—into four sub-concepts, organized according to the categories. While Kant does not draw this out, each of these 'concepts of nothing' is actually a different distinction between concepts of impossible and possible objects. For instance, the first 'concept of nothing' is the concept of an *ens rationis*, which Kant glosses as "the concept of an object to which no intuition corresponds" (A290/B347). This corresponds to the distinction between concepts for which corresponding objects cannot be given in intuition (*ens rationis*, nothing), and concepts for which a corresponding object can be given (something). Kant organizes his 'table of nothings' according to the four moments of the table of categories (quantity, quality, relation, and modality),[45] but for our purposes a different principle of organization is more relevant: the different distinctions between possible and impossible objects are logically subordinate to one another. For instance, a concept of an object that can be given in intuition (something that is not a mere *ens rationis*) is by definition logically consistent, so it is not a *nihil negativum* (the object of a logically inconsistent concept), but the converse does not hold; some 'logical' somethings (objects of logically inconsistent concepts) are merely *entia rationis* (e.g., Kant's example of a fictional force) for which objects in intuition cannot be given. If we unwind these logical subordination relations among different possible contents we arrive at the following table of divisions of possible and impossible objects (summarized in Figure 6.1):

1. *Logical something vs. logical nothing (nihil negativum).* This is the distinction between concepts that are logically possibly instantiated and those that are not.
2. *Real something vs. real nothing (ens rationis).* This is a distinction within the 'logical something,' and I discuss it in detail below.
3. *Material something vs. formal nothing (ens imaginarium).* This is the distinction, drawn within the real something (objects that are not *entia rationis*), between material objects that can be given in space and time through sensory experience, and the *a priori* forms of space and time themselves.

[45] A290/B346.

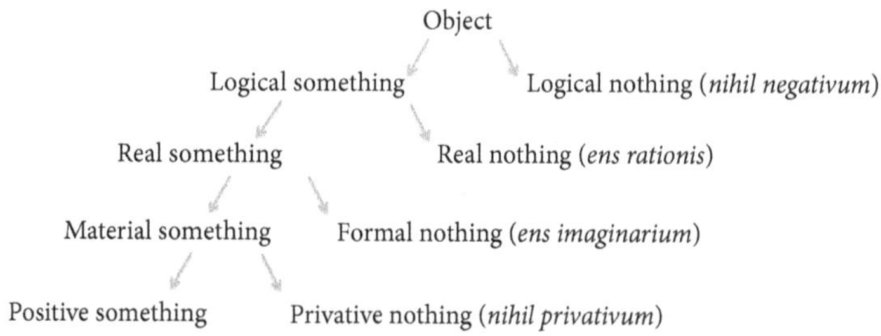

Fig. 6.1. The table of nothings (organized by logical precedence)

4. *Positive something vs. privative nothing (nihil privativum).* This is the distinction, drawn within material objects in space and time (objects that are not *entia imaginaria*), between positive determinations (light, heat) and a mere absence of positive determination (darkness, cold).

I have not attempted to characterize the second pair of distinctions in explicitly modal terms, because to do so would presuppose the details of Kant's Critical epistemology, and because it is not crucial for my purposes here. Instead, I want to focus on the first two something/nothing distinctions.

In his explanation of the distinction between an *ens rationis* and what I have called a 'real something,' Kant writes:

> thus the object [*Gegenstand*] of a concept, to which no corresponding intuition can be given [*gar keine anzugebende Anschauung correspondiert*], = nothing, i.e. a concept without an object, like the noumena, which cannot be counted among the possibilities, although also for that very reason must not be assumed to be impossible (*ens rationis*); or like certain new forces, that, although they are thought without contradiction, lack examples in experience, and therefore must not be counted among the possibilities. (A290/B347)

Two points are important here. First, this is a distinction that concerns *real* possibility; the mere logical possibility of a concept does not entail that its object is possible. Secondly, it is not the distinction between what is really possible and what is not really possible; it is the distinction between what we can know to be really possible and what we cannot know to be really possible.[46] The concept of a noumenon, like the invented concept of a force that has never been encountered, is a concept of 'nothing' because we cannot prove that its object is really possible. We should not, Kant tells us, conclude that such objects are therefore impossible.

This explains why the highest genus of transcendental philosophy is *<object (of representation)>*. Kant defines 'transcendental' at the beginning of the Transcendental Logic as follows: "I call all cognition transcendental that is occupied not so much with objects but rather with our mode of cognition of objects, insofar as this is to be

[46] Cf. MK_3 (Ak. 29: 961).

possible *a priori*" (B25).[47] The most general question for transcendental philosophy is, how can we represent any object whatsoever *a priori*? To answer that question we must distinguish representing a logically possible object from representing a logically impossible one (hence the first division of the Table of Nothings, when it is organized logically: logical nothing versus logical something), and ask, what is the ground of the possibility of our representing *a priori* a logically possible object? The answer to this first transcendental question is relatively straightforward: by representing something that obeys the principle of non-contradiction (PNC) we represent a logically possible object.

The second question of transcendental philosophy, according to my reorganization of the Table of Nothings, would be, what is the ground of the possibility of our representing something really possible that we can know to be really possible (a real something)? Because the answer to the first transcendental question is relatively trivial (PNC), this has a claim to being the first *substantive* question of transcendental philosophy. This characterization of transcendental philosophy may seem unfamiliar to many readers, but recall Kant's characterization of cognition in explicitly modal terms from Bxxvi n. (quoted in section 4): to cognize an object we must be able to prove that the object is really possible. But that is equivalent to representing a real something (in the Table of Nothings). So if this is correct, the first substantive question of transcendental philosophy is, how is *a priori* cognition possible?

Ontology, as conceived by Wolff and Baumgarten, is, the science of all possible beings insofar as they are possible. Transcendental philosophy, the Kantian successor to ontology, differs in two crucial respects. First, it is not a first-order science of some domain of objects, but a second-order inquiry into how representation of objects is possible. Secondly, it distinguishes between representing a (knowably) really possible object and representing a (knowably) logically possible object, and inquires specifically into the grounds of possibility of the former. Note that this distinction does not rest on assuming that logical possibility and real possibility differ in extension, i.e. that there is some p such that it is logically possible that p but not really possible that p. For even if logical possibility and real possibility are the same in *extension* they differ in *intension*. The content of the concepts *<logically possible>* and *<possible>* are different; otherwise, the logicist view that they are co-extensive would be *uninformative*. Consequently, representing an object as really possible is distinct from

[47] This is a modification from Kant's original definition of transcendental cognition at A11: "I call all cognition transcendental that is occupied not so much with objects but rather with our *a priori* concepts of objects in general." However, this modification in the B edition merely brings Kant's definition of *transcendental* cognition into line with his definition at A57/B80. The *CPR*, Kant remarks in both the A and the B Introductions (A12/B25–6), does not contain the complete system of transcendental philosophy because it gives neither a complete inventory, nor an exhaustive analysis, of *a priori* concepts; instead it contains a 'critique' of our rational faculty, an analysis of the fundamental principles and modes of operation which make it possible for us to cognize objects *a priori*, as well as an evaluation of their legitimate scope and use.

representing it as logically possible, and proving that it is really possible is different from proving it to be really possible. By answering the transcendental question about logical possibility, the logicists have not *yet* answered the transcendental question about real possibility.

The question elicited, but not explicitly raised, by *Beweisgrund*—how are real possibilities given to the mind?—becomes the principal question of transcendental philosophy in the *CPR*: how is it possible for us to represent *a priori* knowably really possible objects? This encompasses three interrelated questions, each of which will receive its answer in the course of the *CPR*:

(1) What is it to represent an object *as really possible*? That is, what is the content of the representation of an object *as* really possible?
(2) What is it for an object to be really possible? What are the grounds of real possibility? (At least for objects whose real possibility we can represent (1) and prove (3).)
(3) Of a really possible object we can represent as really possible, how can we prove that it is really possible? And, what's more, prove this *a priori*?

These are questions, respectively, about the cognitive semantics, the metaphysics, and the epistemology of modality. In the next section I am going to reconstruct Kant's positive Critical theory of cognition of really possible objects (his answer to (1) and part of his answer to (3)), so before continuing I want to make a brief remark about the concept of an object.

To relate this back to section 5, the most general concept of object in transcendental philosophy is not 'object of absolute positing.' For us, there are two kinds of representations: concepts and intuitions. Intuitions give us objects of absolute positing (though we may not be consciously aware of every object we intuit, and hence we may not engage in an act of absolute positing for each of them). When we say that a concept C is the concept of an impossible object we are not saying *there is* an object (of absolute positing) that instantiates C but that object is impossible.[48] Instead we are saying something about the content of C: because of this content it is impossible for C to be instantiated by an object (of absolute positing). So we have a notion of object that is, so to speak, internal to representational content, sometimes called the 'intentional' object of the representation. I will call this notion of object 'object of representation' (r-object, for short) and distinguish it from the notion of an 'object of absolute positing' (q-object). Talk of r-objects is really shorthand for talk of

[48] Interpreting the idiom 'C is a concept of a really impossible object' in this way is unattractive for at least three reasons. First, it would populate our theory of what there is (our 'ontology' in the Quinean sense) with a host of absurd beings: the largest prime number, bodies that think, round square, etc. Secondly, by parity of reasoning, it would entail that '<C & ~C> is the concept of a logically impossible object' entails that there is an object, that is both C and ~C, which would require that the PNC be restricted to logically possible objects. Thirdly, there is a more natural interpretation available, given in the body of the text.

conceptual contents. This means that we can more perspicuously formulate questions (1)–(3) in terms of concepts, as follows:

(1*) What is it to represent a concept as really possibly *instantiated by a q-object*?
(2*) What is it for a concept to be really possibly *instantiated by a q-object*? What are the grounds of real possibility?
(3*) How can we prove a concept to be really possibly *instantiated by a q-object*? And, what's more, prove this *a priori*?

Sometimes I will talk about the possibility of the object of a concept; in every case, this means: the possibility that the concept is instantiated by a q-object.[49]

6.7. *A priori* Cognition of Phenomena

In this section I reconstruct Kant's reasons for thinking that we can have *a priori* (modal) cognition of phenomena, objects that can be given to us in sensible intuition. Contrary to the method of the rest of this book, in this section I am going to be somewhat naive about what exactly (real) possibility means; my aim is to start with some stock of clear Kantian principles about possibility (e.g., that logical possibility is not sufficient for real possibility) and construct, over the course of this section, a clearer idea of what notion of real possibility is at work in Kant's positive theory of *a priori* cognition (which I call *formal* possibility). In Chapters 7 and 8 I distinguish formal possibility from several other notions of real possibility in Kant's Critical system.

Cognition [*Erkenntnis*] is a technical term and, as Kant scholars are increasingly appreciating, does not correspond very well to our concept 'knowledge.'[50] First of all, knowledge is standardly taken to be propositional, while Kant regards intuitions and concepts (under the right conditions) as cognitions, but neither of them has a propositional object.[51] Secondly, Kant appears to countenance the possibility of *false* cognition,[52] but the notion of *false* knowledge is prima facie incoherent. Thirdly, Kant's definition of cognition is very different from his definition of knowledge [*Wissen*]. Kant gives two kinds of characterizations of cognition. First, a modal one we have already seen: a representation is a cognition just in case we can prove the real possibility of its object. Secondly, a 'semantic' one: to cognize an object is to think a given object under a (empirical or *a priori*) concept.[53] By contrast, Kant defines

[49] For more on Kant's concept of an object, and the distinction between r- and q-objects, see my unpublished paper "Kant and the Concept of an Object," available on my website (see Notes on the Text).

[50] I have been especially influenced here by Schafer (unpublished) and (forthcoming); the following remarks on the difference between cognition and knowledge are deeply indebted to him. Related ideas are developed by Smit (2000) and Hanna (2006).

[51] The construction '*anschauen daß*' does not (to my knowledge) appear anywhere in Kant's writings.

[52] A58/B83.

[53] E.g., A50/B74. I call this the 'semantic' characterization of cognition because it describes cognition in terms of the interplay of two kinds of representations (concept, intuition), which are themselves defined in terms of their different manner of 'relating to their object' (their content).

knowledge [*Wissen*] in terms of an epistemic relation between a subject's grounds for 'holding [a judgment] to be true' [*Fürwahrhalten*] and the objective grounds of the judgment's truth.[54] This is very different from the definition of cognition, which is not specified in terms of the holding-to-be-true of a judgment at all. There are even passages where Kant explicitly contrasts *Erkenntnis* and *Wissen*, for instance, "there can be a science of that of which our cognition [*Erkenntnis*] is not knowledge [*Wissen*]" (*JL*, Ak. 9: 72).

On this interpretation, what defines cognition is not an epistemic warrant but the way in which it represents its content; cognition is not a purely epistemic but a partly semantic kind. Focusing for the moment on the modal definition, cognition of an object represents it as the really possible object it is; this is why Kant imposes the additional requirement that cognizing an object involves being able to differentiate it from other possible objects.[55] Cognition requires an epistemic warrant for knowing the object to be really possible, but it is not itself that epistemic warrant or knowledge or proof.

As I noted above, Kant gives two different kinds of characterizations of cognition, a modal one (m-cognition, for short) and a semantic one (s-cognition, for short). We saw m-cognition in the previous section, and we have been discussing s-cognition here: thinking a given object under a concept. In some contexts Kant emphasizes m-cognition,[56] in other contexts, s-cognition,[57] and this naturally leads to the question of which one is fundamental and which one is derivative. I think that m-cognition is fundamental and s-cognition (thinking a given object under a concept) is logically downstream of it. I have three main reasons for thinking this.

First, the dialectical role that the notion of cognition plays in Kant's critique of traditional metaphysics. Kant will argue that we cannot cognize non-sensible objects (noumena) and hence that we cannot have knowledge (*Wissen*) of them.[58] Consequently, metaphysics, conceived as a scientifically organized body of knowledge (a science, *Wissenschaft*) of noumena, is impossible for us. This makes perfect

[54] A822/B850. I examine the different modes of '*Fürwahrhalten*' in detail in Ch. 9.

[55] A58/B83, A242 n. Thanks to Karl Schafer for bringing this important point to my attention. Cf. Smit (2000).

[56] The *locus classicus* here is Bxxvi *n*. But I also think that a concept having 'objective validity' is equivalent to it being a cognition, for Kant, so all of the passages that explicitly connect 'objective validity' with real possibility also count as 'modal characterizations' of cognition. See A239/B298.

[57] E.g., A50–1/B74–5, B146.

[58] There is a complication here. Kant takes himself to have sufficient warrant to make certain claims about noumena: that there are noumena, that they are non-spatiotemporal, and that they causally affect us (see *Prol.* 4: 315, A251–2, Bxxii *n.*, A26/B42, *Disc.* Ak. 8: 215). So either these claims do not count as *Wissen* in Kant's sense (although we do possess sufficient rational warrant to justify asserting them) or Kant is willing to allow some minimal *Wissen* of noumena in the Critical period. While I favor the former view, the latter view is compatible with the larger point in the text: even if we *know* that noumena exist and are therefore really possible, this leaves no room for the highly determinate and systematic knowledge of what is possible for noumena required for a general metaphysics (ontology) of noumena. See Smit (2000), Schafer (forthcoming), and Chignell (2014*b*) for further discussion.

sense with m-cognition, because Kant's rationalist predecessors agree that to *know* about some objects in a domain we must be able to prove that those objects are *possible*.[59] So Leibniz, Wolff, and Baumgarten would agree that if m-cognition of noumena is impossible, then noumenal knowledge is impossible for us. What is more, if m-cognition of noumena is impossible, then metaphysics, the science of all possible beings *qua* possible, is impossible for us, since those beings include noumena. By contrast, it would be question-begging for Kant to assume that s-cognition is a necessary condition for knowledge. Leibniz, for instance, did not think that we could have immediate singular thoughts about particular monads (except perhaps in the case of the monad that I myself am), but did not think that was a barrier to knowing about them. So m-cognition is dialectically better suited to Kant's critique of noumenal metaphysics.

Secondly, the concept of s-cognition is implicitly modal to begin with. Kant is not claiming that for a concept to be an s-cognition we must be given an *actual* instance of the concept in intuition. Mathematical concepts, for instance, are cognitions, whether or not they have actual instances (as long as it is really possible that they have instances).[60] Even in passages where Kant is working with the notion of s-cognition, he often makes clear that he means the concept relates to an object *possibly* given in intuition.[61] If s-cognition is implicitly modal then we should ask, what is important about concepts of objects that are *possibly* given in intuition? As I will argue in this chapter and the next, they are the very concepts we can know to be really possibly instantiated.

This leads to my third point: concepts of objects that are possibly given in intuition are concepts whose objects we can prove to be really possible (objects of m-cognition). The converse also holds: objects of m-cognition are objects that are possibly given in intuition. M- and s-cognition are co-extensive, or so I will argue. While this might be thought to undermine the importance of the distinction, I think, on the contrary, that it shows that the semantic notion is logically downstream of the modal notion. In the rest of this section I will reconstruct Kant's reasoning from the notion of m-cognition to the notion of s-cognition. He begins with the problem of m-cognition and *arrives* at the problem of s-cognition. This will take two steps. In the first I reconstruct Kant's reasoning that *if* a concept is the concept of an object that can be given in intuition (s-cognition) then that concept is an m-cognition. In the next section I reconstruct Kant's argument that a concept is an m-cognition (a concept of a knowably really possible object) *only if* it is an s-cognition (a concept of an object that can be given in intuition). Kant begins

[59] Recall Leibniz's objection to the Cartesian ontological argument: we must first prove that <*God*> is the concept of a possible object. See Ch. 1.3 for more on Leibniz's idea that knowledge in metaphysics rests on *a priori*$_G$ knowledge of the possible instantiation of its concepts.

[60] See B147, A156/B195, and esp. A224/B271.

[61] E.g., Bxxvi, A3/B6, A4/B8, A63/B88, B146, A239/B298, A245, A247/B304, B383, B391, and B479.

with the modal problem—how do we represent knowably really possible objects *a priori*?—and proves that it is equivalent to a problem in cognitive semantics—how can *a priori* concepts represent objects that can be given in sensible intuition?

If we begin by assuming the modal characterization of cognition, and we assume further that our basic representations of objects are either concepts or intuitions,[62] then it is relatively straightforward to explain how empirical cognition is possible. We intuit a given object and we abstract a concept from the object by generalizing from one of its features (the object is red and we form the general concept <*red*>). It is actually the case, hence really possible, that an object instantiates the concept; to know its real possibility we need only know that it instantiates the concept, and know that actuality entails real possibility. There might be some difficult work to do in spelling out exactly how empirical concept formation works, or how intuition of an object puts one in a position to know that the object instantiates the concept, but Kant does not regard it as mysterious how any of this is possible in the first place. But this empirical model will not account for *a priori* cognition or for cognition of unactualized possibility.

Kant's theory of *a priori* intuition gives him some purchase on the problem of *a priori* cognition and cognition of unactualized possibility, but not enough. We intuit space and time *a priori*, which I take to mean: independently of how we are sensorily affected by objects (hence *a priori*) we are immediately presented with (intuit) the spatiotemporal structure in which any object given to us must appear. So we have *a priori* cognition of the real possibility of at least two objects: space and time. Furthermore, we can use this model to account for *a priori* mathematical cognition: a mathematical concept is an *a priori* cognition just in case I can construct its object in pure intuition and thereby prove that such an object can be given in space and time.[63] While I cannot go into the details of Kant's theory of construction in pure intuition, I want to point out that what 'can and cannot be constructed' in pure intuition is itself a modal notion, so we need some account of it. Let us say that it is *intuitively possible* that a given concept is instantiated just in case it is compatible with the forms of intuition, space and time, that we intuit an instance of that concept. In the next chapter I will have more to say about the nature of this compatibility relation.

A priori constructibility in intuition *by itself* has serious limitations, though, as a guide to real possibility because instances of the fundamental concepts of metaphysics (e.g., substances, forces, etc.) cannot be presented in pure intuition.[64] The fundamental concepts of metaphysics are concepts of *concrete* beings with causal powers, not of abstract formal mathematical objects that lack causal efficacy; consequently, no mathematical construction is a construction of an instance of such a

[62] Judgments are not basic because they are composed of concepts.
[63] See esp. B147, A224/B271, A714/B742, and A141/B180.
[64] A715/B743, A722/B750.

concept. In the Prize Essay Kant had already pointed out that the concepts of metaphysics cannot be constructed, but in the context of the *CPR* this has an additional methodological consequence: *a priori* intuition alone will not explain how cognition through the categories is possible.

This is the point which Kant's thinking about the problem of *a priori* cognition had reached by the time of his 1772 letter to Herz. He does not find empirical cognition problematic, and he has the rudiments of an explanation of *a priori* cognition of pure quantities in mathematics,[65] but he cannot explain (or is unwilling to explain in his correspondence with Herz) how *a priori* cognition of concrete non-mathematical objects using the fundamental concepts of metaphysics—what he will come to call the categories or 'pure concepts of the understanding'—is possible. But if we cannot cognize objects using categories (the fundamental concepts of metaphysics) then metaphysics is impossible for us, because cognition, in the modal sense, is a precondition for knowledge. We cannot, for instance, know the nature of substance (e.g., whether all substances are simple) unless we can prove that substances are really (not merely logically) possible. Kant's ingenious solution in the *CPR* is to come up with (i) an account of what it is to represent an object given in sensible intuition as falling under a category, that (ii) specifies this in terms of the *a priori* forms of intuition, space and time, such that (iii) representation of such objects is really possible, according to a notion of real possibility that will be articulated in the course of executing tasks (i) and (ii).

Kant's answer to (i) is provided by his view of the relation between the logical functions of judging and the categories. Judgments, for Kant, are hylomorphic complexes of matter (their constituent concepts) and form (their principle of unity).[66] The logical functions of judging are principles of unity of judgments, the manner in which concepts can be arranged to make judgments. The table of the logical functions of judging (A70/B95) lists these different principles of unity, organized according to four different moments: quantity, quality, relation, and modality. By specifying the quantity, quality, relation, and modality we specify the principle of unity of a judgment; judgments that agree in quantity, quality, relation, and modality can only differ in their *matter*: the concepts that are combined into a unified judgment according to the same form.[67]

Kant gives the logical functions a double duty, not merely uniting concepts into judgments, but also uniting the manifold of representations in an intuition: "the very same function that gives unity to different representations **in a judgment** also gives unity to the different representations **in an intuition**, which, expressed generally, is

[65] Corr., Ak. 10: 131.

[66] For more on Kant's 'hylomorphic' conception of judgment see Stang (2014*a*).

[67] As Strawson (1966), 78 points out, hypothetical judgments (if *p* then *q*) do not have a quantity or quality (but *p* and *q* do, assuming they are not further hypothetical judgments), so the specification of the form of a judgment will be slightly more complicated than this.

called the pure concept of the understanding" (A79/B104–5).[68] This brings in two crucial Kantian notions: the manifold and synthesis. An intuition represents a *single* object but contains in itself a manifold (plurality) of parts, which are themselves representations of the *parts* of the object.[69] To intuit an object we do not need to become conscious of the manifold of its parts, or to combine them in any way; we can be given, in intuition, objects of which we are not conscious.[70] But if we consciously intuit an object then we are conscious of the manifold of representations in our intuition of it and it is *possible* for us to think of that manifold as composing an intuition of a single object, i.e. to represent that intuition as a complex whole composed of those parts. Representing a complex whole of parts as a complex whole is what Kant calls 'synthesis' or combination.[71] To represent a complex whole of parts as such, to 'synthesize' it, we also need to represent the principle of unity of that complex, the relation that holds among its parts in virtue of which that complex is the complex it is.[72] So for any intuition of an object of which we are conscious, we can be conscious of the manifold of representations in that intuition, and it is possible for us to combine or synthesize that manifold and think of that intuition as a complex but unified intuition of a single object. For each logical function of judging there is a corresponding principle of unity for synthesis of manifolds in intuition. Just as logical functions are principles of unity for judgments, these 'schemata' are principles of unity of conscious acts of combination of the manifold of parts in an intuition. To think of a manifold of intuition as unified according to a given schema (as having that principle of unity) is to think of the object of that intuition as falling under the corresponding category. For instance, I have a temporally extended intuition of an object, whose parts are temporal parts or 'time slices' of the whole intuition. If I represent those 'time slices' as constituting a unified intuition of a single object that persists through time while its appearance changes then I am representing that intuition as having a certain principle of unity: representation of the changing accidents in a persisting object. This corresponds, according to Kant, to the logical form of categorical judgment and the category <*substance-accident*>. If I combine that manifold according to that principle of unity then I am thinking of the object of that intuition as a substance, that which persists while its accidents change.

I interpret this to mean that (a) for each logical function of judging there is a corresponding rule for synthesizing manifolds (pluralities) of representations in

[68] Cf. "I will merely precede with this explanation of the categories. They are concepts of an object in general, by means of which its intuition is regarded as determined with regard to one of the logical functions for judgments" (B128). My interpretation here is deeply indebted to Longuenesse (1998).

[69] B136 n. [70] See *JL* (Ak. 9: 35) and *Anthr.* (Ak. 7: 135). [71] B129–131.

[72] The qualification "to represent a complex whole of parts *as such*" is crucial—we do not need to synthesize every whole, for, given that space and time are infinite given wholes, this is an infinite, hence impossible, task. We are given space and time as infinite wholes. To *think* of their parts as parts of a whole, though, we need to engage in an activity of *synthesis*. Cf. McLear (2015).

intuitions, (b) we synthesize manifolds of representation in intuitions by applying these rules, and (c) because of this, the object of that intuition is represented as instantiating the corresponding category. These rules for synthesis are what Kant calls the 'schemata' of the categories. This gives us an account of what it is to represent an object as falling under a pure concept of the understanding, Kant's answer to question (i) from above: to represent an object as falling under a category is to synthesize the manifold of representations in an intuition of that object according to the schema of the corresponding logical function of judging.

Kant's answer to (ii) and (iii) follows almost immediately from this initial move. The schemata of the categories are given in *temporal* terms; I interpret this to mean that the manifolds of representation that are being synthesized are temporally extended manifolds, that is, manifolds of outer intuition that occur at different times, and the rules are specified in terms of the temporal properties of these manifolds. To return to the example of <*substance*>, the schema for this concept is the rule of organizing a temporally extended series of outer representations as representations of the *changing* accidents of a *persisting* substance.

Finally, we need an answer to question (iii) from earlier: we need to show that it is really possible for us to synthesize a manifold in an intuition according to these logical functions of judging. This raises, of course, the pressing question of what exactly real possibility means here. I will examine that question in detail in the next chapter, but for now it suffices to point out that the very minimal notion of real possibility articulated earlier—compatibility with the *a priori* forms of intuition, space and time—will not suffice, for the simple reason that this combination of a manifold of representations is a spontaneous activity of the understanding, not of the passive faculty of sensibility. If all we have done is demonstrate that it is compatible with the spatiotemporal structure of intuition that a temporally extended manifold of intuition be represented as successive apprehensions of the changing accidents of a persisting object, we will not have shown that it is really possible for the understanding to so unify that manifold. Kant needs a more complete notion of real possibility to answer this question.

The answer lies in Kant's doctrine of the unity of apperception, one of the most puzzling and complex ideas in the whole Transcendental Deduction, which, by Kant's own admission, is both the most difficult (and most important) part of the *Critique* itself.[73] I cannot offer a complete interpretation of the Deduction here, for it would quickly engulf this entire book.[74] For our purposes, a few points will have to suffice. The unity of apperception is the unity of consciousness, the unity that obtains

[73] Axvi.

[74] Nor, for the same reason, can I engage with the extensive secondary literature on the Deduction. My discussion here is deeply indebted to Longuenesse (1998), as well as Strawson (1966), Henrich (1968–9) and (1976), Allison (1983) and (2004), Hanna (2001), and Allais (2015).

among all of my conscious representations in virtue of being the conscious representations of a single conscious subject. Kant claims for the unity of apperception in relation to the understanding the very status he had earlier claimed for space and time in relation to intuition, that it is a 'highest' ground of (real) possibility:

> The supreme principle of all intuition in relation to sensibility was, according to the Transcendental Aesthetic, that all the manifold of sensibility stand under the formal conditions of space and time. The supreme principle of all intuition in relation to the understanding is that all the manifold of intuition stand under conditions of the original synthetic unity of apperception. (B136)[75]

Just as the 'highest' grounds of the possibility of objects of intuition are space and time, the 'highest' ground of the possibility of those objects insofar as they are thought by the understanding is what Kant here calls the "original synthetic unity of apperception."

Let me first explain what that term refers to. The analytic unity of apperception is Kant's term for the unity that obtains among my conscious states in virtue of their being *mine*: "the 'I think' must be able to accompany all of my representations" (B131). He argues further that the analytic unity among the manifold of representations in an intuition is only possible if a further *synthetic* unity holds among them, and this synthetic unity is itself defined in terms of the possibility of a *synthesis*: "only because I can comprehend their manifold in one consciousness do I call them all together **my** representations; for otherwise I would have as multicolored, diverse a self as I have representations of which I am conscious" (B134).

I interpret this to mean for any intuition whose object I can *think* about (consciously subsume its object under a concept), the manifold of parts in that intuition must have synthetic unity of consciousness, which means not that I *actually* consciously synthesize that manifold, but that it is *possible* that I do so. The synthetic unity of apperception refers to a necessary possibility: necessarily, if a sensibly given object is thinkable by me, then possibly I synthesize (combine) the manifold of the intuition in which that object is given.

That the synthetic unity of apperception is the 'highest' ground of the possibility of objects of the understanding means, I take it, that the synthetic unity of apperception is the ultimate ground of the possibility of using the understanding to *think* about given objects. This means two things: anything that is a condition of the possibility of thinking about given objects is *either* among the grounds of the synthetic unity of apperception, or among its consequences. Everything that conditions the possibility of *thinking* sensibly given objects must 'go through' the synthetic unity of apperception: it must be either explanatorily downstream or upstream of it. But notice there are also conditions of the possibility of being *given* objects: space and time. So we

[75] Cf. A107: "the numerical unity of this apperception grounds all concepts *a priori* just as the manifoldness of space and time grounds the intuitions of sensibility."

have two distinct sets of conditions of possibility: the conditions of being given objects (space and time) and the conditions of thinking about them (the synthetic unity of apperception, and its conditions). To experience an object is to think about a sensibly given object using a concept. So we need to supplement our former conception of real possibility—compatibility with the forms of intuition—with the conditions of the possibility of thinking about objects to form a conception of what we might call *formal* (real) possibility:

> (*Formal*) It is *formally possible* that we represent a certain (conceptual or intuitional) content just in case it is compatible with the forms of intuition (space and time) and the highest principle of the understanding (the synthetic unity of apperception) that we represent a sensibly given object using that (conceptual or intuitional) content.[76]

That the synthetic unity of apperception is the highest or 'first' ground of real possibility of thinking about given objects means that any thought that is compatible with the synthetic unity of apperception is formally possible, provided it also agrees with the forms of intuition (i.e. it is not a thought of an object that cannot be given in intuition). Since the synthetic unity of apperception just *is* the possibility of synthesizing manifolds of intuition, it follows that if the synthetic unity of apperception is itself *possible* (i.e. does not have inconsistent conditions of possibility) then it is formally really possible that we synthesize intuited manifolds. I take it that Kant assumes in the Deduction not only that the analytic unity of apperception is possible, but that it is *actual* (I am aware of myself as the numerically identical subject of my representations). It follows that the synthetic unity of apperception (which makes analytic unity possible) is itself possible, from which we can conclude that synthesis of manifolds of intuition is possible. Now we have already seen that the synthesis of a manifold of intuition is synthesis according to the logical functions of judging and their schemata, and this just is representing (specifically, thinking) the object of intuition under the corresponding category. So if the former is really possible, the latter is really possible as well. We can conclude, finally, that it is really possible to think a sensibly given object under the categories, and we have given a preliminary characterization of what the relevant notion of real possibility is (compatibility with the intuitional and intellectual conditions of experience of objects—respectively, space and time, and the synthetic unity of apperception).

This is an explanation of what grounds the real possibility of representing sensibly given objects under the pure concepts of the understanding. It constitutes an answer to *one* question Kant is raising in the Analytic of Concepts: how can we so much as represent sensibly given objects as falling under the pure concepts of the understanding?

[76] Cf. Chignell (2014*b*), which argues that the notion of real possibility involved in the 'real possibility constraint' on knowledge is compatibility with background knowledge of nature and its laws. See Ch. 7.3 for discussion of the textual evidence for this claim.

This is ultimately a question about *representational content*: what representational content do pure concepts of the understanding have when they are applied to sensibly given objects and how can that content be applied to them? This does not by itself answer another question: how can we know *a priori* that it is really possible that objects instantiate the categories? The account I have sketched above—which essentially summarizes some key points in the Metaphysical Deduction, the first half of the Transcendental Deduction (roughly, up to §23),[77] and the Schematism—leaves open the possibility that, while we can *represent* sensibly given objects under the categories, it is impossible for them to actually *instantiate* those concepts. For instance, a Leibnizian might allow, although we can *think* of objects of sensation as substances in an unqualified sense by representing them as bearers of absolutely intrinsic properties (underlying their perceptible relational-dispositional properties), it is impossible that sensibly given objects could *be* substances because they are merely *phaenomena bene fundata*, the well-founded phenomena of perception. Likewise, a Humean (though probably not David Hume himself) might allow that we can *think* of sensibly given objects using the concept <*necessary connection*> without thereby admitting that it is really possible that objects could instantiate such a concept.[78]

To fill this gap between, on the one hand, the real possibility that we think of sensibly given objects using the categories, and, on the other, the real possibility that they actually instantiate the categories, Kant must invoke his transcendental idealist thesis that the objects given in intuition are appearances, not things in themselves. The meaning of Kant's transcendental idealism and the distinction between appearances and things in themselves has been controversial since the publication of the A edition in 1781, and that controversy has continued down to the present day.[79] It is not my intention to enter into that controversy here, for we can understand how transcendental idealism fills the gap in Kant's account while remaining neutral among competing interpretations of transcendental idealism. The key idea in closing this gap is that *the real possibility of thinking a given object under the categories* entails *the real possibility of an object falling under the categories*.[80] The first terms refers to the real possibility of a certain representational content (thinking of a given object that it instantiates a category); the second term refers to the real possibility of given objects instantiating categories.

[77] Which naturally raises the question, what is left to do in the second half of the B Deduction? My own view, which I do not defend here, is that the second half shows the real possibility of sensibly given objects *instantiating* the categories. The solution that I offer to that problem in the rest of this section is a more modest solution than, on my reading, Kant actually gives in §§24–7.

[78] I would not attribute this view to Hume himself for, on my reading, Hume denies that we have any idea *of* (what Kant would consider) a necessary connection (except insofar as the idea of a necessary connection is just the idea of a constant succession accompanied by a feeling of certainty), so, in Kantian terms, we cannot even *think* of objects under (what Kant considers) the category <*cause-effect*>.

[79] Stang (forthcoming) provides an overview of the current state of the debate; Beiser (2002) covers the historical background.

[80] See A111.

The reason why this entailment holds is that the objects of experience are appearances, not things in themselves. While there are different ways of understanding this distinction, they all agree that this entailment holds. For instance, on the phenomenalist reading of this distinction 'appearances' and 'things in themselves' refer to distinct classes of entities. Appearances are merely the 'intentional objects' of experience, objects whose 'whole being' is grounded in the contents of experience; if it is really possible that we experience appearances a certain way (e.g., falling under the categories) then it is really possible that they do fall under the categories.[81] On the 'one object' reading 'appearances' and 'things in themselves' do not refer to different classes of objects but to different properties possessed by one and the same class of objects.[82] There are many different interpretations of what this distinction amounts to, but I think it should be clear that on any such reading the entailment holds. For instance, if we take this distinction to be the distinction between the properties objects can appear to have in experience and the properties that can never be revealed in experience, the entailment holds: if it is really possible for us to experience objects as having certain properties (e.g., falling under the categories) then it follows that those properties are among their 'empirical' properties, which just means it is really possible that appearances (or objects *qua* experienced by us) have those properties.

The third class of interpretations of transcendental idealism differ from the first two in not taking the distinction between appearances and things in themselves to be a metaphysical distinction (either between two kinds of objects, in the case of the phenomenalist, or between two kinds of properties, in the second case) but a methodological distinction between two standpoints we can take on objects: we can consider them as objects of experience for spatiotemporal discursive intellects (the 'empirical standpoint') or we can consider them as objects of an intellect in general (the 'transcendental standpoint').[83] This interpretation also upholds the entailment, for if it is really possible (compatible with the conditions that have to be in place for us to experience objects in the first place) for us to represent objects under the categories, then it is really possible for objects to fall under the categories, insofar as objects are considered from the empirical standpoint. So even on the 'methodological' reading of Kant's transcendental idealism, the entailment holds.

This is a reconstruction of Kant's account of how it is possible for us to cognize the real possibility of sensibly given objects falling under the categories. Note that it will

[81] The phenomenalist interpretation of Kant was inaugurated by the (now infamous) Feder-Garve review, the text of which can be found in Karl Vorländer's edition of the *Prolegomena* (Kant 1976), 167–74; Sassen (2000), 54–8 contains a translation. Sellars (1968) and (1976), Aquila (1979) and (1983), ch. 3, Van Cleve (1999), and Stang (forthcoming) attempt, in different ways, to rehabilitate the phenomenalist reading of transcendental idealism, while rejecting the Feder-Garve view.

[82] Recent defenders of this view include Allais (2004), (2006), and (2007); Rosefeldt (2007) and (forthcoming); Marshall (2013); and McDaniel (2013).

[83] I have in mind mainly Allison (1983) and (2004).

only succeed for objects that are *appearances*, not for things in themselves. For things in themselves, objects that do not depend upon our manner of sensibly cognizing them, the entailment does not hold. There is a 'modal gap' in the case of things in themselves, between the real possibility of our representing them under the categories and the real possibility of their falling under the categories. More fundamentally, though, the account of *what it is* to represent an object as falling under the categories does not apply to things in themselves, because (given Kant's transcendental idealist thesis that all objects given to us in intuition are appearances) things in themselves cannot be given to us in intuition. So this account does not issue in an explanation *either* of (a) what it is to represent things in themselves under categories or (b) *whether* we can cognize (knowingly represent the real possibility of) things in themselves instantiating the categories. In section 8 I reconstruct Kant's argument that (b) we *cannot* cognize things in themselves using the categories. In Chapters 9 and 10 I explore his account of (a) what we are doing in representing things in themselves under the modal categories.

Before continuing, I want to remark on the difference between two distinctions Kant uses to articulate his transcendental idealism: appearance/thing in itself, and phenomena/noumena. The appearance/thing in itself distinction is a distinction between things that depend upon being sensibly intuited (appearances) and things that do not so depend (things in themselves).[84] The phenomena/noumena distinction is a *cognitive* distinction between objects that can be given in sensible intuition (phenomena) and objects that cannot be so given (noumena).[85] This distinction assumes nothing about the metaphysical status of phenomena or of noumena; even a transcendental realist could accept it. Since all of our intuition is sensible, we intuit only phenomena, by definition. All objects we intuit are either in space or time, and Kant argues in the Transcendental Aesthetic that space and time present only appearances, not the things in themselves that appear. Consequently, <*appearance*> and <*phenomenon*> are co-extensive concepts, though they differ in intension (meaning).[86] By the same reasoning, all things in themselves are noumena, for they cannot be sensibly intuited. Conversely, a noumenon that is not a thing in itself would be a being that cannot be sensibly intuited (noumenon) but depends upon being sensibly intuited (an appearance to sensible intuition), which is absurd. <*Noumena*> and <*things in themselves*> [*Dinge an sich selbst*] are co-extensive

84 See esp. Kant's definition of transcendental idealism at A369; cf. A42/B59.

85 A252; this corresponds in the B edition to the 'negative' concept of a noumenon (B307).

86 Kant does distinguish appearances from phenomena at A248–9, but I think the dominant meaning of 'phenomena' in this section is: possible object of a sensible intuition. One might also wonder whether there could be phenomena (objects of sensible intuition) for intellects other than ours, but which are not appearances (whether some discursive intellects intuit things in themselves). Kant claims that sensible intellects, in virtue of being passive, cannot intuit things in themselves (B68, B152–3, B156, B159), but does not base his case for transcendental idealism on this 'short' argument. See Ameriks (2003) for discussion.

concepts, though they differ in intension.[87] Hence, if we cannot cognize noumena, *a fortiori* we cannot cognize things in themselves.

Bear in mind, though, that by drawing the distinction between *<phenomena=possible object of sensible intuition>* and *<noumena=not a possible object of sensible intuition>* Kant is not thereby committing himself to there being noumena. This is his point when he writes: "the division of objects [*Gegenstände*] into *phenomena* and *noumena*, and of the world into a world of sense and a world of understanding, can therefore not be permitted at all in a positive sense, although concepts certainly permit of division into sensible and intellectual ones" (A255/B311, underlined portion added in the B edition).[88] But we should not, for that reason alone, conclude that noumena are impossible, for "one cannot assert of sensibility that it is the only possible kind of intuition" (A255/B310).

6.8. No *a priori* Cognition of Noumena

In this section I will reconstruct Kant's reasons for denying that we can (modally) cognize objects other than phenomena; since cognition is a precondition for knowledge [*Wissen*] of an object, it will follow both that we cannot have knowledge of noumena (objects that cannot be given in sensible intuition) and thus that we cannot know all possible objects *überhaupt*.[89] Metaphysics is possible for us *only* if it is restricted to phenomena.

Kant typically argues that we cannot cognize noumena by claiming that we can only cognize objects that we can intuit. However, this strategy of argument relies on a premise that would not be granted by all of Kant's interlocutors: that we can only cognize (know the real possibility of) objects that can be given to us in intuition. This premise becomes especially problematic when it is paired with Kant's assumption, for which he never explicitly argues in the *CPR*, that all of our intuition is sensible (that we have discursive intellects). While many of Kant's commentators find these claims

[87] I know these claims will strike some readers as misguided, but there is strong textual evidence that Kant regards '*Dinge an sich selbst*' and 'noumena' as co-extensive terms: Cf. A254/B310, A256/B312, A259/B315; *Prol.* §30 (Ak. 4: 312), §32 (314–15), §33 (315), and §59 (360); *Disc.* (Ak. 8: 208); and *Prog.* (Ak. 20: 292, 308).

[88] Cf. his comment in the previous sentence that the concept *<noumenon>* does not allow us to "posit [read: posit absolutely] anything positive outside" the domain of the senses.

[89] The exact interplay between real possibility, cognition, and knowledge is a matter of scholarly debate; I do not have the space to discuss it fully here. All parties agree that cognition limits in some fashion what we can know; the question is how to spell this out in detail. Chignell (2014*b*) points out that Kant seems to allow some very general knowledge about noumena (e.g., that they are non-spatiotemporal, that they affect us) while denying that cognition of noumena is possible (see discussion in section 7 of this chapter). Schafer (forthcoming) argues that, while this entails something very much like modal knowledge of noumena (e.g., if we know that there are noumena we know *a fortiori* that they are really possible), it fails a further requirement on cognition: that we cognize objects determinately (e.g., we know that there are noumena but nothing about their determinate positive properties). If Schafer is correct that we 'know' (in some sense, though not perhaps Kant's technical notion of *Wissen*) some minimal real possibilities for noumena, my main point still stands: this minimal knowledge of real possibility would not allow for the robust modal knowledge of noumena required for noumenal metaphysics. See also Smit (2010).

prima facie plausible they would find universal assent *neither* among Kant's early modern predecessors *nor* among later philosophers (even those influenced by Kant).[90] Since I am attempting to reconstruct Kant's best argument that cognition of noumena is impossible, I will focus on arguments that do not rely on contentious premises about the intuition-dependence of cognition or on the assumption that our faculty of intuition as such is sensible.

The closest thing Kant gives to a general argument that cognition of noumena is impossible, without relying on controversial premises about the intuition dependence of cognition, is the passage at the beginning of the Transcendental Deduction, from which I quoted a small section at the beginning of section 4:

> There are only two possible cases in which synthetic representations and its objects can come together, necessarily relate to each other, and, as it were, meet each other: either if [1] the object alone makes the representation possible, or if [2] the representation alone makes the object possible. If it is the first [1b], then this relation is only empirical, and the representation is never possible *a priori*. And this is the case with appearances in respect of that in it which belongs to sensation. But if it is the second [2], then since representation in itself (for we are not here talking about its causality by means of the will) does not [2a] produce its object as far as its **existence** is concerned, the representation is still determinant of the object *a priori* if [2b] it is possible through it alone to **cognize something as an object**. (A92/B124–5—bracketed numbers refer to my analysis of the argument below)

This parallels the following passage from the letter to Herz:

> [O]n what grounds rests the relation [*Beziehung*] of what in us is called representation [*Vorstellung*] to the object? If [1a] the representation contains only the way in which the subject is affected by the object, then it is easy to see how the representation corresponds to the object, as an effect to its cause, and how this determination of our mind can *represent* something, i.e. how it can have an object. The passive or sensuous representations thus have a graspable relation to objects, and the principles that are derived from the nature of our soul have a graspable validity for all things insofar as they might be objects of the senses. Likewise, if [2a] that in us which is called representation were active with regard to the object, i.e. if the object were produced by the representation itself (as one thinks of divine cognitions as archetypes of things), then the conformity of the representations with the objects would also be able to be understood. And so one can at least understand the possibility of both an archetypal intellect [*intellectus archetypus*], upon whose intuition the things themselves are grounded, as well as ectypal intellect [*intellectus ectypus*], which attains the data of its logical activity from the sensuous intuitions of things. But neither is [2a] our understanding by means of its representations the cause of the object (except in morals, the cause of good ends), nor is [1a] the object the cause of the representations of the understanding in the real sense (*in sensu reali*). The pure concepts of the understanding must, therefore, not be [1a] abstracted from the sensation of the senses, nor must those concepts express the receptivity of representations

[90] E.g., Hegel, who criticizes Kant on exactly this point in his 1802 work, *Faith and Knowledge*: H.*Werke* 2: 325–6.

through sense; but they must, to be sure, have their sources in the nature of the soul, though not [1a] insofar as they are produced by the object nor [2a] insofar as they bring forth the object itself. (Corr., Ak. 10: 130)

I argued above, in section 4, that the relation between representation and object with which Kant is concerned in both passages is *cognition*: representing a knowably really possible object. Although his own positive solution depends upon intuition, Kant raises the question, in both passages, primarily about the pure concepts of the understanding (categories): how can we cognize objects through pure concepts of the understanding?[91] His reasoning, in both passages, takes the form of a dilemma: either (1) the object is the ground of the concept, or (2) the concept is the ground of the object. Whereas, in the Herz letter, Kant concludes pessimistically that *a priori* cognition using pure concepts of the understanding seems to be impossible on either horn of the dilemma, in the *CPR* he thinks this possibility can be understood through the second horn (2): the pure concepts of the understanding make their objects possible. We have already seen Kant's positive solution, so now I want to reconstruct his reasons for thinking this dilemma is *exhaustive* and his reasons for excluding the other option(s).

Kant's argument relies on a disjunctive premise:

> (*Disjunction*) Where C is a concept, if C is a cognition then either (1) the object of C grounds C, or (2) the concept C grounds its object.

This disjunctive premise needs to be fleshed out further, in several respects.[92] First of all, it is not clear what kind of grounding relation is involved. On the one hand, Kant seems to assert a causal relation between an object and its empirical concept in the letter to Herz ("the representation corresponds to the object, as an effect to its cause"); causal grounding is also relevant in the practical case, where the representing of a particular action as my end is among the causes of the existence of that action ("neither is our understanding by means of its representations the cause of the object (except in morals, the cause of good ends)"). On the other hand, the non-causal grounding of possibility also clearly has a role to play, since Kant's ultimate conclusion will be that the categories are non-causal grounds of the possibility of objects. Consequently I propose that we divide each horn of the dilemma into a causal and a non-causal case, and represent Kant's disjunctive premise as a table of possible views (Table 6.1).

Regarding (1b), note that, while in the Herz letter Kant says that the object of an empirical concept is the cause of the concept, in the CPR he says that the object makes the concept possible (and this is why the concept is "not possible *a priori*"). I explain what this difference might amount to below.

[91] This is because he possessed a rudimentary account of *a priori* intuition and mathematical cognition, since at least 1770 in ID. See §3 of ID (Ak. 2: 398–406).

[92] This section is heavily indebted to Eli Chudnoff, who helped me think through Kant's options.

Table 6.1. Kant's disjunctive premise

	(1) *The object grounds the concept*	(2) *The concept grounds the object*
(a) *Causal grounding* (ground of existence)	Empirical concepts (Herz letter)	Practical representation
(b) *Non-causal grounding* (ground of possibility)	Empirical concepts (*CPR*)	Kant's positive view: cognition of phenomena

However, further refinements are needed before we can reconstruct Kant's arguments for excluding the other options. In this study I have typically formulated grounding relations as relations between *facts* (or between an object and a fact, e.g., between God and the fact that some predicate is really possible), so we must determine what fact about the concept is being grounded by what fact about the object (and vice versa). The easiest case to consider is the empirical one: an empirical concept is a cognition because we experience an actual instance, so we know *a fortiori* that it is really possibly instantiated. In the empirical case it is only possible for us to represent the relevant content if there is an actual instance in our environment and we experience it. This is a point that Kant constantly makes about concepts of particular forces: we cannot have a concept of a particular force unless we actually encounter it.[93] In the empirical case (horn (1)) we can identify the relata of the grounding relation as the fact that the concept is instantiated and the fact that we have a concept with that representational content (that we possess *that* concept, assuming concepts are individuated by their content). Assuming the relata of the grounding relation are parallel in the two horns of the dilemma, then in horn (2) the relata will be the same (the fact that we possess a concept, and the fact that the concept is instantiated). In the case of non-causal grounding of possibility (b), what is grounded is the possibility of the second fact (e.g., that we possess a certain concept, (1a)). If this is correct, we can reformulate the disjunctive premise as the chart of alternatives in Table 6.2, where bracketed expressions denote facts.

Kant of course, wants to exclude (1) entirely and conclude that, in the case of *a priori* theoretical cognitions, only (2b) remains, which will lead to the conclusion that the only objects we can cognize are phenomena (objects that are made possible by our representations of them, appearances). But before we examine his reasons for rejecting (1) it should be pointed out that even on Kant's, admittedly idiosyncratic, way of seeing the problems and their potential solutions there is a third alternative that he does not consider explicitly in the quoted passages, but which does come in

[93] Thus there are no *a priori* concepts of particular forces; see A207/B252, A222–3/B269–270, A291/B347.

Table 6.2. Kant's initial alternatives

	(1) [the concept is instantiated] *grounds* [we possess the concept]	(2) [we possess the concept] *grounds* [the concept is instantiated]
(a) *Causal grounding* (ground of existence)	Empirical concepts (Herz letter)	Practical cognition
(b) *Non-causal grounding* (ground of possibility)	Empirical concepts (*CPR*)	Kant's positive view

Table 6.3. Complete set of alternatives

	(1) [the concept is instantiated] *grounds* [we possess the concept]	(2) [we possess the concept] *grounds* [the concept is instantiated]	(3) *Common source grounds both* [the concept is instantiated] *and* [we possess the concept]
(a) *Causal grounding* (ground of existence)	Empirical concepts (Herz letter)	Practical cognition	Crusius (see below)
(b) *Non-causal grounding* (ground of possibility)	Empirical concepts (*CPR*)	Kant's positive view	Malebranche (see below)

for discussion later, both in the Herz letter and the *CPR*.[94] The third alternative is that our concepts and their objects both depend upon a common source; in the context of the eighteenth century, the natural candidate is God. So we should really add a third column of possible alternatives, as in Table 6.3.

Option (3) actually involves two different grounding relations: one between the 'common source' (God) and our concept, and one between the 'common source' (God) and objects. We might want to divide (3) into *four* rather than merely *two* sub-cases, but, for the sake of simplicity, I shall make the Kantian assumption that the 'common' source is the non-causal ground of the very possibility of objects instantiating our concepts, so we only need to distinguish whether God is the cause of the existence of our concepts with their representational content or their non-causal ground. We can quickly dispense with (2a), for, while this is the relation between concepts and objects in practical cognition, in theoretical cognition our concepts do not causally bring about the existence of their objects. Kant's strategy is to exclude

[94] Ak. 10: 131 f. and B167–8. I take Kant's mention of a "preformation system of pure reason" to be a reference to Crusius; see discussion below.

each of the options represented in this table as a potential explanation of *a priori* cognition, until he is left only with (2b), his own view.

By representing these options in a table I do not mean to be committing Kant to the assumption that they are *exclusive*. To take the empirical case (1), it might be that the presence of gold in my environment is both a non-causal ground of the possibility that I have the concept *<gold> and* the causal ground of my actual possession of that concept; Kant's argument will be that, if either of them is true for a given concept, that concept is not an *a priori* cognition. What is more, Kant's reasons for rejecting option (1) and its sub-options are different in kind than his reasons for rejecting option (3) and its sub-options. Kant's objection to (1) is that if objects were the grounds of our possessing the categories then *a priori* cognition through the categories would be impossible. His objection to (3), by contrast, is that even if there is some common source, both of our possession of category-concepts and of their instantiation by objects, this does not *explain* why *a priori* cognition is possible. This does not allow him to conclude that (3) and its sub-options are *false*, but merely that they have no role to play in a transcendental critique of our cognitive faculty: an inquiry into the grounds of possibility of our *a priori* representation of knowably really possible objects.

(i) Kant's reasons for rejecting object-dependence

Kant's reasons for excluding (1a)—the instantiation of the concept is (one of) the *causal* grounds of our possessing the concept—are fairly straightforward. If my possession of the concept causally depends upon the concept being instantiated by an object, then I have experienced an instance and abstracted the concept. This model might be acceptable for empirical concepts, but it cannot be correct for *a priori* concepts like the categories; if my possession of the categories depends causally on their objects, then the fact that I possess these concepts depends upon the particular course of my experience, so these concepts would not count as *a priori* concepts in Kant's sense: concepts, the possession of which does not depend upon the particular course of our experience.[95]

This option, (1a), corresponds to the causal model Kant rejects in the Herz letter: "if the representation contains only the way in which the subject is affected by the object, then it is easy to see how the representation corresponds to the object, as an effect to its cause" (Ak. 10: 131). However, in the *CPR* Kant suggests a different model: "if it is the first [the object alone makes the representation possible], then this relation is only empirical, and the representation is never possible *a priori*. And this is the case with appearances in respect of that in it which belongs to sensation" (A92/B124–5). Kant here claims that the object of an empirical concept makes that concept *possible*; this means that, if I do not experience an instance of the empirical concept,

[95] See Chapter 7.3 for more on this notion of the *a priori*.

then that empirical concept is not so much as thinkable for me. This is plausible, for instance, about certain very basic sensory concepts; if one has not yet experienced the taste of pineapple then one cannot have the concept <*taste of pineapple*>. Kant makes the analogous claim about concepts of particular physical forces: unless the force in question is instantiated by some object, you cannot so much as have the concept of that force (think about that force in particular).[96] Consequently, we cannot so much as think about *particular* uninstantiated forces (we can only think about *force* in general, and particular instantiated forces).

This is a similar thought to what is now called 'semantic externalism.' For instance, if the clear drinkable stuff in lakes and rivers were XYZ rather than water, we could not have thoughts about water, but only about XYZ.[97] This is relevant to Kant, because if we apply the idea of semantic externalism back to the categories we obtain the following possible alternative picture: possessing the concept <*substance*>, that is, being able to think about substances, requires being in an environment where there are substances (likewise for the other categories). This might be thought to face the same problem that (1a) did—how can we know *a priori* that we are thinking about substances, and thus are in a substance-laden environment?—but the externalist may be able to resolve this problem by appealing to another Kantian doctrine: to think about objects we must think using the categories. However, on the externalist view I am sketching, this principle receives a different interpretation: to think about objects is to use certain concepts (e.g., <*substance*>), but to use those concepts requires being in a world where those concepts are instantiated. Now the principle that "to think about objects we must think using the categories" is a principle that Kant regards as *a priori*, and what is more, as *a priori* whether or not the objects in question are phenomena or noumena. So the externalist defender of metaphysical realism could elaborate the following alternative: to think at all is to be intentionally related to certain conceptual contents, the fundamental concepts of all thought (the categories), and to possess those concepts requires there actually being objects (noumena) instantiating them in one's world. This promises to give us an *a priori* proof that the categories are instantiated, but a proof that rests on the opposite of the ground Kant wants it to: the instantiation of the categories is the ground of the possibility of our possessing these concepts and, indeed, of thinking at all.[98]

However, in appealing to the principle that the possibility of category-thoughts non-causally depends upon the instantiation of the categories in one's environment, the 'metaphysical externalist' is making a claim about objects *überhaupt*, including both phenomena and noumena. Our ability to cognize noumena is precisely what is at issue, so a circularity worry might arise: this is only an explanation of our ability to

[96] E.g., B252, *MFNS* (Ak. 4 :486–7). [97] See Kripke (1972) and Putnam (1975).

[98] This 'externalist' response to Kant is closely related to the issue of externalism and 'armchair' knowledge; see Burge (1988). I think it is also part of Hegel's idealism in the *Science of Logic*, but I do not have space to defend that claim here. Thanks to Eli Chudnoff for helpful discussions on this point.

cognize noumena if we can cognize (the concepts involved in) the 'metaphysical externalist' principle, but that principle is itself a principle (in part) about noumena so it is circular to appeal to it. It is hard to know how problematic circularity of this kind is; I will postpone further discussion of this issue until the final subsection.

(ii) A priori *cognition of phenomena*

Kant dismisses option (2a) out of hand as an account of *theoretical* cognition because our concepts of objects do not *cause* the *existence* of their objects. Not even the categories make it the case that, for instance, substances *exist*. Kant adamantly maintains, in the face of those who would assimilate his view to Berkeley's, that our representations are among the grounds of the *possibility* of objects but do not suffice to ground their *existence*. For objects of experience to *exist* there must be objects affecting us, providing the matter of experience and its objects. Option (2b) represents Kant's own positive view of *a priori* cognition, reconstructed in the previous section. If the categories are among the grounds of the possibility of their objects (objects instantiating them), as they are on Kant's transcendental idealist theory, then *a priori* cognition of those objects is possible. But this only explains how *a priori* cognition of phenomena, objects that can be given in sensible intuition, is possible; it does not explain *a priori* cognition of noumena.

(iii) Kant's reasons for rejecting 'common source' models

The idea behind (3) is that the categories are concepts of really possible objects because God is the common source of both our concepts and real possibility. Kant distinguishes two different versions of such a view: a 'pre-established harmony' view (which elsewhere he refers to as the 'preformation system' of pure reason—B167) on which God has created our intellects so that they track real possibility,[99] and a 'hyper-physical' influx view he associates with Malebranche, on which we currently stand in connection with God's mind.[100] I will begin by explaining the former view, which Kant associates with Crusius, and why it should be understood as option (3a) in my schema; then I'll explain why the Malebranchean view is an instance of (3b). I conclude by explaining Kant's common objection to both views as explanations of the possibility of *a priori* cognition: they are circular.[101]

The highest principle of Crusius's modal epistemology, which figures in his explanation of why (to use Kant's terms) the categories (the basic concepts of metaphysics) are *a priori* cognitions, is that conceivability and possibility are co-extensive: "the very highest distinguishing feature of possible and actual things is the

[99] Cf. Kant's discussion of Crusius in *Disc.* (Ak. 8: 245 f.), *Refl.* 4275 (Ak. 17: 492), 4446 (Ak. 17: 554), 4866 (Ak. 18: 14), as well as *LP* (Ak. 24: 335), *MH* (Ak. 28: 10–11, 156), *MV* (Ak. 28: 372), and *MvS* (Ak. 28: 467–8).

[100] Corr., Ak. 10: 131. Cf. *Refl.* 4275 (Ak. 17: 492), 4446 (Ak. 17: 554), and 4866 (Ak. 18: 14).

[101] Kant also includes Plato as a representative of the 'preformation' view in the Herz letter, but I will focus on the early moderns, Crusius and Malebranche.

essence of the understanding, namely that what cannot be thought as such is not possible or actual, and that, by contrast, what can be thought is possible."[102] The scope of what is 'thinkable' by our intellects is specified by three fundamental laws: the principle of contradiction, the principle of the inseparable, and the principle of the uncombinable. The law of non-contradiction provides the 'formal' criterion of conceivability; anything that violates it is inconceivable, hence impossible. The second two principles—the principle of the inseparable and the principle of the uncombinable—are 'material' principles of thought because they do not pertain merely to the logical form of a putative possibility. Some concepts are uncombinable in thought, even though they are not logically contradictory. Consequently, it is impossible for them to be co-instantiated. For instance, it is inconceivable, hence impossible, that a person be in two locations simultaneously, although this is not logically contradictory.[103] Likewise, some concepts are inseparable in thought, even though there is no contradiction in affirming one, while denying the other. Consequently, they are necessarily co-instantiated, but this necessity is not a logical necessity. For instance, it is inconceivable, hence impossible, that an alteration in a substance lack a sufficient ground, although there is nothing contradictory in the concept of an uncaused event.[104] Since not all logically consistent concepts are conceivable (given these principles), not all logically consistent concepts are possibly instantiated. Real possibility, for Crusius, as for Kant, is a restriction on logical possibility: not all logical possibilities are really possible.[105]

Conceivability (defined in terms of the three laws above) is a guide to possibility because God has given us intellects that cannot combine in thought really incompatible concepts, and can combine in thought really compatible concepts.[106] In Kantian terminology, Crusius's view is that we know that the categories are really possibly instantiated because it is *conceivable* (in the sense described above) that they are; thinking of a substance with accidents, or a cause that is sufficient to bring about the existence of an alteration, does not involve combining anything that cannot be thought together, or separating anything that cannot be thought separately.[107]

Crusius's view is most naturally understood as the view that God is the *causal ground* of our having the intellects we do, with certain laws of conceivability. In the context of Kant's argument, this means that God is the causal ground of our possession of category concepts, our ability to think about objects as, for instance, substances (3a). On Malebranche's view, though, our mind's dependence on God takes a more radical form: we represent objects (other than our own mental states) by

[102] *Ent.* §15; translation from Watkins (2009), 140. Cf. *Weg* §261.

[103] *De Usu* §27. [104] *De Usu* §27.

[105] In fact, it is plausible that Kant borrowed his distinction of logical from real possibility from Crusius.

[106] See *De Usu* §29; *Ent.* §237, 287. Cf. Heimsoeth (1956), 163.

[107] In the letter to Herz, Kant writes of this solution: "Crusius believed in certain implanted rules for the purpose of forming judgments and ready-made concepts that God implanted in the human soul just as they had to be in order to harmonize with things" (Corr., Ak. 10: 131).

perceiving ideas of them in the divine intellect. Without God's intellect it would not be so much as *possible* for any created mind to represent outer objects, much less represent them as substances.[108] "Through his presence God is in close union with our minds," writes Malebranche in *The Search after Truth*, "such that He might be said to be the place of minds as space is, in a sense, the place of bodies."[109] If we take this metaphor in a Kantian spirit, then Malebranche's view would be that the divine intellect is the complete space of possible thinkable content; the ideas presented to our minds are merely a limited subset of that space. In the context of Kant's argument, this means not only does God cause it to actually be the case that we have category-concepts (that we perceive those ideas) by creating us and willing that we do; God's mind is the non-causal ground of the mere possibility of those concepts (those ideas). In my way of dividing the options, Malebranche represents (3b): the possibility of our possessing category-concepts (the possibility of our 'seeing' those ideas) non-causally depends upon God.[110] To apply Malebranche's view to Kant's specific problem, the reason our category-concepts are possibly instantiated is that they are ideas in God's mind, and divine ideas are essences of possible things.[111]

Despite this difference, though, Kant levels the same objections against both the 'preformation' system (Crusius) and the 'hyper-physical influx theory' (Malebranche): he makes a 'bad company' objection, and he accuses them of circularity. In the Herz letter he says of both views that they "encourage all sorts of wild notions and every pious and speculative barbarism" (Ak. 10: 131), a point to which he returns in the *Prolegomena*: "with the lack of sure criteria for distinguishing an authentic origin from a spurious one, the use of such a principle [either Crusian divine preformation or Malebranchean immediate presence to God's mind] looks very precarious, since one can never know what the spirit of truth or the father of lies may have put into us" (Ak. 4: 320 n.). Kant is raising an essentially Cartesian objection to Crusius and Malebranche: perhaps an evil demon has implanted the 'laws of thought' in our minds, or is revealing to us ideas of impossibilities.

However, I am going to focus on Kant's objection in the Herz letter that the theories of Malebranche and Crusius are *circular*: "the *deus ex machina* [...] has—besides its vicious circularity in drawing conclusions concerning our cognitions—also this additional disadvantage [...]" (Corr., Ak. 10: 131). Recall the dialectical context in which Kant discusses these views: we want to explain how (and whether) *a priori* cognition of noumena is possible, and cognition is a prerequisite for

[108] Malebranche makes the un-Kantian assumption that representation of bodies (outer sense) is in principle different than representation of states of the mind (inner sense). Awareness of the states of the mind, according to Malebranche, requires no mediating idea. See *Search*, 1.2.3–7.

[109] *Search*, 3.2.6 (Malebranche 1980, 230).

[110] It should be noted that Crusius *also* thinks that God is the ground of all (real) possibility, so *a fortiori* he is the ground of the possibility of our being able to think of objects as substances. While this is part of Crusius's overall metaphysical theory it is not part of his account of how we can know that our thoughts track real possibility, or in Kant's terms, his theory of how we can cognize objects *a priori*.

[111] Malebranche (1997), 74.

knowledge. To prove the real possibility of noumena by appealing to some metaphysical theory requires, presumably, that that metaphysical theory itself is an object of knowledge and, therefore, of cognition. But the metaphysical theory to which Crusius and Malebranche appeal is a theory about noumena themselves: God and his intellect. Therefore, in arguing that cognition of noumena is possible, Crusius and Malebranche must first assume that it is. This is the alleged circularity in their theories.[112]

However, as anyone familiar with the 'Cartesian circle' and the debates surrounding it knows, the issue of circularity in epistemological theories is a complex one. We might think of Malebranche and Crusius as providing sufficient conditions on knowledge of possibility (cognition), namely:

(3a) An agent S knows that a concept C is really possibly instantiated if (i) C is logically consistent and contains no elements that cannot be combined by S in thought, and does not involve separating any elements that cannot be thought separately by S, (ii) God has created S's intellect so that if (i) is true then C is really possibly instantiated, and (iii) S bases her judgment that C is really possibly instantiated on (i). [Crusius]

(3b) An agent S knows that a concept C is really possibly instantiated if (i) S perceives C, (ii) C is in God's mind and (iii) S bases her judgment that C is really possibly instantiated on (i). [Malebranche][113]

These principles do not entail that, if S knows that some concept of noumena is really possibly instantiated, then S has some antecedent knowledge of God (or about any other noumena), in particular. In this sense, they are not circular. To borrow a slogan from Van Cleve (1979), to know real possibility I do not need to *know* (3a) (or (3b)), I just need to *fall under* it (in particular, I need to satisfy the right-hand side). In particular, to satisfy the cognition requirement with respect to some metaphysical theory about noumena, I do not need to *know* the metaphysical theory of God that underlies (3a) (or (3b)); I just need to *satisfy* it, in particular, the right-hand side of (3a) (or (3b)).[114]

But a problem will arise if we think of (3a) or (3b) as providing a *justification* for claims of knowledge of real possibility for noumena. By using (3a) or (3b), and their right sides, to infer, for some concept C, that I know that C is really possibly

[112] Kant's objection is thus very similar to the problem of the 'Cartesian circle.' My argument in the rest of this section is indebted to Van Cleve (1979), who argues, convincingly, that the problem of the Cartesian circle is the problem of the criterion.

[113] I have formulated both (3a) and (3b) as externalist principles on which an agent need not base her judgment of possibility on the grounds that make her concept the concept of a possible object: divine preformation/vision in God. Presumably, Crusius and Malebranche would both agree that agents had knowledge of possibility before they ever offered their theories. This does not affect, I think, the point I go on to make about epistemic principles.

[114] Van Cleve (1979), 70.

instantiated, I do not confer any *justification* on the conclusion of that logically valid inference unless the premises themselves are justified. This becomes even more problematic when we recognize that the right-hand sides of (3a) and (3b) are claims about noumena (in particular, God); since knowledge of a theory of noumena depends on antecedent cognition of the concepts involved, it would seem that we must be in a position to know that <*God*> is really possibly instantiated before we can be in a position to know (3a) or (3b) (or their right-hand sides).

Recall that the context of Kant's objection is his attempt to explain how *a priori* cognition (knowledge of the real possibility) of noumena is possible, a problem, he claims, "in my long metaphysical labors, I, as well as others, had failed to consider" (Ak. 10:131). In *sensu stricto*, though, this is clearly false, and Kant would have known it to be false; we have just seen Crusius's and Malebranche's explanations of how *a priori* cognition of noumena (the relation of intellectual concepts to their objects, in the language of the Herz letter) is possible and the conditions under which we actually have such cognition, (3a) and (3b). Kant's claim to innovation is strengthened in the *CPR*, however, when he distinguishes between the dogmatic and the critical procedure in metaphysics: "dogmatism is therefore the dogmatic procedure of pure reason, without an antecedent critique of its own capacity" (Bxxxv). I take 'antecedent' to mean that the critique of our capacity of pure reason will not presuppose any substantive metaphysical theory about noumena. It is not enough to require of a 'critique of pure reason' that we refrain from claims to *know* a metaphysical theory about noumena, for Crusius and Malebranche can use (3a) and (3b) and the supporting metaphysics to generate knowledge of real noumenal possibility *without* assuming that (3a) or (3b) are *known* to be true. Kant must be setting up the project of a critique of pure reason as follows: suspend any positive *claims* about noumena and explain our putative *a priori* cognition of them. And within a project of that kind it is appropriate to exclude Crusius's (3a) and Malebranche's (3b) not because they are 'circular' (they are not) but because they presuppose a substantive metaphysical theory of noumena (in particular, of God).[115]

This vindicates Kant's reasons for setting aside Crusius's and Malebranche's *deus ex machina* theories within his own project: a transcendental critique into our faculty of pure reason. But this does not undermine the coherence of Crusius's or Malebranche's project, which in very general terms we might characterize as: starting with prima facie plausible metaphysical and epistemic assumptions, find the best overall theory of metaphysics and our epistemic access to it.[116] What is more, it might cast

[115] They are also circular if Crusius and Malebranche take (3a) and (3b) to be our source of justification for claiming that the concepts that figure in those very principles (e.g., <*God*> and <*cause*>) are really possibly instantiated.

[116] Obviously, both Crusius and Malebranche have more specific philosophical/theological projects in their writings, e.g., to defend libertarian freedom against the excessive rationalism of Leibniz and Wolff (Crusius), or to perfect the Cartesian system and make it safe against atheism and skepticism (Malebranche).

some suspicion on the coherence of Kant's *own* project. First of all, the negative result of that critique—that there is no explanation of how *a priori* cognition of noumena is possible—is unsurprising if we are required at the outset to bracket any positive theory of noumena, for presumably any such explanation will consist in a theory of noumena and how they are accessed by our minds (as Crusius's and Malebranche's do). One might reasonably suspect that Kant has built the result of that critique (noumenal ignorance) into its foundational assumptions. Secondly, there are reasons to be suspicious whether Kant respects his own methodological scruples. Kant's transcendental theory of our cognitive capacity begins by assuming that we have a sensible (receptive, non-spontaneous) form of intuition and distinguishing two 'stems' of cognition: sensibility and understanding.[117] There are reasons to think that, in knowing that sensibility and understanding are distinct faculties, we do not merely know phenomena; after all, phenomena are possible objects of experience, and experience itself depends upon our faculty of sensibility (intuitions) and understanding (concepts), so presumably those faculties themselves cannot be phenomena (cannot be sensibly intuited). So there are grounds to wonder whether Kant's own answer to the transcendental question of how cognition *a priori* is possible does not, in fact, begin with a (controversial) assumption about the non-phenomenal noumenal nature of our minds.

Fully adjudicating these issues would require addressing some of the thorniest issues in the whole Critical philosophy, principal among them the epistemic status of Kant's own transcendental theory of experience, and whether it is compatible with his denial of synthetic *a priori* knowledge of noumena.[118] I do not intend to address those issue here. I want to propose merely that the best way of understanding Kant's circularity objection to Crusius and Malebranche is as motivating a project in transcendental philosophy (bracketing our positive metaphysical theories of noumena, explain how *a priori* cognition of real possibility is even possible in the first place), since it is far from clear that it can succeed as a non-question-begging objection to their own metaphysical projects.

Ontology, for the logicists, was the science of all possible beings *qua* possible. In Kant's pre-Critical works he argued that logical possibility is not sufficient for possibility, which he would later refer to as *real* possibility (to distinguish it from mere logical possibility). However, his doctrine that all real possibilities are grounded in God raised pressing questions in the metaphysics, epistemology, and cognitive semantics of modality that he did not answer:

[117] A16/B29–30. Kant's concession that "they may perhaps raise from a common but to us unknown root" does not, however, ameliorate the point; the question is what ground Kant could have for claiming that sensibility and understanding are distinct faculties with distinct contents and distinct principles of operation that "cannot exchange their function" (A51 f./B75 f.), if this is a claim about how they are *in themselves*.

[118] See e.g., Pereboom (1991).

(1) What is it to represent an object *as really possible*? That is, what is the content of the representation of an object *as* really possible?
(2) What is it for an object to be really possible? What are the grounds of real possibility?
(3) Of a really possible object we can represent as really possible, how can we prove that it is really possible? And, what's more, prove this *a priori*?

The outline of Kantian answers to these questions have now emerged, and they depend upon the crucial distinction of all objects *überhaupt* into phenomena (objects that can be given to us in sensible intuition) and noumena (objects that cannot be so given):

(1) To represent a phenomenon as (formally) really possible is to represent it as compatible with the formal intuitional (space and time) and intellectual (unity of apperception) conditions of experience.
(2) For phenomena, the first grounds of (formal) real possibility are space, time, and the unity of apperception.
(3) We prove the (formal) real possibility of phenomena *a priori* through a transcendental critique of our capacity for cognition, which reveals the first real grounds of the possibility of experience: space, time, and the unity of apperception. We cannot prove *a priori* what is really possible for noumena; consequently, a metaphysics of noumena is impossible for us.

This leaves open, however, whether there are notions of real possibility other than formal real possibility (compatibility with the forms of experience); in the next two chapters I will explore other Kantian notions of real possibility for phenomena. It also leaves unanswered questions (1) and (2) about real possibility in relation to noumena, and leaves open the possibility (3) that we may have some epistemic warrant falling short of theoretical knowledge with respect to real possibility for noumena. The final two chapters of this study concern Kant's Critical theory of the cognitive semantics, metaphysics, and epistemology of real possibility for noumena.

7

Three Kinds of Real Possibility

7.1. Introduction

In the *Critique of Pure Reason* (*CPR*) Kant answers two questions that had arisen originally, albeit implicitly, in *Beweisgrund*: how are real possibilities 'given' to the mind? How can we know *a priori* what is really possible? However, Kant cannot retain his pre-Critical theory of real possibility wholly unchanged within the Critical philosophy. Pre-Critical real possibility concerns what is possible for all objects *überhaupt*, including objects that cannot be sensibly intuited (noumena). Given Kant's Critical doctrine that we cannot cognize noumena and hence cannot cognize all objects *überhaupt*, his pre-Critical modal metaphysics cannot constitute an object of knowledge (*Wissen*); cognition of real possibility for noumena is impossible. The domain of real possibility shifts in the Critical period to real possibility we can cognize: real possibility for *phenomena*.

In this chapter I begin to explore Kant's Critical theory of real possibility, although, in one form or another, it will occupy us for the rest of this study. In this and subsequent chapters I distinguish various different kinds of real possibility in the Critical system, but in section 2 I begin by taking a step back and considering what all of these notions have in common that make them kinds of *real* possibility in the Critical system. In the remainder of the chapter I focus on three different kinds of real possibility for phenomena, which I call *formal* real possibility (section 3), *empirical-causal* real possibility (section 4), and *noumenal-causal* real possibility (section 5).

7.2. Critical Real Possibility

Recall that Kant's pre-Critical theory of real possibility had these elements:[1]

(*Non-Logical*) *Real possibility is distinct from logical possibility*. That some proposition[2] p is logically consistent ($\neg p$ does not entail a contradiction) does not entail

[1] In *Beweisgrund* Kant focused on the possibility of *predicates*; I have formulated these claims in terms of the possibility of propositions being true, to accommodate the wider range of things whose possibility Kant will investigate in the Critical period: the possibility of experience, the possibility of objects in experience, the possibility of those objects undergoing change, etc. Kant typically defines possibility in terms of the possibility of an *object*—see his metaphysics lectures (*MV*, Ak. 28: 410; ML_2, Ak. 28: 543; *MM*, Ak. 29: 81; MK_3, Ak. 29: 960). These are interconvertible definitions: an object is possible just in case the proposition that there is such an object is possible. Likewise, a concept *C* is possible just in case the proposition that the concept is instantiated is possible.

[2] By 'proposition' I just mean the content of a judgment; I introduce this terminology to make clear I am not talking about the modal status of *token* acts of judging. Kant himself, however, defines the term

that p is really possible. So logical possibility and real possibility are at least *conceptually distinct*; it is not a conceptual truth that if $\Diamond_L p$ then $\Diamond_R p$ (equivalently, it is not a conceptual truth that if $\Box_R p$ then $\Box_L p$).

(*Ground*) *Real possibilities have real grounds in actuality*. If it is really possible that p then there is some actual object or principle that grounds the fact that it is really possible that p. The relation between *grounds of real possibility* and the *real possibilities they ground* is not a logical relation; if x grounds the real possibility that p, this is not because of the conceptual containment of p within the concept of x.

(*Worldly*) *Real possibility is a form of metaphysical, or world-based, possibility*.[3] Real possibilities in general are grounded in facts about the world that do not depend upon how our minds are constituted, or how we experience or conceptualize the world.[4]

In the Critical philosophy Kant deploys several different *kinds* of real possibility without explicitly distinguishing them; in this chapter I will distinguish three of them and, in the next chapter, a fourth. What makes all of these different kinds of possibility kinds of *real* possibility is that they obey Non-Logicality and Groundedness (with modifications to be noted below). However, some of these kinds of real possibility violate Worldliness; Kant grounds some kinds of real possibility in our cognitive faculties and in the causal powers of phenomena,[5] rather than in objects that do not depends upon a discursive intellect, noumena.[6] For now this will constitute an interpretive hypothesis; it will take me the rest of this study to argue for it fully.

This following is a provisional characterization of the concept of real possibility in the Critical philosophy:

(*Real Possibility*) For any kind of possibility $\Diamond_X p$ (and its associated kind of necessity $\Box_X p$, where $\Box_X p \leftrightarrow \neg\Diamond_X \neg p$), $\Diamond_X$ is a kind of *real possibility* (and $\Box_X p$ is a kind of *real necessity*) only if

'proposition' [*Satz*] to refer to an assertoric rather than merely problematic judgment; see *Disc.* (Ak. 8: 193 n.) and Kant's handwritten note at A74 (E XXXVIII, Ak. 23: 25).

[3] This is not strictly true for all notions of real possibility in *all* of Kant's pre-Critical writings. In ID Kant already holds that objects in space and time are phenomena whose possibility depends on our spatiotemporal forms of intuition. So, while the pre-Critical writings do contain some notion of real possibility for non-noumenal objects, the dominant strand in Kant's pre-Critical theorizing about modality is noumenal.

[4] Another element of Kant's pre-Critical modal metaphysics (*GARP*—see Ch. 5.7) was the doctrine that there is a unique being that grounds all real possibilities and exists with absolute necessity. I discuss the fate of this doctrine in the Critical period in Ch. 9.

[5] Phenomena are mind-dependent in this sense: their possibility is grounded in the possibility of our experience of them. See A111 and Ch. 6.7.

[6] The role of God is maintained in a particular kind of real possibility, which I will call 'noumenal real possibility'; see Ch. 9 for more on noumenal possibility.

(i) *Non-logicality*: it is not a conceptual truth that $\Diamond_L p \supset \Diamond_x p$ (equivalently, it is not a conceptual truth that $\Box_x p \supset \Box_L p$), and
(ii) *Groundedness*: if $\Diamond_x p$ then the fact that $\Diamond_x p$ has a real ground in some actual object or principle.

Every aspect of this definition requires further philosophical analysis, but let me begin by pointing out that clause (i) means that every concept of real possibility (real possibility$_x$) is a *different concept* than the concept of logical possibility. This does not entail that they have different *extensions* (it does not entail that there are logical possibilities that are not real possibilities of the relevant kind), only that they might.[7] In some cases, though, we know *a priori* they do have different extensions. For instance, we know *a priori* that it is logically possible that a triangle have internal angles that sum to less than 180 degrees but this is not really possible. While there are kinds of real possibility that are extensionally distinct from logical possibility, extensional distinctness is not a conceptual requirement of real possibility.

The other aspect of this definition of real possibility that stands in need of clarification is the notion of 'ground' involved in the second clause. The question that I raised at the beginning of this chapter—what is the relation between pre-Critical real possibility and Critical real possibility?—has its counterpart in the question, what is the relation between Kant's pre-Critical conception of real grounds, and his Critical conception of real grounds? Fully answering that question would require a book of its own. Here I can only sketch an answer. The most important aspect of Kant's pre-Critical conception of real grounds is retained within the Critical system, namely:

(*Non-Logicality*) If x is the real ground of y then the concept of x does not contain the concept of y, and the judgment that x exists does not logically entail the judgment that y exists.[8]

In other words, real grounding relations do not obtain in virtue of relations of logical entailment or conceptual containment. Kant's Critical system involves at least two very different kinds of real grounds: formal grounds (e.g., triangularity grounds the possession of internal angles that sum to two right angles)[9] and efficient causal grounds (e.g., the sun shining on the stone grounds the stone's warming). Both of these kinds of real ground obey the Non-Logicality requirement. For each kind of real possibility I distinguish below, I will also explain the grounds of this kind of real possibility, as well as the nature of the relevant non-logical grounding relation.

[7] Because I am treating possibility and necessity as modal operators on judgments, rather than predicates of objects, their extension is the class of judgments to which they apply, rather than a class of objects; e.g., the extension of <*possible*> is the set of judgments which are possibly the case.

[8] In the metaphysics lectures, see *MM* (Ak. 29: 807), *ML*$_2$ (Ak. 28: 549), *MV* (Ak. 28: 402), and *MvS* (Ak. 28: 486).

[9] What Crusius would call 'real existential grounds.' See Ch. 3 and *Ent.* §79 (p. 136), as well as §3 of this chapter.

Whereas Kant's pre-Critical view required that real possibilities be grounded in an actually existing substance, his Critical view broadens the candidates for real grounds of possibility to include grounds that are not substances (e.g., the forms of experience themselves) and, indeed, in some cases are not even objects at all (e.g., the unity of apperception). Precisely articulating the 'actuality' requirement on grounds would require investigating the details of Kant's Critical theory of actuality [*Wirklichkeit*] and existence [*Dasein*], which I postpone until Chapter 10; in the meantime, though, I will require merely that some actual object or principle grounds real possibilities: the *actual* forms of experience, the *actual* natural laws, the *actual* past, etc.

This introduces an important distinction, however, among the kinds of real possibility. We are concerned in this chapter with real possibilities for phenomena, but phenomena have two kinds of grounds in Kant's Critical system: 'immanent' ones and 'transcendent' ones. By 'immanent' grounds I mean grounds that are either themselves phenomena or are subjective conditions of phenomena, such as the forms of experience. Phenomena have immanent grounds insofar they are conditioned (causally) by other phenomena and (non-causally) by subjective forms of experience. By 'transcendent' grounds I mean grounds that are not immanent to experience, namely, noumena. Empirical objects have noumenal grounds in that they are the appearances of those noumena; noumena causally affect subjects' minds, giving rise to the sensory matter (but not the form) of experience.[10] Of the kinds of real possibility I distinguish in this chapter, two have immanent grounds (formal possibility, empirical-causal possibility) and one (noumenal-causal possibility) has a noumenal ground.

Figure 7.1 represents the distinctions I have sketched so far and indicates provisionally where these different kinds of real possibility fit. In the penultimate chapter of this study I will expand our focus to include kinds of real possibility for noumena as well. The different kinds of real possibility are distinguished both by the range of propositions to which they apply, as well as by the kinds of grounds they have.

7.3. Formal Possibility

In the previous chapter we explored Kant's answer to the question, how can we know *a priori* what is really possible? His answer to this question, on my reconstruction, invoked a notion of possibility that I there dubbed 'formal' possibility. My discussion of that concept was, to borrow a Kantian trope, a reconstruction of it 'from above,' from the philosophical problem it is supposed to solve. In this section, to continue the Kantian trope, I will reconstruct it 'from below,' from the basic elements of Kant's theory of experience, and show that the two reconstructions arrive at the same point. In the second half of the section I explore the notion of grounding involved in formal possibility.

[10] For noumena as grounds of the matter of phenomena see A190/B235, A387, A494/B522, A614/B642, *Prol.* (Ak. 4: 289, 314, 318), *CPrR* (Ak. 4: 451), and esp. *Disc.* (Ak. 8: 215).

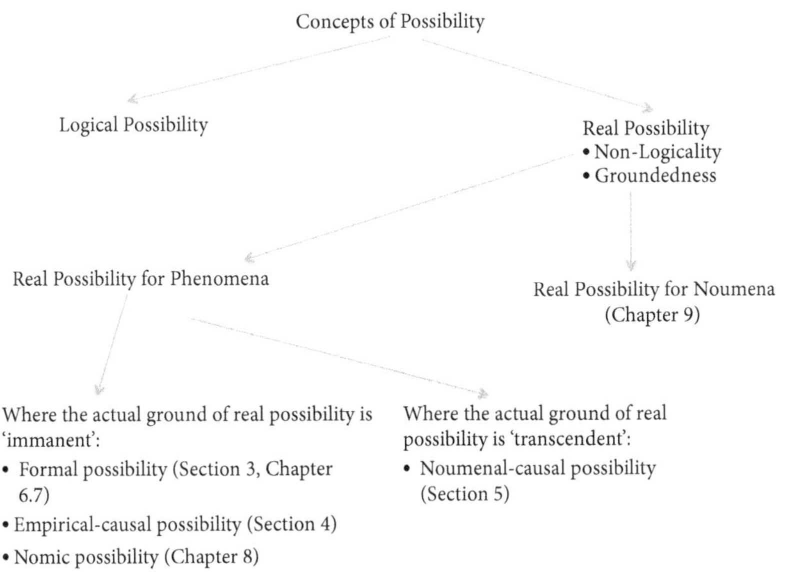

Fig. 7.1. Kinds of real possibility in Kant's Critical system

Kant defines experience at B147 as "empirical cognition": thinking a sensibly given object under a concept.[11] This is a 'semantic' characterization of experience in terms of the kind of representation it is and the kind of content it has. Experience is also, according to Kant, the joint product of affection by mind-independent objects—noumena or 'things in themselves'—and processing by the subject's mind (synthesis).[12] This characterizes experience in terms of its causal origin. Noumena affect the subject's mind, creating low-level representations, sensations.[13] The subject's cognitive faculties process these sensations by synthesizing them into representations of spatiotemporal objects, experience.[14]

Some of the representations generated by this cognitive process depend upon the particular sensations had by the subject. For instance, the determinate shape and size of my desk, as represented in my intuition of it, depends upon the sensations I have. Other representations do not depend in this way upon the particular sensory content.

[11] "To think of an object and to cognize an object are thus not the same. For two components belong to cognition; first, the concept through which an object is thought at all (the category), and second, the intuition, through which it is given [...] [the categories] serve only for the possibility of empirical cognition. This, however, is called experience [...]" (B146–8). This is a quite minimal notion of experience. In this section I build up from this minimal sense of experience to a stronger sense (the sense in which Kant usually uses the term *Erfahrung*): a unity of perceptions that are connected according to laws. See further discussion in text.

[12] B1.

[13] Kant asserts the noumenal affection of our sensibility by things in themselves at A190/B235, A387, A494/B522, *Prol.* (Ak. 4: 289, 314, 318), *GMM* (Ak. 4: 451), and esp. *Disc.* (Ak. 8: 215).

[14] B1–2.

Regardless of the sensations I have, these representations will have the same content. A representation is said to be *a priori* just in case its content does not depend upon the particular sensations had by the subject.[15] An *a posteriori* representation is a representation whose content does depend upon particular sensations. For instance, regardless of the sensory content I receive, I intuit outer objects (i.e. objects that are not my inner psychological states) in space. Similarly, regardless of the sensory content I receive, I intuit space as Euclidean, according to Kant. While the determinate spatial properties of objects depend upon the particular sensory content I receive and are therefore *a posteriori*, the representation of space and its Euclidean character are *a priori*.[16] The forms of experience (forms of intuition and forms of understanding) are *a priori* representations: determinable intuitional and conceptual structures of which any particular experience is a determinate filling-out.[17] The Euclidean structure of space is an *a priori* form of intuition; it is shared by all intuitions of outer objects, and it does not depend upon the particular course of my experience.[18]

Experiencing an object, by definition, requires thinking of it under a concept and, Kant argues, this requires thinking of objects under the categories. They are *a priori* concepts, because they provide a determinable conceptual structure for all experience of objects. For instance, the category <*substance–accident*> requires that we experience each object as either a substance or a modification of a substance, and the category <*cause–effect*> requires that we experience alterations in substances (changes in their accidents) as governed by deterministic causal laws. The conceptual forms of experience (categories) do not dictate which objects we experience as substances, and which as modifications, nor which causal laws govern objects,[19] just as the intuitional forms (space, time) do not dictate the determinate sizes, shapes, locations, and durations of objects. Each particular experience is a combination of

[15] This conception of the *a priori* is equivalent to what Kitcher calls '*a priori*$_O$' in Kitcher (1993), 15–16.

[16] This concept of *a priority* is not defined in terms of our experience-independent justification for making a judgment about objects (e.g., that they are in space), nor in terms of deriving something from its ground. On this conception of the *a priori*, see *Disc.* (Ak. 8: 221) where Kant distinguishes his conception of the *a priori* from innatism. *A priori* representations are not innate, for they require experience for their activation. What is innate is the capacity to represent objects as, for instance, spatial, regardless of the course of experience. See also Kant's discussion of the origin of the categories at B167. Cf. the discussion of both passages in Longuenesse (1997), 221.

[17] See A50–1/B75: "Thus pure intuition contains merely the form under which something is intuited, and the pure concept only the form of thinking in general. Only pure intuitions or concepts alone are possible *a priori*, empirical ones only *a posteriori*."

[18] The representation of space is determinable in the sense that it does not determine the spatial properties of particular objects. However, the structure of space is determinate. My *a priori* representation of space represents space as Euclidean. Thus, in Kantian terms, our form of intuition of outer objects is Euclidean. Similar remarks apply for time: the form of time dictates the structure of time, but not the determinate temporal properties of objects.

[19] Kant claims that which particular causal laws govern empirical objects is not wholly determined by the forms of experience (concepts and intuitions) alone at A125, B167, and in the *CJ* (Ak. 5: 180 and 184). See Ch. 8 for more discussion.

determinable *a priori* intuitional and conceptual forms (space, time, the categories) and more determinate *a posteriori* content or 'matter' (size, shape, location, particular causal laws).

The idea that experience has *a priori* forms corresponds naturally to the concept of what is necessary *because* of the forms (formally necessary), or, more precisely:

> (*Formal necessity*) It is *formally necessary* that p if and only if the fact that p is wholly grounded in facts about the actual intuitional form (space and time) and conceptual form (categories) of experience

where 'p' ranges over synthetic propositions about phenomena.[20] The word 'actual' is there to remind us that it is at least conceivable that there could be discursive cognizers with different intuitional forms than ours; formal possibility is defined in terms of our actual forms of intuition, space and time, not the idea of forms of intuition in general (or the requirement that there be some form of intuition or other). This definition of formal necessity, according to the interdefinability of possibility and necessity, corresponds to the following definition of formal possibility:

> (*Formal possibility*) It is *formally possible* that p if and only if it is not the case that facts about the actual intuitional form (space and time) and conceptual form (categories) of experience wholly ground the fact that $\neg p$.

For instance, it is formally possible that we experience unicorns; the fact that we do not experience such objects is partly grounded in how we are affected.

Kant's standard formula for expressing the (formal) necessity of space, time, the categories, the principles of experience, etc. is to say that they are 'conditions' [*Bedingungen*] or 'grounds' [*Gründe*] of the possibility of experience,[21] and readers may wonder about the relation of that famous Kantian formula to my definitions of formal necessity and possibility. Before answering that question directly, I want to make two observations about this formula. First, it is not clear in what sense of 'possibility' these forms are conditions of possibility; they are not, for instance, conditions of the *logical* possibility of experience. Nor is it sufficient to say that they are conditions of the *formal* possibility of experience, for Kant defines what I am calling 'formal' possibility in terms of the forms of experience *themselves*: "whatever agrees with the formal conditions of experience (in accordance with intuition and concepts) is **possible**" (A218/B265 f.). Secondly, not all principles that are necessary in the relevant sense are conditions of the possibility of experience, for some of them are consequences of those conditions. For instance, no possible object of experience can violate a theorem of geometry, but it is implausible that *every* theorem is a

[20] Since in this chapter I am focusing only on kinds of modality that apply to phenomena, I will not always make this restriction explicit.

[21] E.g. A94/B126, A95, B161, A158/B197, and A237/B296; cf. *Prol.* (Ak. 4: 297, 319, 351).

condition of experience. The basic Euclidean structure of space is a *condition* of the possibility of experience, while a theorem is a necessary consequence of that condition; both apply to every possible object of experience, but this is *because* of the Euclidean structure of space.

However, I do not think that the necessity of the forms of experience is defined in terms of some antecedently intelligible notion of possibility (of which they are conditions); I take these forms to be *constitutive* of the relevant notion of (formal) possibility and (formal) necessity. When Kant says that the forms are 'conditions' or 'grounds' of the possibility of experience he means: they are the grounds that are constitutive of the formal possibility of experience in the first place. With respect to the second issue, my definition of formal necessity has built into it a distinction between the forms of experience (which ground formal necessity and possibility) and the consequences of those forms. So we can say that the basic structure of space is a *ground* of the formal possibility of outer objects, the theorems of geometry are *grounded* in that structure, and both are equally formally necessary (they apply to all formally possible objects). This repeats a familiar structure from Kant's pre-Critical theory of absolute necessity (see Chapter 5.2): there is a hyperintensional relation of *grounding* among the formal necessities.

In some texts Kant talks about formal possibility in terms of *compatibility* with our forms of experience, for instance, "whatever agrees with [*mit … übereinkommt*] the formal conditions of experience (in accordance with concepts and intuitions) is **possible**" (A218/B265 f.).[22] I take 'agreement with' the forms of experience to mean 'compatibility with' the forms of experience. If A218/B265 f. and other passages that characterize possibility in terms of compatibility with the forms of experience are discussing what I am calling formal (real) possibility, this raises the question: what is

[22] Abaci (2013) and Chignell (2014*b*) dissent from this reading of the definition of possibility at A218/B265 f. In the body of the Postulates themselves Kant discusses a series of concepts "the possibility of which is entirely groundless, because it cannot be grounded in experience and its known laws" (A223/B270), namely: a "substance persistently present in space yet without filling it […], or a special power of our mind to intuit the future (not merely, say, to deduce it), or, finally, a capacity of our mind to stand in community with the thoughts of other people (no matter how distant they may be)" (A222–3/B270; cf. A290–1/B347). Abaci and Chignell argue that Kant does not mean that the assertion that these objects are formally possible is groundless, but that we have no reason to think they are actually instantiated (Abaci) or compatible with our background knowledge of natural laws (Chignell). In each case, though, there is a serious question as to whether these are concepts of objects that can be given in experience. To take the first case, Kant argues in *MFNS* that matter present in space without filling it through dynamic force is incompatible with our forms of experience (Ak. 4: 496–7). Likewise, in the second case, it is questionable whether intuition of the future is compatible with the forms of experience, given that our sensible faculty must be affected by its object, and 'backwards causation' is ruled out by the Second and Third Analogy. In the third case, it is questionable whether the unity of apperception allows me to attach the 'I think' to a thought of another person; arguably, what I can attach the 'I think' to determines the range of what thoughts are mine, so apperceptive access to the thoughts of another person is impossible (they would thereby be my thoughts as well). In each case there are reasons for Kant to assert that the possibility of these objects is "entirely groundless because it is not grounded in experience and its known *a priori* formal laws." See also A144/B184, *MvS* (Ak. 28: 494), *ML*$_2$ (Ak. 28: 556), *MM* (Ak. 29: 815, 821 f.), *Refl.* 4081 (Ak. 17: 732), and *Refl.* 6360 (Ak. 18: 688).

the relevant relation of compatibility? The definition in terms of compatibility is equivalent to the definition in terms of grounding if the compatibility relation is defined as follows:

(*Defn.*) *p* is *compatible with our forms of experience* if and only if it is not the case that facts about our forms of experience wholly ground the fact that $\neg p$.

This expresses an intuitive notion of compatibility; for instance, it is compatible with the forms of experience that I experience motions that violate Newton's laws just in case the forms of experience alone do not make it the case that I do not experience such objects. My definitions of formal possibility and necessity in terms of grounding explain and unify two standard Kantian formulas for necessity and possibility, respectively: "conditions of the possibility of experience" (necessity) and "agreement (compatibility) with the forms of experience" (possibility).[23] I take this to be indirect evidence of their correctness.

Experience is thinking of sensibly given objects under concepts, and thus has two kinds of conditions: sensible conditions and intellectual conditions. In the intellectual case, I have focused on the categories, but Kant's ultimate view is that the categories are derivative of a more fundamental intellectual condition, the unity of apperception.[24] I take this to mean that any intellectual ground of the possibility of experience is ultimately grounded in the unity of apperception. For instance, the categories are intellectual grounds of experience of objects because they are grounded in the unity of apperception. So we can reformulate our definition of possibility as follows:

(*Formal possibility*) It is *formally possible* that *p* if and only if it is not the case that facts about the actual intuitional form (space and time) and intellectual form (unity of apperception) wholly ground the fact that $\neg p$.

In the previous chapter I reconstructed 'from above' a notion of formal possibility:

(*Formal*) It is *formally possible* that we represent a certain (conceptual or intuitional) content just in case it is compatible with the forms of intuition (space and time) and the highest principle of the understanding (the synthetic unity of apperception) that we represent a sensibly given object using that (conceptual or intuitional) content.[25]

[23] I depart here from Robert Hanna, who attributes to Kant the familiar 'possible worlds' analysis of possibility and necessity, and then defines what an 'experientially possible world' is in terms of the conceptual and intuitional form of experience (Hanna (2001), 239–45). However, an experientially possible world is going to be one that does not have features that are incompatible (in the sense defined above) with our forms of experience, so we can define formal possibility directly in terms of those forms of experience, and skip the intermediate step of constructing a set of possible worlds for the modal operators-cum-quantifiers to range over. Furthermore, a possible worlds analysis will not help in accounting for the hyperintensional grounding relation among formal necessities (see Ch. 5.2), because, by definition, both relata of the grounding relation will be true in all 'experientially possible worlds.'

[24] A111, B136. See Ch. 6.6.

[25] See Ch. 6.7.

If we understand compatibility here in terms of grounding, via the definition of compatibility above, I think it is clear that this notion of formal possibility is subsumed by the more general definition of formal possibility I have just constructed 'from below' (that we represent a certain content is just a value of p).

Since all objects of possible experience are appearances (since all sensibly given objects are appearances),[26] if F is a content that figures in the content of experience (e.g., the property of being a substance), we can infer from claims of the form:

(1) It is formally possible that we experience objects as being F

to conclusions of the form:

(2) ∴ It is formally possible that objects of experience are F.

This inference would be invalid if the objects of experience were things in themselves (noumena), rather than appearances (phenomena). This entailment holds because facts of the form of (2) are grounded in facts of the form of (1); if it is formally possible for objects to have property F, this is in virtue of the fact that it is formally possible to experience objects as having property F. Since the forms of experience (space and time, the unity of apperception) are the grounds of formal possibility, they are the grounds of facts of the form of (1). It follows that the forms of experience are grounds of the formal possibility of objects of experience (phenomena) themselves, as Kant states in the A Deduction:

> The *a priori* conditions of a possible experience in general are at the same time conditions of the possibility of the objects of experience. Now I assert that the categories that have just been adduced are nothing other than the conditions of thinking in a possible experience, just as space and time contain the conditions of the intuition for the very same thing. They are therefore fundamental concepts for thinking objects in general for the appearances, and they therefore have *a priori* objective validity, which was just what we really wanted to know. (A111)[27]

This means that the forms of experience (Kant here focuses on the categories, but the same point applies to space and time as well) apply to all possible objects of experience. The forms of experience alone make it the case that all objects of experience obey these forms. It follows, by definition, that it is formally necessary that the objects of experience obey the forms of experience (e.g., they instantiate the categories).

It is clear that formal possibility (and necessity) satisfy the two conditions on real possibility (and necessity) from section 2:

(*Real Possibility*$_F$) Formal possibility $\Diamond_F p$ (and its associated kind of necessity $\Box_F p$, where $\Box_F p \leftrightarrow \neg\Diamond_F \neg p$) is a kind of *real possibility* (and $\Box_F p$ is a kind of *real necessity*)[28] only if

[26] A30/B45. [27] Cf. A158/B197.

[28] Abaci (2013, 25 n. 8) objects to my assertion that Kant has a notion of formal necessity, not made explicit in the Postulates itself, but derived from formal possibility in the standard fashion. While I admit

(i) *Non-logicality*: it is not a conceptual truth that $\Diamond_L p \supset \Diamond_F p$ (equivalently, it is not a conceptual truth that $\Box_F p \supset \Box_L p$), and
(ii) *Groundedness*: if $\Diamond_F p$ then the fact that $\Diamond_F p$ has a real ground in some actual object or principle.

Formal possibility is not conceptually equivalent to logical possibility because it is not even extensionally equivalent: there are logically possibilities (e.g., that we experience non-spatial outer objects) that are not formally possible. Formal possibility is grounded in actuality because, if it is formally possible that *p*, then this is so in virtue of a fact about our *actual* forms of experience (they do not make it the case that $\neg p$). If we, *per impossibile formaliter*, had different forms of intuition, the facts about what is formally possible might be different. It is not grounded in an actual object, because the forms of intuition, much less the unity of apperception, are not *objects*.

Readers might wonder what notion of 'ground' is involved in the definition of formal possibility and necessity. It should be clear that formal necessity and possibility cannot be defined in terms of logical grounding (entailment). The relation between the axioms and theorems of geometry parallels the relation between the forms of outer experience and the formally necessary principles of experience that follow from them. If we analyzed formal possibility and necessity in terms of logical grounding, it would follow that geometrical axioms logically entail geometrical theorems, i.e. that all formally impossible propositions are logically incompatible with the axioms. But Kant holds that geometric reasoning is irreducibly synthetic because it relies on construction in pure intuition.[29] Consequently, the relation between forms of experience and what is formally necessary cannot be assimilated to the relation of logical entailment; the formal necessities do not follow *logically* from the forms of experience. Consequently, formal possibility satisfies the requirement that kinds of real possibility have *real* grounds in actuality.

What, then, is the nature of this real grounding relation? It is very hard to extract a fully satisfying answer to that question from Kant's writings. However, certain broad features of Kant's Critical theory of real grounds of possibility can be noted. As noted in section 2, the Critical Kant continues to distinguish logical from real grounds,[30] of which causal grounds are a species.[31] Further, he follows Wolff, Baumgarten, and Crusius[32] in distinguishing between grounds of becoming (*rationes*

that Kant could have been more explicit about making all of these modal distinctions, he clearly needs a notion of formal necessity to make sense of the modal status of the *a priori* formal principles of experience themselves. This cannot be empirical necessity (see below) because empirical necessity only applies to alterations in substances and is only knowable *a posteriori*$_J$.

[29] For more on Kant's views on construction and the irreducibly synthetic character of geometric reasoning, see Friedman (1992*a*), 80–7, Parsons (1992), 78–9, and Shabel (2007), 97–107.

[30] *MM* (Ak. 29: 807), *ML*$_2$ (Ak. 28: 549), *MV* (Ak. 28: 402), and *MvS* (Ak. 28: 486).

[31] *MV* (Ak. 28: 403)and *MvS* (Ak. 28: 486).

[32] *Ont.* § 874, *Meta.* § 311, and *Ent.* § 34. Thanks to Smit (2009), 200–2 for drawing my attention to this important distinction.

fiendi) and grounds of being or possibility (*rationes essendi*).[33] A ground of becoming (*ratio fiendi*), as its name suggests, is the ground of a change, either an alteration in, or the generation or corruption of, a substance (its arising or perishing). Kant defines a ground of being (*ratio essendi*) as follows: "the *ratio essendi* is the ground of that which pertains [*zukömmt*] to a thing considered according to its possibility" (*MM*, Ak. 29: 809). The *ratio essendi* of a being is what makes it possible, while its *ratio fiendi* accounts for its actuality. The forms of experience are real *rationes essendi* of objects of possible experience but this does not tell us what this real grounding relation is.

While I do not think we can find a reductive analysis of the real grounding relation between the forms of experience and their consequences, I think we can make some progress towards determining the *structure* of this relation. I will begin by focusing on Kant's example of a *ratio essendi* in the Mrongovius lectures: "the three sides [*Seiten*] in the triangle are the ground of the three corners [*Winkel*]" (Ak. 29: 809).[34] The example is potentially confusing because we refer to triangles as 'tri*angles*' [*Dreiecke*] but we typically define them as three-*sided* [*dreiseitig*] plane figures. Kant is assuming (not implausibly, I think) that their trilaterality is more fundamental than their triangularity; triangles are triangular *because* they are trilateral. This is not a logical grounding relation (that *T* is trilateral does not logically entail that it is triangular), nor is it a causal relation. Furthermore, the relation is asymmetric: trilaterality grounds triangularity, and not vice versa. Nonetheless, the two properties are (formally) necessarily co-instantiated: necessarily, a figure is trilateral if and only if it is triangular.[35] So the relevant grounding relation is a

[33] *ND* (Ak. 1: 392), *MH* (Ak. 28: 37, 54), *MM* (29: 809, 810), ML_2 (Ak. 28: 571), *MD* (Ak. 28: 648), MK_2 (Ak. 28: 802), and *Refl.* 5182 (Ak. 18: 111). Cf. Kant's claim in the *CPrR* that "freedom is the *ratio essendi* of the moral law" (Ak. 5: 14). Strictly speaking, Kant's example in the quoted passage is one in which the essence of a possible thing is the ground of that very (numerically identical) thing's properties; in the rest of the section I am applying that notion to cases where one thing is the ground of the possibility of a numerically distinct thing (e.g., the form of experience and an object of experience). But we have seen that Kant's system requires a real grounding relation between a ground and the possibility of a numerically distinct thing; this, I take it, is sufficient reason to generalize the notion of *ratio essendi* beyond the case of grounding of properties 'within' one individual (e.g., a triangle).

[34] The *MM* transcript does not use the term '*Dreieck*' but merely a drawing of a triangle. This example is sufficiently similar to Crusius's example of a real-existential ground (the three sides of a triangle and their relation ground the size of its angles—*Ent.* §36) that it is quite likely that Kant is borrowing both the idea of an inefficacious ground of possibility, and the example, from Crusius (see Watkins (2005), 83–4). However, Crusius's discussion of real-existential grounds in *Ent.* §36 does not cast much light on what exactly this relation is; his discussion in §79 is even less helpful for understanding Kant, for his examples are concrete beings (a wedge and a lever). I think it is also descended from the notion of a 'formal cause' in Descartes, which he takes from Aristotle (see CSM II: 169/AT VII: 238–42 where he cites *Posterior Analytics*, bk. II, ch. 11).

[35] In his lectures on metaphysics Kant attempts to reconcile the asymmetry of grounding with the fact that the positing of grounds and consequences mutually (though not *logically*) entail one another: if you posit a ground you posit a consequence, and if you posit a consequence you posit a ground (*MV*, Ak. 28: 486; *MM*, Ak. 29: 808). Kant's solution is to say that the positing of a consequence does not involve positing a determinate or specific ground but merely some ground in general (*MV*, Ak. 28: 408; ML_2, Ak. 28: 549; *MD*, Ak. 28: 625, 628), but all of his cases involve causal preemption. It stands to reason, then, in the case of

(i) non-causal, (ii) non-logical, (iii) asymmetric relation between (iv) (formally) necessarily co-instantiated properties. If we were to attempt to analyze this notion we might say that trilaterality grounds triangularity because we can prove the latter from the former; however, we can also prove the former from the latter, violating the asymmetry condition. We might supplement this with the requirement that the ground (e.g., trilaterality) *explains* the consequence (e.g., triangularity) and not vice versa. But this is just to appeal to an unanalyzed notion of one fact asymmetrically (non-logically, non-causally) *explaining* another, which is the notion of grounding in another guise. Clearly, a reductive analysis will be hard to come by, and I will not further pursue one here.[36]

In keeping with the Critical shift away from ontology (which studies the highest grounds of all possible beings) to transcendental philosophy (which studies the highest grounds of possible objects of representation), Kant internalizes the grounding relation within experience. As I will argue in the rest of this section, there are really two basic grounding relations involved in Kant's definition of the formal modalities: first, a grounding relation between the *forms* of experience and its *contents* (which explains the formal possibility of *representing* objects as having certain properties), and secondly, a grounding relation between those contents and the properties of objects (which explains the formal possibility of objects *having* those properties). I will refer to these as *form-content* grounding and *content-property* grounding, respectively, and will discuss them in that order. There will also be various derivative relations of real grounding, for instance, causal grounding between phenomenal substances and their accidents; these relations are derivative, because the obtaining of them depends upon the grounding relation between experience and its objects (content-property).

The first grounding relation holds between the form of experience and the content of experience. I will start with the case of (specifically outer) intuition. Our spatial form of outer intuition grounds the fact that any outer intuition has spatial content (represents its object in space).[37] This grounding relation has the same structure from earlier: our spatial form of intuition (i) non-causally, (ii) non-logically makes it the case that objects of outer intuition are represented in space, and this relation is (iii) asymmetric because the spatial content of intuition does not explain why we have the form of intuition we do. Finally, (iv) both our form of intuition and the space in which we intuit objects are formally necessary. But this grounding relation is not

non-causal grounding that the positing of the ground and the consequence mutually necessarily entail one another.

[36] Asymmetric non-logical, non-causal explanatory relations between mutually necessarily entailing truths (e.g., that Socrates exists and that {Socrates} exists) have recently become a major topic of metaphysical research, and even the eighteenth-century name for this relation has been retained: grounding. See the essays in Correia and Schnieder (2012), as well as Bliss and Trogdon (2014) and Raven (forthcoming), for a survey of the recent discussions.

[37] A26/B42, A42/B59, A48/B65.

limited to the contents of intuition; objects of (formally) possible intuition have the spatial properties they are (formally) necessarily intuited as having. So the next grounding relation holds between the *contents* of intuition and the *properties* of the objects of intuition; objects of (formally) possible intuition have spatial properties because of the content of intuition (which is itself grounded in the form of intuition). The same structure repeats itself: the content of possible intuition is a (i) non-logical, (ii) non-causal, (iii) asymmetric ground of the possible spatial properties of objects, even though (iv) both are formally necessary.

This applies at the level of the pure form of space (and time) and objects that can be presented (constructed) in pure intuition, but we want to understand the structure of the grounding relation within experience generally. To bring out this further structure it will help to think about the structure of experience itself. The most minimal level of experience,[38] according to Kant, is merely intuiting and conceptualizing sensory manifolds, without relating those manifolds to intersubjectively accessible 'public' objects.[39] I will refer to this as *perception*[40] to distinguish it from more robust conceptions of experience (see later discussion). Perception of sensory manifolds requires thinking them under the mathematical categories of quantity and quality, but not (yet) the categories of relation (substance, cause, community). Thinking them under the categories of quantity requires representing them as spatiotemporal wholes composed of parts (extensive magnitudes),[41] while thinking them under the categories of quality requires representing their sensory qualities as continuously gradable (intensive magnitudes).[42] This means that the grounding relations that obtain between space and objects of pure intuition holds for perception and its sensory manifolds: the forms of experience (space and time, categories of quantity and quality) ground the fact that they are perceived as extensive and intensive magnitudes, and this content grounds the fact that they have the properties of extensive and intensive magnitudes. The first relation holds between the form of experience (in this case, the forms operative in perception) and its content, and the

[38] There is a more minimal notion of (s-)cognition available: thinking about the pure manifolds of space and time. It would be inappropriate to call this *experience* because it is *a priori*. In the main body of the text I start with perception of sensory manifolds.

[39] This is how I interpret Kant's distinction between 'mathematical' (quantity, quality) and 'dynamical' (relation, modality) categories (B110, A160/B199, A176/B218): the former are involved in any cognition (thinking an intuited object under a concept) of sensory manifolds as such, while the latter are required to think about sensory manifolds as 'appearances' (in the empirical sense—A30/B45, A45–6/B63) of objects in an intersubjective space that do not depend upon perceptions of them (the 'empirical thing in itself'—see previous passages). To cognize sensory manifolds as such we need to think them under categories of quantity (which brings them under the Axioms of Intuition) and quality (which brings them under the Anticipations of Perception); to cognize sensory manifolds as appearances of intersubjective objects we need to think them under categories of relation (which brings them under the Analogies of Experience). My thinking about these issues has benefited greatly from conversations with Clinton Tolley and from his (forthcoming).

[40] "Perception is empirical consciousness, i.e. one in which there is at the same time sensation" (B207).

[41] A162/B203. [42] A168/B210.

second relation holds between the content of experience (perception) and the properties of its objects.[43]

This describes the structure of the most rudimentary notion of experience, mere perception of sensory manifolds. Let us now consider experience of objects in some intersubjectively accessible 'public' space, objects which we think of as existing independently of our perceptions of them and as being accessible to other cognitive subjects.[44] Kant refers to this as 'experience' in the System of Principles to distinguish it from mere 'perception' of sensory manifolds.[45] To experience these objects, Kant argues, we must think of them using the relational categories: "inherence and subsistence, causality and dependence, community [*Gemeinschaft*] (reciprocal action [*Wechselwirkung*] between agent and patient)" (A90/B106).[46] Although these are often referred to by Kant's own parenthetical explanations in the *CPR* ("substance

[43] The titles of the Axioms of Intuition ("all intuitions are extensive magnitudes"—B202) and the Anticipations of Perception ("in all appearances the real, which is an object of the sensation, has intensive magnitude, i.e. a degree"—B207) announce the outcome of this line of reasoning (the properties of intuited manifolds), while the body of those sections gives the argument for the first grounding relation (form-content): the forms of intuition plus the categories of quantity (Axioms of Intuition) and the categories of quality (Anticipations of Perception) ground the fact that intuition presents manifolds with extensive and intensive magnitudes. Kant takes the second grounding relation (content-property) to be clear to the reader by that point in the *CPR*, for he announced it earlier as the supreme principle of synthetic *a priori* judgment (A158/B197).

[44] This is my gloss on what Kant means by distinguishing intuition of an object from cognition of its *existence* (A160/B190) and why the Analogies of Experience are requirements on cognition of *existence* specifically (A176, B219, A182, B225, B233, B257, A215–16/B262–3). Existence cannot here mean 'absolute positing' (see Ch. 1) for we can absolutely posit mere sensory manifolds (there is such-and-such a sensory manifold) without using the categories of relation. Likewise, 'existence' here cannot mean causal efficacy for that would render at least the Second Analogy tautologous. Instead, I propose, 'existence' of an object here means: its being intersubjectively available to multiple cognizers and possessing properties independently of any particular discursive cognizer's perceptions of it. To use another Kantian distinction, the Axioms of Intuition and Anticipations of Perception state the necessary conditions on representing appearances *in the empirical sense* while the Analogies state the necessary conditions on representing the thing in itself *in the empirical sense*. See A29–30/B45, A44–5/B60, and what I take to be the closely related notions of 'necessary universal validity' and the distinction between judgments of perception and judgments of experience in *Prol.* §§18–19 (Ak. 4: 298–301).

[45] This corresponds to one of Kant's dominant uses of '*Erfahrung*': a synthetic unity of perceptions (A110, A124 f., A156/B195, B161, B218, A213/B260, A183/B226) that can, consequently, represent 'existing' objects (see previous note). I have been using 'experience' in a more minimal sense to refer to 'empirical cognition,' thinking a sensibly given object under a concept (B146, B165), in order to reserve 'cognition' for the specifically modal notion of representing an object one can know to be possible (Bxxvi n.).

[46] Given the topic of this book, readers might wonder where the modal categories belong, which Kant includes, with the categories of relation, among the dynamical categories (A162/B201). I think Kant is oversimplifying matters by including the modal categories among the dynamical categories. Even in the case of cognition of pure manifolds, we can distinguish the possible, that which agrees with the form of intuition and the categories of quantity, from that which does not so agree; in the case of perception of sensory manifolds we can distinguish, additionally, between what is *actually* present to the senses from what is not. I think that at each 'stage' of cognition (intuition, perception, experience) we can apply modal categories. Only at the final level (experience in the full sense) do we apply the modal categories according to the principles given in the Postulates; Kant is correct, then, in counting the Postulates *as he formulates them* as dynamical principles rather than mathematical principles. In the note he added to the B edition to explain the mathematical/dynamical distinction (B201*n*.), Kant does not mention modal categories.

and accident, cause and effect, reciprocal action"), or by the shortened names he gives them in the *Prolegomena* (substance, cause, community),[47] it is clear from his initial naming of them in the *CPR* that they are concepts of *relations* between objects. At the form-content level, the forms of experience (space, time, and the categories of relation) (i) non-logically, (ii) non-causally, (iii) asymmetrically explain why the content of experience represents its objects as standing in relations of inherence (accidents to substances),[48] causation (substances, objects which do not stand in inherence relations to other objects, to accidents of substances), and community (substances to substances that cause one another's accidents), while (iv) it is formally necessary both that we have these forms of experience and that experience has this content. At the content-property level, though, this grounds *relations* among objects of possible experience: the content of experience (i) non-logically, (ii) non-causally, (iii) asymmetrically explains why objects of experience stand in the relations of inherence, causation, and community.[49]

Notice, though, that these three relations among objects of experience are themselves real grounding relations: a substance is a *ratio essendi* of the possibility of its accidents, for an accident would not be possible without the very substance it inheres in; and substances are *rationes fiendi* of one another's accidents (grounds of alteration). These are *real* relations because the ground of a substance's accidents is not its concept, but the other substances with which it interacts, and substances are not *logical* grounds of the accidents they cause to inhere in one another. In Chapter 3 we saw Kant, in *Negative Magnitudes*, grappling with the problem of how we understand logically irreducible relations of real grounding; in that work he settled on the—admittedly not very satisfying—view that we simply possess a primitive stock of unanalyzable concepts of kinds of real grounding relations.[50] Kant's pre-Critical project was not to logically analyze the real grounding relation (this, by definition, is impossible) but to acquire 'insight' [*Einsicht*] into such relations, in Kant's technical sense of that term: *a priori*$_G$ knowledge.[51] So Kant's pre-Critical problem was to explain relations of real grounding among finite things, that is, to acquire knowledge of the *ground* that makes real grounding relations among finite things possible in the first place.[52] His pre-Critical solution to this problem (see Chapter 5.5) was to

[47] Ak. 4: 303.

[48] Kant defines an accident as something that inheres in something else, and a substance as that which does not inhere in anything else (nothing can inhere in itself). See A187/B230, Ak. 8: 225 n., Ak. 28: 562–3, Ak. 28: 639, and Ak. 28: 1104 f.

[49] The titles of the three Analogies announce the relations that obtain among objects of experience, while the body of the text of each Analogy is an argument for the corresponding form-content grounding claim. Kant takes the second grounding relation (content-property grounding) to be clear to the reader by that point in the *CPR*, for he announced it earlier as the supreme principle of synthetic *a priori* judgment (A158/B197).

[50] *NG*, Ak. 2:204.

[51] See *LPö* (Ak. 24: 539), *LDW* (Ak. 24: 730), *Refl.* 1866 (Ak. 16: 141), *Refl.* 1955 (Ak. 16: 169), and *Refl.* 2394 (Ak. 16: 342), *CPrR* (Ak. 5: 27, 46, 47), and *CJ* (Ak. 5: 83).

[52] For a clear statement of the problem, see *NG* (Ak. 2: 203–4) and *MH* (Ak. 28: 12, 102).

appeal to the common source of the existence of a real ground and its consequence, God; finite things can be the real grounds of one another's alterations because they share a common real ground of their existence.[53] Although this may merely push the question back one step (how can we have insight into the grounding relationship between God and finite substances?),[54] there is a deeper reason why Kant must reject it in the Critical period: it violates his restriction on cognition of noumena (God).

But now we have seen Kant's solution to at least part of this problem within the *CPR*: we can have insight (knowledge from the ground) into the real grounding relations *inherence* and *causation* among phenomena because we know that their ground (the ground of the possibility of these relations obtaining) is the grounding relation between the content of experience and the objects of experience. We also have insight into why this grounding relation obtains: objects of experience are phenomena, not noumena, so the content of experience grounds their properties. But our insight goes yet further: we have insight into the possibility of experience having this content because we know the *a priori* ground of its possibility, the forms of experience, and we have *a priori* insight into how the forms of experience make it possible for experience to have this content, synthesis of sensory manifolds according to categorial schemata (see Chapter 6.7). This explains one aspect of a typical Kantian formulation in the *CPR*; he does not merely describe the form of experience as *grounds* or *conditions* of possible experience but as *a priori* grounds.[55] Read in its full historical context, though, this does not merely mean that the forms of experience can be known *a priori*$_J$ (although they can) but that they are the antecedently determining *rationes essendi* (grounds of possibility) from which we can have *a priori*$_G$ knowledge of possible experience and, thus, of all possible objects of experience.[56] Kant has solved several of the problems that arose for him in the 1760s regarding our ability to have insight (knowledge from the ground of possibility) into real grounding relations by internalizing the grounding relation, and its relata, within experience itself.[57]

[53] *ND* (Ak. 1: 413), *MH* (Ak. 28: 51–2, 132), ID (2: 407–8).

[54] Kant implicitly acknowledges that his own view (see Ch. 5) explains the possibility of finite substances standing in real grounding relations only through an appeal to the unexplained real grounding relation between God's will and those very substances: "the will of God is something. The world which exists is *something completely different*" (*NM*, Ak. 2: 202).

[55] A30/B46, A116, B150, A136/B175, A156–7/B196–7, B202, A177/B220, A268/B324.

[56] See Chs. 1.3 and 3.2 for more on eighteenth-century theories of grounding.

[57] Abaci (2013) attributes to Kant a view on which his modal metaphysics (within experience) is oddly divorced from his modal epistemology: the grounds of real possibility are the forms of experience but we cannot know that an object is really possible just by proving *a priori* its compatibility with the forms of experience (we must actually experience the object or prove what I call its 'formal necessity'). My interpretation delivers a more unified account of Kant's modal epistemology and metaphysics.

7.4. Empirical-Causal Possibility

We have already seen that the "Postulates of Empirical Thinking in General", the section of the System of Principles in the *CPR* devoted to the modal categories (possibility, existence, and necessity), contains a definition of formal possibility.[58] However, the Postulates section also contains the following discussion of necessity:

> From this it follows that the criterion of necessity lies solely in the law of possible experience that everything that happens is determined *a priori* through its cause in appearance. Hence we cognize only the necessity of effects in nature, the causes of which are given to us, and the mark of necessity in existence does not reach beyond the field of possible experience, and even in this it does not hold of the existence of things, as substances, since these can never be regarded as empirical effects, or as something that happens and arises. Necessity therefore concerns only the relations of appearances in accordance with the dynamical law of causality, and the possibility grounded upon it of inferring *a priori* from some given existence (a cause) to another existence (the effect). Everything that happens is hypothetically necessary; that is a principle that subjects alteration in the world to a law, i.e. a rule of necessary existence, without which not even nature itself would obtain. (A227–8/B280)

That Kant here writes that we "cognize the necessity of effects in nature" shows that he does not have formal necessity in mind, for particular events are not formally necessary. In this section I develop an alternate conception of necessity, which I call *empirical-causal* necessity, and argue that, in this passage and in others, Kant invokes this kind of necessity. However, he does not explicitly distinguish this conception of empirical-causal necessity from formal necessity. The distinctions I am drawing between kinds of modality are made implicitly, not explicitly, by Kant.

Empirical substances are in time, so they undergo alteration [*Veränderung*], which Kant defines as successive change [*Wechsel*] of accidents.[59] In the Second Analogy of Experience, Kant argues that the form of time and the category of cause–effect ground (see section 3) the fact that every empirical alteration has a determining cause,[60] a

[58] A218/B265.

[59] A188/B232. My discussion of Kant on causation is informed by the meticulous analyses of Watkins (2005).

[60] A determining cause is to be distinguished from a merely sufficient cause. A sufficient cause is opposed to a merely *partial* or insufficient cause, which is insufficient, by itself, to bring about its effect. A determining cause, on the other hand, operates according to a rule: any substance of the same kind, under the same conditions, would cause the same effect. The crucial distinction between determining and merely sufficient causes derives from Crusius (*De Usu* §§I, III, XX); a determining ground is one that, under the same conditions, would always produce the same effect. Crusius introduces this distinction to save freedom from the necessitarian consequences of the Leibnizian and Wolffian versions of the PSR: the free actions of rational agents have *sufficient*, but not *determining* grounds, according to Crusius. Kant praises this distinction in *ND* (Ak. 1: 398) but denies that free actions lack determining grounds, a position he will later reverse (see section 5). It is clear, both from the schema of cause–effect (A144/B183), and from the second Analogy itself, that Kant takes himself to have proved that empirical alterations have determining and not merely sufficient grounds: "in accordance with such a rule there must therefore lie in that which in general precedes an occurrence the condition for a rule, in accordance with which this occurrence always and necessarily follows" (A193/B238). Cf. Hogan (2005) and Hogan (2009a).

cause from which the effect follows according to a law.[61] This means that it is formally necessary that every empirical event has a determining cause and that cause–effect relations are governed by laws. In other words, the principle of causation is formally necessary, but which causal laws govern the actual world, and which empirical events occur, are formally contingent.[62] Let p be the proposition that alteration E occurs at time t. Consider the following definition:

> (*Empirical-causal possibility*) It is *empirically-causally possible* that p if and only if it is compatible with actual natural laws, and the past history of the empirical world up until time t, that p.[63]

For instance, it is empirically-causally possible (empirically possible, for short) that I refrain from making a lying promise at time t just in case it is compatible with the entire history of the empirical world up to that moment, and the natural laws, that I so refrain. As a merely heuristic device, we might introduce the notion of 'possible worlds' and say: it is empirically possible that E occur at t if and only if there is a possible empirical world (series of empirical alterations) that is qualitatively indistinguishable from the actual world before t, and has the same natural laws, in which E occurs at t.[64] According to Kant, it is formally necessary that every event has a determining cause. That causes are determining, for Kant, means that the laws that govern the operation of these causes are deterministic. This means that if C is the cause of E, and C occurs, and the background conditions are held fixed, it is incompatible with the natural laws that E not occur. To invoke the heuristic of possible worlds once again, we might define a 'divergent world' as one that is qualitatively identical to the actual world up to time t, in which the actual laws obtain, but is qualitatively different from the actual world after time t. The natural laws governing the actual world are deterministic if and only if there are no divergent possible worlds. The history of the empirical world up to a moment, plus the natural laws, determine a unique future history. From the definition of empirical-causal

[61] Some scholars have questioned whether the Analogies of Experience are supposed to demonstrate that there are causal laws, or merely that every alteration has a cause. I am going to assume that Kant took himself to have established the stronger conclusion; see section 4, and the first footnote in Ch. 8, for more discussion.

[62] See A766/B795: "Thus if wax that was previously firm melts, I can cognize *a priori* that something must have preceded (e.g., the warmth of the sun) on which this has followed in accordance with a constant law, though without experience, to be sure, I could determinately cognize neither the cause from the effect nor the effect from the cause *a priori* and without instruction from experience."

[63] In Stang (2011) I called this *empirical* possibility. There is a slight complication here, though: on Kant's view 'the entire history of the empirical world up to time t' does not refer to any object of possible experience so there is no such object (see the resolution of the first antinomy, A517–23/B545–51). See later discussion for how to deal with this problem.

[64] By introducing this heuristic device, I am not claiming that Kant accepts a 'possible worlds' *analysis* of modality, something I have consistently rejected. On the possible worlds analysis of determinism, see Lewis (1983), 31–3.

possibility above,[65] we can define a companion notion of empirical-causal necessity (letting p be as before):

(*Empirical-causal necessity*) It is *empirically-causally necessary* that p if and only if it is incompatible with actual natural laws, and the past history of the empirical world up until time t, that $\neg p$.

That natural laws are deterministic entails that every empirical event is empirically-causally necessary (empirically necessary, for short).[66] No non-actual event is empirically possible, because the prior history of the empirical world and the natural laws ensure that it does not occur. Consequently, the actual, the empirically possible, and the empirically necessary are co-extensive.

I have defined empirical possibility and necessity in terms of the relation of being compatible with the past and the laws. As with the corresponding definition of formal possibility, the question arises: what does 'compatibility' mean here? I propose that we define compatibility in terms of *grounding*:

(*Defn.*) *p is compatible with actual natural laws and the past history of the empirical world up to time t* if and only if the facts about the actual natural laws and the past history of the empirical world up until t and the laws do not wholly ground the fact that $\neg p$.

This, in turn, allows us to define empirical-causal possibility and necessity in terms of grounding:

(*Empirical-causal possibility*) It is *empirically-causally possible* that p if and only if it is not the case that the fact that $\neg p$ is wholly grounded in facts about actual natural laws, and the past history of the empirical world up until time t.
(*Empirical-causal necessity*) It is *empirically-causally necessary* that p if and only if the fact that p is wholly grounded in facts about actual natural laws, and the past history of the empirical world up until time t.

The same problem that arose in the case of defining formal possibility in terms of grounding—precisely delineating the relevant notion of *ground*—arises again in the case of empirical possibility and necessity, but it is more tractable in this case. The laws describe the causal powers of phenomenal substances,[67] so the ultimate grounds of empirical-causal possibility are those powers, which are both *rationes fiendi* of the *existence* of alterations in phenomenal substances and *rationes essendi* of the empirical-causal *possibility* of those very alterations. In section 3 I discussed at length

[65] Combined with the principle that $\Box p \leftrightarrow \neg\Diamond\neg p$; see *MV* (Ak. 28: 418), *MvS* (Ak. 28: 498), *ML_2* (Ak. 28: 556, 557), and *MD* (Ak. 28: 633).
[66] As I have set up the definitions, it would be more precise to say: for any event E, the proposition that E occurs is empirically necessary. However, I am treating the two formulations as identical, for ease of exposition.
[67] See Ch. 8.

how the content of possible experience grounds the possibility of causal relations among phenomena, but now we have a relation of possibility-grounding among phenomena themselves. The powers of phenomenal substances are grounds of the empirical-causal possibility of alterations in one another, and this grounding relation is itself made possible by the content of experience (see section 3). So we have a relation of real grounding between, on the one hand, powers of phenomenal substances, and, on the other hand, possible alterations in substances, and although this relation is not logically intelligible (the concept of the former does not contain the concept of the latter), we can have *a priori*$_G$ insight into it: the ground of its possibility is the content of experience itself.

However, there is a problem with this definition of empirical-causal possibility. According to Kant's transcendental idealism, alterations in empirical substances, and temporally ordered series of them, are phenomena: objects of a (formally) possible experience. A formally possible experience of a series of empirical alterations is a unity synthesized from experiences of its individual elements. Since we can only ever synthesize finitely many elements (a completed infinite series is formally impossible) it follows that no infinite series of empirical alterations is a possible object of experience for us. Consequently, "the past history of the empirical world up until time *t*" does not refer to any possible object of experience. For any finite empirical series S there is a 'larger' finite empirical series S* that includes S as a proper part; in other words, finite empirical series leading up to time *t* can be indefinitely 'extended' to include alterations further in the past, but the "complete past before *t*" refers to no (formally) possible phenomenon. Since it clearly refers to no noumenon (it is the concept of a temporal series, and noumena are not in time) I conclude that it refers to nothing: there is no such thing, according to Kant, as the complete past before a given time.[68]

Fortunately, this problem can be resolved. Kant's deterministic conception of natural laws entails, not that the natural laws plus the *past history of the empirical world up until time t* determine a unique future, but that the natural laws plus *the past history of the empirical world for an arbitrary interval of non-zero duration before time t* determines a unique future.[69] To use, once again, the heuristic of possible empirical worlds (series of empirical alterations) from previously, a world *w* is said to *diverge* from the actual world just in case *w* is qualitatively identical to the actual world for a non-zero interval before *t*, but is qualitatively different after *t*. Kant's determinism entails that there are no worlds governed by the actual natural laws that diverge from the actual world after a period of being qualitatively identical. So if we

[68] This is not to deny that we have an *idea* of the "complete series of alterations before a given time," a concept of the unconditioned series of temporal conditions of a given alteration. This idea is the subject of the first Antinomy. But this does not mean there is any object answering to this idea (a complete series of past times).

[69] This is a consequence of the *continuity* of the relation between cause and effect. See A207–8/B253–4. This means that if empirical series A is the same as empirical series A* for a non-zero interval before time *t*, they are the same after *t*.

fix a non-zero interval of the empirical series before *t*, given the natural laws, only one future is possible. This requires a slight change in the definitions of empirical-causal possibility and necessity:

(*Empirical-causal possibility*) It is *empirically-causally possible* that *p* if and only if it is not the case that the fact that $\neg p$ is wholly grounded in facts about the actual natural laws and a non-zero interval of the past before time *t*.
(*Empirical-causal necessity*) It is *empirically-causally necessary* that *p* if and only if the fact that *p* is wholly grounded in facts about the actual natural laws and a non-zero interval of the past before time *t*.

In what follows I will sometimes speak of empirical-causal possibility as being grounded in the natural laws and "the past." In every case this means: the natural laws and an arbitrary non-zero interval of the past.

It should also be clear that empirical possibility and necessity are kinds of *real* possibility and necessity, respectively. Recall the criteria on real modality from section 1:

(*Real Possibility*$_E$) Empirical-causal possibility $\Diamond_E p$ (and its associated kind of necessity $\Box_E p$, where $\Box_E p \leftrightarrow \neg\Diamond_E \neg p$) is a kind of *real possibility* (and $\Box_E p$ is a kind of *real necessity*) only if
(i) *Non-logicality*: it is not a conceptual truth that $\Diamond_L p \supset \Diamond_E p$ (equivalently, it is not a conceptual truth that $\Box_E p \supset \Box_L p$), and
(ii) *Groundedness*: if $\Diamond_E p$ then the fact that $\Diamond_E p$ has a real ground in some actual object or principle.

Empirical possibility and necessity satisfy these criteria. It is logically possible that, for instance, some non-actual empirical alteration occurs (e.g., no contradiction results from the proposition that it occurs), but it is empirically-causally impossible for any non-actual alteration to occur. Similarly, if it is empirically-causally possible that some alteration occurs at some designated time, this alteration has real grounds in actual objects: the actual natural laws, and a non-zero interval of the past before that time. Similar remarks apply to empirical necessity. In section 2 I noted that one difference between Kant's pre-Critical conception of real possibility and his Critical conception of real possibility is that pre-Critical real possibility is grounded in mind-independent reality (noumena). Empirical possibility (and necessity) is grounded in facts about phenomena: powers of phenomenal substances and the past before a given time. As such, it is not a kind of 'noumenal' possibility.

When Kant writes, in the long passage from the Postulates section quoted at the beginning of this section (A227–8/B280), that "it follows that the criterion of necessity lies solely in the law of possible experience that everything that happens is determined *a priori* through its cause in appearance," he means that it follows from the principle that every alteration has a determining cause that every alteration is empirically necessary. He writes that "everything that happens is hypothetically necessary" because empirical necessity is a form of hypothetical necessity. Alterations are necessary *given the hypothesis* of the prior history of the empirical world, and the natural laws. He contrasts substances with accidents because, on Kant's view, substances are permanent; they do not arise or perish, unlike alterations. Thus, restricting ourselves to causes within the empirical world, substances are not caused—their *alterations* are caused—and the existence of a given set of substances is neither empirically-causally possible nor necessary; empirical-causal modality is simply not defined for substances themselves.[70]

Before concluding my discussion of 'immanent' modality, I want to comment briefly on Kant's crucial claim that "the categories have this special feature: they do not augment the concept to which they are ascribed in the least, as a determination of the object, but express only the relation to the faculty of cognition" (A219/B266). I have already explained at length what it means that <*existence*> is not a determination or real predicate. In this chapter I have focused on thinking of formal possibility and necessity as operators on propositions (judgment-contents), because many questions in formal possibility are most naturally formulated this way, but we can also define these modal concepts as predicates of objects:

(*FP-objects*) A phenomenon *x* is *formally possible* if and only if it is compatible with the forms of experience that we experience *x*.

(*EN-objects*) An alteration *x* in a phenomenal substance is *empirically-causally necessary* if and only if it is incompatible with the natural laws and a non-zero interval of the past relative to *x* that *x* not occur.

But it follows from this that these object-level predicates are not determinations of the domain of objects for which they are defined: it is (formally) impossible that there is a phenomenon that is not formally possible, and it is (formally) impossible that there is an alteration that is not empirically necessary.[71] To say of a phenomenon that it is formally possible is not to further determine that object (ascribe to it a predicate it might lack) but to express the compatibility of the experience of it with the forms of experience. To say of an alteration that it is empirically necessary is not to determine it (ascribe to it a predicate it might lack) but to express its relation to the past and the laws, which, given your forms of experience, make it necessary.

7.5. Noumenal-Causal Possibility

We have already seen that the Postulates of Empirical Thought, the section of the *CPR* devoted to the modal categories (possibility, actuality, and necessity), contains

[70] This is not the only passage in which Kant deploys this conception of empirical necessity. See *MV* (Ak. 28: 417) and *MM* (Ak. 29: 814), as well as *Refl.* 4298 (Ak. 17: 499) and 5177 (Ak. 18: 109).

[71] In the technical terminology of Ch. 1.6, <*formally possible*> does not determine <*phenomenon*> and <*empirically necessary*> does not determine <*alteration in phenomenal substance*>.

discussions of formal possibility and of empirical-causal possibility, although they are not explicitly distinguished. The conclusion of the Postulates, however, is harder to characterize. Kant raises a pair of questions: "whether the field of possibility is greater than the field that contains everything actual, and whether the latter is in turn greater than the set of that which is necessary" (A230/B282). Given the interdefinability of possibility and necessity, though, these are really one question: are there non-actual possibilities?[72] Kant's subsequent discussion of non-actual possibilities is quite intricate, so I will analyze this text in some detail.[73]

The first case Kant discusses is the possibility of a non-actual experience:

> [a] [The two questions from above] mean, roughly, to ask whether all things, as appearances, belong together in the sum total [*Inbegriff*] and the context [*Kontext*] of a single experience, of which each given perception [*Wahrnehmung*] is a part and which [*der*] therefore could not be combined with any other appearances, or whether my perceptions could belong to more than one possible experience (in their general connection). (A230–1/B283)[74]

"A single experience, of which each given perception is a part" is a reference to Kant's *strongest* conception of experience, experience as the totality of all perceptions combined according to natural laws, which he characterizes as follows in the A Deduction: "there is only **one** experience, in which all perceptions are represented as in thoroughgoing and lawlike connection [...] if one speaks of different experiences, they are only so many perceptions insofar as they belong to one and the same universal experience" (A110).[75] I will refer to this concept of a "single all-encompassing experience" (A582/B610) as 'universal' experience. Universal experience, as I understand it, is the idealized limit of a process of combining individual perceptions according to *a priori* principles and natural laws; this is a rough characterization, but for our purposes it will have to suffice.[76] Since alternate individual experiences are possible, according to A110, in virtue of belong to a 'universal' experience, Kant focuses on the question of whether a 'universal' experience, other than the actual one, is possible. Possible experiences are playing a role analogous to possible worlds here, and just as 'possible worlds' accounts of possibility standardly require that worlds be *maximal* (if *p* is consistent with world *w*, then *p* is true in *w*) Kant is concerned with *maximal* experience, or 'universal' experience.

[72] If possibility and necessity are interdefinable in the standard way—necessarily *p* if and only if not possibly not-*p*—the questions *is the field of possibility greater than the field of the actual?* (i.e. are there non-actual possibilities?) and *is the field of the actual greater than that of the necessary?* (i.e. is anything contingent?) receive the same answer.

[73] Cf. Stang (2011), 454–8, and Abaci (2013), 4–7.

[74] Note that, in the second relative clause ("which [*der*] therefore could not be combined with any other appearances [...]"), the masculine relative pronoun '*der*' must take sum total (*Inbegriff*) or context (*Kontext*) as its referent, not perception (*Wahrnehmung*), which is not clear in Guyer–Wood (I have modified their translation accordingly).

[75] Cf. A492–3/B520–1 and A582/B610.

[76] I say more about 'universal experience' in Stang (2012) and (forthcoming).

This passage can be read as raising either of two modal questions:

(1) Could my actual perceptions be part of more than one possible universal experience?
(2) Is there more than one possible universal experience?

On the first reading, in the final clause of the sentence (underlined above) Kant is referring to the *de re* possibility that my actual perceptions be part of a non-actual single, universal experience. On the second reading, Kant is referring to the *de dicto* possibility that I have some perceptions or other that are part of a non-actual single universal experience. To continue with the possible worlds analogy, these are analogous to asking, of some proper part of the actual world, (1*) would this exist in another possible world? or (2*) are there any possible worlds other than the actual one? While this passage alone does not settle the interpretive question, the continuation of Kant's discussion in the rest of the paragraph does. He goes on to discuss the possibility of a non-actual *form* of experience, and then the possibility of a non-actual *matter* of experience; I take this to mean that the original question posed at the beginning of the paragraph is about the possibility of a non-actual experience, (2). The most urgent question raised by this passage, though, is what sense of possibility is involved in these questions, and my argument, that it is a kind of possibility grounded in noumena, will still apply if the original question raised at A230–1/B283 is (1) rather than (2).

Kant goes on to discuss two modal issues, the first of which is the possibility of non-actual forms of experience:

> [b] Even were they possible, we could still not conceive of and make comprehensible [*erdenken und faßlich machen*] other forms of intuition (than space and time) or other forms of understanding (than the discursive form of thinking, or that of cognition through concepts); and even if we could, they would still not belong to experience, as the sole cognition in which objects are given to us. (A230 f./B283)

Kant remains agnostic about the possibility of forms of intuition other than space and time, and about the possibility of a non-discursive (intellectual—see Chapter 10) understanding.[77] We cannot cognize the real possibility of such non-actual forms of experience because the actual forms of experience ground the only space of real possibilities we can cognize: formal possibility and empirical-causal possibility. Since (by definition) no non-actual form is either an object in that space or a form constitutive of such an object, we cannot cognize their real possibility. Immediately after this, he writes:

> [c] Whether other perceptions than those which in general belong to our entire possible experience, and therefore an entirely different field of matter, can obtain cannot be decided

[77] Kant expresses agnosticism about the possibility of a non-spatiotemporal form of intuition elsewhere. See A42/B59, A231/B283 and *Prol.* §57 (Ak. 4: 351).

by the understanding, which has to do only with the synthesis of that which is given. (A231/B283).

Coming, as it does, immediately after Kant's discussion of non-actual forms in [b], this is most plausibly read as a discussion of whether a non-actual *matter* of experience is possible. Retrospectively, it sheds light on [a]. Kant's question there was (2), and he then goes on to separate this into two subordinate questions: [b] is a universal experience with a non-actual form possible?; and [c] is a universal experience with a non-actual matter possible?

We can conclude our line-by-line reading of the Postulates discussion of non-actual possibilities with this passage:[78]

[d] All that can be added to my understanding is something beyond agreement with the formal conditions of experience, namely connection with some perception or other; but whatever is connected with this in accordance with empirical laws is actual, even if it is not immediately perceived. However, that another series of appearances in thoroughgoing connection with that which is given to me in perception, thus more than a single all-encompassing experience, is possible, cannot be inferred from that which is given, and even less without anything being given at all. (A231/B284)

I take the first sentence to mean that what is left undetermined by the forms of experience is the particular objects of experience, which are either perceived directly or inferred from their causal connection with directly perceived objects.[79] However, Kant reminds us, "whatever is connected with" particular perceived objects agrees with empirical laws.[80] In other words, whatever formally contingent empirical objects we experience, these objects are governed by causal laws. Consequently, their alterations are empirically necessary because they follow from previous alterations by natural laws. So no non-actual empirical alterations are empirically possible. In the second sentence Kant expresses agnosticism about the possibility of "more than a single all-encompassing experience" which I read as a reference back to "the sum total [*Inbegriff*] and the context [*Kontext*] of a single experience" in [a]. Kant is returning to the original question with which he began in [a] (is a universal experience other than the actual one possible?) and claiming that what is given to us (actual empirical objects) does not allow us to infer such a possibility.

[78] I have so far ignored Kant's comment that these are "appropriate [*artige*] questions, and are to be resolved synthetically [*von synthetischer Auflösung*], though they also fall under the jurisdiction of reason" (A230/B282). I take this to mean that the questions I am investigating in the body of the text (whether a non-actual matter or form of experience is possible) are not settled by analysis of our concepts <*possible*> and <*actual*> but must be referred to the faculty of reason, which seeks *conditions* (grounds) for the *conditioned*; in this case, reason seeks a ground of non-actual possibilities outside the realm of possible experience, in noumena. Cf. Ch. 9.6.

[79] Cf. the postulate of actuality, A218/B266 and A225/B272.

[80] It is possible to read this passage as concerning the possibility of non-actual empirical *laws*. However, the same questions arise: non-actual empirical laws are clearly formally possible and empirically impossible, so why does Kant express agnosticism? Cf. Guyer (1998), 304–8.

I have been analyzing these passages in detail to determine exactly what question about non-actual possibility Kant is investigating, but the deeper issue is: what kind of *possibility* is involved here? As Kant himself points out, in [d], the *matter* of a non-actual universal experience is formally possible, but the alterations that would be part of that matter are empirically-causally impossible. So if Kant has either formal or empirical-causal possibility in mind, it is unclear why he expresses agnosticism, since his own theory straightforwardly delivers a determinate answer. The same point holds with respect to the question of the possibility of a non-actual *form* of experience in [b]; non-actual forms of experience are trivially formally impossible, and causal-empirical possibility is not defined for them (they are not alterations of phenomenal substances). Consequently, Kant must have a distinct kind of modality in mind in these passages. Both formal possibility and empirical-causal possibility have what I have called 'immanent' grounds, in the sense that they are either objects of experience (empirical possibility) or the subjective forms that condition those objects (formal possibility). However, phenomena and their subjective forms are not the only domain in Kant's metaphysics; there are also the noumena. Although we cannot cognize noumena (know what is really possible for them) this does not necessarily preclude us from raising questions about noumena and whether they ground non-actual possibilities.[81]

When Kant expresses agnosticism about whether the matter or form of a non-actual experience are possible, he might be expressing agnosticism about whether there is a noumenal ground of the possibility of a non-actual matter and form for experience. While it is somewhat harder to say what role noumena might play in the possibility of a non-actual *form* of experience, it is relatively straightforward to understand the role they might play in the possibility of non-actual *matter*. The *matter* of experience is given to us through affection of our sensibility by noumena, as Kant makes abundantly clear in *On a Discovery*:

> Having raised the question 'what gives sensibility its matter, namely sensations?' [Herr Eberhard] believes himself to have pronounced against the *Critique* when he says "we may choose what we will—we nevertheless never arrive at *things-in-themselves*." Now, that, of course, is the constant contention of the *Critique*; save that it posits this ground of the matter of sensory representations not once again in things, as objects of the senses, but in something super-sensible, which *grounds* the latter, and of which we can have no cognition. It says that objects as things-in-themselves *give* the matter to empirical intuitions (they contain the ground by which to determine the faculty of representation in accordance with its sensibility), but they *are* not the matter thereof. (*Disc.*, Ak. 8: 215)

Not only here, but in numerous other passages in the *CPR* and other Critical writings, Kant maintains that things in themselves (noumena) affect our sensibility, producing

[81] On the restriction of formal and empirical possibility to empirical objects, and the question of whether there is a kind of modality that applies beyond the bounds of experience, see Schneeberger (1952), 32–3.

the matter of experience.[82] This, in turn, allows us to understand how noumena might ground the possibility of a non-actual *matter* for experience: by having the causal power to produce in us a matter distinct from the one they actually produce. I will provisionally label this concept of possibility *noumenal-causal possibility* and define it as follows: where it is formally contingent that *p*,[83] it is *noumenally-causally possible* that *p* if and only if noumena have the causal power to so affect our sensibility that the resulting experience is one of which *p* is true. Obviously, this is only defined for values of *p* that are about experience and its objects. Since causal powers are real grounds, this kind of noumenal possibility is a kind of real possibility.

But this is not the only place where the notion of noumenal-causal possibility arises in Kant's system. His theory of freedom, one of the foundations of the entire Critical system, rests on a notion of possibility grounded in the causal powers of noumena (of agents' wills considered as noumena). According to Kant, *ought* implies *can*, so if I ought not to perform a given act, then I can omit that act. If the moral law (which has the form of an imperative or 'ought' in relation to finite wills, like ours,[84] which are afflicted with sensible inclination) applies to me then it must be possible that I omit the actions I ought to omit. But every act I commit is an empirical alteration (the changing of accidents in empirical substances, e.g., my body) and is therefore a necessary consequence of prior history and natural laws. In the case of a morally wrong act, how can we reconcile the empirical-causal necessity of that act with the possibility that I omit it?[85] Kant thinks that if we restrict our account to the empirical world, freedom, and therefore the moral law, is impossible:

> For if appearances are things in themselves, then freedom cannot be saved. Then nature is the completely determining cause, sufficient in itself, of every occurrence, and the condition for an occurrence is always contained only in the series of appearances that, along with their effect, are necessary under the law of nature. (A536/B564)

If appearances were things in themselves, i.e. if we could not appeal to anything other than appearances and their properties, then freedom would be destroyed. Freedom

[82] In the third chapter of Adickes (1924), Erich Adickes assembles an impressive array of textual evidence for this claim. See esp. A190/B235, A387, A494/B522, *Prol.* (Ak. 4: 289, 314, 318), and *GMM* (Ak. 4: 451).

[83] If we want to extend the notion of noumenal-causal possibility to include the noumenal-causal possibility of non-actual forms of experience, we would need to remove this clause.

[84] *GMM*, Ak. 4: 413.

[85] Kant confronts precisely this question in the *CPrR*: "If I say of a human being who commits a theft that this deed is, in accordance with the natural law of causality, a necessary result of determining grounds in preceding time, then it was impossible that it could have been left undone; how, then, can appraisal in accordance with the moral law make any change in it and suppose that it could have been omitted because the law says that it ought to have been omitted? That is, how can that man be called quite free at the same point of time and in regard to the same action in which and in regard to which he is nevertheless subject to an unavoidable natural necessity?" (Ak. 5: 95–6). Kant is trying to reconcile the empirical necessity of all of my acts with the possibility of my omitting them. He thinks that my acts are normatively subject to the moral law only if they are free, and freedom requires that it be possible for me to act otherwise.

would be destroyed because each of my actions would be empirically necessary, and there would be no possibility of my omitting any of my acts.[86]

Kant resolves the conflict between our freedom and the empirical necessity of our actions by appeal to a non-empirical domain: noumena and their causal powers. Self-conscious rational subjects, for Kant, are noumena (things in themselves)[87] that appear to themselves and other subjects as objects in space and time, subject to causal laws. Therefore, as a rational subject, I can be considered both as a noumenon (outside of space and time) and as the appearance of that noumenon (in space and time, subject to causal determination). In *CPrR* Kant writes:

> Every action [of a subject] ... is to be regarded in the consciousness of his intelligible existence as nothing but the consequence and never as the determining ground of his causality as a noumenon. So considered, a rational being can now rightly say of every unlawful action he performed that he could have omitted it even though as appearance it is sufficiently determined in the past, and so far, is inevitably necessary; for this action, with all the past which determines it, belongs to a single phenomenon of his character, which he gives to himself and in accordance with which he imputes to himself, as a cause independent of all sensibility, the causality of those appearances. (*CPrR*, Ak. 5: 97–8)

Kant resolves the conflict between freedom and necessity by appealing to the causal powers of the subject, considered as a noumenon, "a cause independent of all sensibility." When Kant writes that "a rational being can now rightly say of every unlawful action he performed that he could have omitted it," I take him to be claiming that for any unlawful action an agent performs, the omission of that action is possible because the agent, considered as a noumenon, has the causal power to choose in such a way that the empirical world does not include that action.[88]

Kant renders compatible the empirical-causal necessity of my actions (as events in time) with the possibility of my omitting them by claiming that, considered as a noumenon, I could have caused those actions not to occur. I will call this kind of possibility *noumenal-causal possibility*, and define it as follows:

> (*Noumenal-causal possibility*) It is *noumenally-causally possible* that *p* if and only if (i) it is formally contingent that *p*,[89] and (ii) there is some noumenon with the causal power to make it the case that *p*, where *p* is a synthetic proposition about phenomena.

[86] Kant assumes throughout that the mere *formal* possibility of my omitting some act is not sufficient for freedom. He appears to assume that freedom requires that I have the causal power to act otherwise than I do. The mere compatibility of acting otherwise with the forms of experience entails nothing about my having the power to act otherwise. Similarly, he passes over *logical* possibility: the mere logical consistency of the proposition that I act otherwise is not sufficient for my being able to do otherwise.

[87] See the end of Ch. 6.7 for an explanation of how I am using these terms.

[88] Exactly how the very same event can be both empirically necessary and noumenally contingent involves Kant's notions of 'empirical character' and 'intelligible character' and lies outside the scope of this chapter. Cf. Wood (1984), 102–12, Watkins (2005), 325–39, and Pereboom (2007), 550–9.

[89] Noumenal-causal possibility is only defined for the *matter* of experience, which includes the actions of rational agents. If we want to inquire into the possibility of non-actual forms of experience, as Kant does at A230–1/B282–3, we will need another notion of possibility grounded in noumena.

Noumenal-causal possibility, like empirical-causal possibility, is grounded in the causal powers of actual things. Unlike empirical-causal possibility, though, it is grounded in the causal powers of things in themselves, noumena. In this chapter I am applying these conceptions of possibility—formal possibility, empirical-causal possibility, and noumenal-causal possibility—to propositions about empirical objects, e.g., the proposition that I omit some action I actually commit. In Chapter 9 I will consider whether we can extend this notion of noumenal possibility—possibility grounded in noumena—to the modal status of propositions about noumena themselves, that is, whether noumena can be grounds of possibilities for other noumena.

It should also be clear that noumenal-causal possibility and necessity satisfy the two conditions on real possibility and necessity, respectively, from section 2:

> (*Real Possibility*$_N$) *Noumenal possibility* $\Diamond_N p$ is a kind of *real possibility* only if
> (i) *Non-logicality*: it is not a conceptual truth that $\Diamond_L p \supset \Diamond_N p$, and
> (ii) *Groundedness*: if $\Diamond_N p$ then the fact that $\Diamond_N p$ has a real ground in some actual object or principle.[90]

In the case of formal and empirical possibility Kant can *prove* that they are *extensionally* distinct from logical possibility: there are logical possibilities that are not formally possible (e.g., that 2+2=5) and there are logical possibilities that are not empirically possible (e.g., that some actual alteration not occur). Within theoretical philosophy he cannot argue in this fashion in the case of noumenal-causal possibility because we do not have sufficient theoretical knowledge of what is (and is not) noumenally-causally possible.[91] However, he can rest this claim on the fact that no analysis of our concepts reveals <*logically possible*> and <*noumenally-causally possible*> to be co-extensive; we cannot determine whether these concepts apply to all and only the same propositions merely by analyzing them. Consequently, they are different concepts; noumenal-causal modality satisfies the Non-Logicality criterion. Furthermore, it is clear that noumenal-causal modality satisfies the Groundedness criterion: facts about what is noumenally-causally possible (and necessary) are grounded in the causal powers of actual noumena, and causal powers are real grounds.

Whereas Kant's pre-Critical conception of real metaphysics incorporates what I called the 'Worldliness' criterion—that real possibility is grounded in the nature of mind-independent reality, what the Critical Kant calls *things in themselves* (noumena)—Kant's Critical modal metaphysics does not. It includes kinds of real possibility that are not 'worldly' because they are grounded in subjective forms of

[90] Using the interdefinability of possibility and necessity it is easy to define a notion of noumenal-causal necessity; for the sake of brevity I omit this.

[91] Kant's account of practical reason, and how it underwrites cognition, and even, potentially, knowledge, lies outside the scope of this study. For more on this notion, see Schafer (unpublished).

experience (formal possibility) or in the properties of mind-dependent objects (empirical possibility). But clearly, noumenal-causal possibility and necessity are 'worldly': they are grounded in the causal powers of mind-independent objects, things in themselves (noumena). Consequently, noumenal-causal possibility and other kinds of real possibility grounded in noumena have some claim to be the closest Critical analogue of pre-Critical real possibility, a theme I explore in greater depth in Chapter 9.

Some readers might object that my definitions of noumenal-causal modality attribute causal and modal properties to noumena. This might be taken to violate Kant's doctrine that categories like *<cause–effect>*, *<possible>*, and *<necessary>* cannot be used to cognize noumena. My response is that I am not *yet* attributing modal properties to noumena.[92] I am attributing modal properties to empirical objects on the basis of the causal powers of noumena.[93] As I argued in Chapter 6, Kant distinguishes between thinking [*Denken*], cognition [*Erkenntnis*], and knowledge [*Wissen*]. All that is required for thinking of an object falling under a concept is that there be no logical inconsistency in the concept. Since there is nothing logically contradictory about noumena instantiating categories, we can think of them as having causal powers.[94] Since Kant expresses agnosticism about the possibility of a non-actual matter and form of experience, it is sufficient that we can *think* of noumena grounding these possibilities.[95] We can *think* about whether non-spatial and non-temporal intuition and a non-actual matter of experience are possibilities grounded in noumena, but we cannot cognize this as a real possibility.

[92] However, Kant does attribute modal properties to things in themselves, and distinguishes, for instance, the possibility of a thing in itself from the possibility of an object of experience. See *Refl.* 5184, 5723, and 5177. See Ch. 9 for a more extensive discussion of modality for things in themselves.

[93] In a variety of passages from the Critical period, Kant describes things in themselves in causal terms. See especially his discussion of the *forces* of things in themselves in his response to Mendelssohn's *Morgenstunden* at Ak. 8: 153–4. See also A42/B49, A44/B61, B72, A190/B235, A358 f., and A372. Cf. Adickes (1924), 28–37.

[94] See "On the ground of the distinction of all objects in general into *phenomena* and noumena" in the 2nd edition (B294–315).

[95] In a crucial passage of the *CPrR* Kant considers his warrant for applying the category of causality to the supersensible world: "But how is it with the application of this category of causality [. . .] to things that are not objects of possible experience but lie beyond its boundaries? For I was able to deduce the objective reality of these concepts only with respect to objects of possible experience. But what gives them a place in the pure understanding, from which they are referred to objects in general (whether sensible or not) is just this: that I saved them only in case I proved that objects may nevertheless be *thought* through them although not determined *a priori*. If anything is still wanting, it is the condition *for the application* of these categories and especially that of causality to objects, namely intuition; where this is not given, application *with a view to theoretical cognition* of an object as noumenon is made impossible" (Ak. 5: 54). We can *think* supersensible objects under categories, but we cannot *cognize* or know them to fall under the categories.

8

Nomic Necessity

8.1. Introduction

In the *Critique of Pure Reason* (*CPR*), Kant argues that the *a priori* forms of experience—space, time, and the categories, especially the category <*cause-effect*>–make it the case that alterations in phenomena are governed by causal laws that are universal (they have no exceptions) and deterministic (the cause necessitates the effect).[1] The forms of experience make it the case that empirical objects are law-governed, but they do not determine which particular laws govern them. In the terminology of the previous chapter this means that the principle of causation itself (that every event has a sufficient cause, and causation is governed by universal, deterministic laws) is formally necessary, while the particular causal laws are formally contingent. However, Kant also holds that these causal laws are *necessary*. The combination of these two views raises an interpretive puzzle: in what sense are particular causal laws *necessary* for Kant? Answering that question will be the aim of this chapter.

Before continuing, it is important to be clear about what this question means. I have attributed two different modal claims to Kant:

(1) Necessarily, there are universal laws governing nature.
(2) For any law *L* that governs nature, necessarily *L*.

The sense of necessity involved in the first claim is 'formal necessity,' as defined in the previous chapter. Here I am interested in the second claim and the sense of necessity involved in it. I will call it 'nomic necessity.'[2]

[1] Some commentators have questioned whether the forms of experience ground the fact that there are universal and necessary causal laws; e.g., Paton (1936); Bird (1962); Beck (1978); Buchdahl (1965), (1969), and (1992); and Allison (1983), 228–34 and (1994), 298. Some commentators claim that the second Analogy proves only that there must be 'rules' (e.g., A-type events always cause B-type events) but not laws; for the basis of the distinction, see A113, A126, *Refl.* 5414, and the discussion in Gloy (1976), 19 f. I do not have space in this book to argue for the traditional view that the forms of experience require that there be laws; see, however, A108, A113, A114, A127–8, B165, A159/B198, A216/B263, and *Prol.* § 36, as well as Watkins (2005), 286–91. I am not claiming that the second Analogy *successfully* proves that there are causal laws, but only that this is what Kant takes himself to have shown there.

[2] My interpretation agrees, in broad outline, with Kreines (2009) and Watkins (2005). While Kreines focuses mainly on arguing against the 'best systems' account of nomic necessity, and arguing for the philosophical appeal of a view on which laws state explanatory grounds, I focus mainly on exploring the

In section 2 I present the textual case that nomic necessity is not formal necessity. I also consider several other candidate conceptions of nomic necessity and argue that none of them is correct. This clears the ground for my own interpretation, that Kant holds what is now called an 'essentialist' view of laws: laws are grounded in the essences of empirical natural kinds.[3] This entails the following definition of nomic necessity:

(*Nomic necessity*) It is *nomically necessary* that p if and only if p and the fact that p is grounded in the real essences of empirical natural kinds (e.g., matter, water, etc.).

I begin my argument in section 3 with an account of the Kantian distinction between real and logical essence. Then, in section 4, I offer a close reading of the Preface to Kant's 1786 work *Metaphysical Foundations of Natural Science* (*MFNS*), and argue that he identifies the essences of empirical objects with their 'formal natures' and that laws, on Kant's view, are grounded in the formal natures (essences) of empirical objects. This entails my account of nomic necessity. I then explain why this 'essentialist' account of laws and nomic necessity is crucial to Kant's philosophical project in *MFNS*. In sections 5 and 6 I diagnose a tension between my 'essentialist' conception of laws (and their necessity), and the *a priori* 'constructive' procedure of *MFNS*: how can *a priori* constructions inform us of the essential properties of matter? In section 7 I explain how this problem can be solved. In section 8 I return to the texts that jump-started my interpretation in section 2 and explain them in light of that interpretation.

First, though, a note about how I am using the term 'law.' In some texts, Kant defines 'law' quite broadly; in some places he even calls the formally necessary principle that all alterations in empirical substances have causes the 'law' of causality.[4] In other texts, though, Kant seems to use 'law' more narrowly, so that transcendental principles of this kind would not count as 'laws.'[5] In this chapter I investigate the modal status of the truths 'intermediate' between contingent empirical generalizations (e.g., all dogs have fleas) and formally necessary principles grounded in the

modal status of such laws, and its relation to other kinds of modality in Kant's system. Watkins (2005) argues, on the basis of a detailed exegesis of the Analogies of Experience, that Kant conceives of causation as the exercise of a power in a substance, where that power is part of the substance's essence (2005), 286–91; this is very much in line with my 'causal essentialist' reading of Kant on laws, although I do not have the space in this book to ground that claim in a detailed interpretation of the Analogies.

[3] See Shoemaker (1980) and (1998), Ellis and Lierse (1994), and Bird (2005). Hawthorne (2001) contains a very helpful overview of this literature.

[4] See esp. the title of the second Analogy in the B edition (B232); also A188/B234, A227–8/B280, A460/B488, A536/B564, and A636/B664.

[5] E.g., B165, A127, and *CJ* (Ak. 5: 180 and 184). Kant also refers to the metaphysical principles of matter as 'laws'; see the 'Metaphysical foundations of mechanics' chapter of *MFNS* where Kant refers to the 'laws' of mechanics (Ak. 4: 541, 543, 544). On the distinction between metaphysical and transcendental principles see the Preface to *MFNS* (Ak. 4: 469–70), *CJ* (Ak. 5: 181), and *Prol.* §15 (Ak. 4: 295).

forms of experience. In some contexts Kant might not call all of these truths laws, but that is not a serious problem for my project. In section 5 I explore the various kinds of laws in Kant's system.

8.2. What Nomic Necessity Is Not

In a variety of texts from the Critical period, Kant makes clear that which causal laws obtain is not determined by the *a priori* forms of experience alone. For instance:

The pure faculty of understanding does not suffice, however, to prescribe to the appearances through mere categories *a priori* laws beyond those on which rests a **nature in general**, as lawfulness of appearances in space and time. Particular laws, because they concern empirically determined appearances, **cannot** be **completely derived** from the categories, although they all stand under them. Experience must be added in order to know particular laws **at all**; but about experience in general, and about what can be cognized as an object of experience, only those *a priori* laws offer instruction. (B165; cf. A127)

But there is such a manifold of forms in nature, as it were so many modifications of the universal transcendental concepts of nature that are left undetermined by those laws that the pure understanding gives *a priori*, since these pertain only to the possibility of a nature (as object of the senses) in general, that there must nevertheless also be laws for it which, as empirical, may seem to be contingent in accordance with the insight of **our** understanding, but which, if they are to be called laws (as is also required by the concept of a nature), must be regarded as necessary on a principle of the unity of the manifold, even if that principle is unknown to us. (*CJ*, Ak. 5: 180)

The understanding is of course in possession *a priori* of universal laws of nature, without which nature could not be an object of experience at all; but still it requires in addition a certain order of nature in its particular rules, which can only be known to it empirically and which from its point of view are contingent. These rules, without which there would be no progress from the general analogy of a possible experience in general to the particular, it must think as laws (i.e. as necessary), because otherwise they would not constitute an order of nature, even though it does not and never can cognize their necessity. (*CJ*, Ak. 5: 184)

There is a lot going on in these passages. I want to focus, though, on one claim that Kant makes in all of them: that the forms of experience underdetermine which particular causal laws govern empirical objects. Kant is not merely claiming that *for all we know a priori*, particular causal laws are not completely determined by the forms of experience. While that reading may be compatible *sensu stricto* with the first passage (from the *CPR*), it is incompatible with the second two texts. In the second passage he claims that these laws are "left underdetermined by the laws that the pure understanding gives *a priori*," and in the third text he claims that they are contingent "from the point of view" of the understanding (a claim he also makes in the second passage). I take this to mean that these laws are contingent given the 'laws' of the understanding (principles of experience) *alone*: the principles of experience *themselves* (in concert with the *a priori* forms of intuition) do not necessitate that empirical objects are governed by the laws that actually obtain. I interpret this to mean that these causal laws are not formally necessary; because they are not wholly grounded in the *a priori* forms of experience, the laws that actually obtain are formally contingent. However, in the underlined portions Kant also claims that these particular causal laws are necessary. This is a claim Kant makes in a wide variety of other texts.[6] This constitutes, I take it, clear evidence that, according to Kant's Critical view, laws are necessary but formally contingent. Consequently, the necessity of laws (nomic necessity) is not formal necessity.

There are two ways this claim—that nomic necessity is not formal necessity—can be taken, namely:

(1) No law is formally necessary.
(2) Not all laws are formally necessary. The necessity that applies to laws as such is not formal necessity.

I take it that the texts above support attributing (2), at least, to Kant and I will proceed on that hypothesis. If 'law' is taken in the broadest sense, (1) is clearly false. Kant sometimes refers to *a priori* formally necessary principles (like the causal principle) as 'laws.'[7] So I interpret the texts quoted earlier, in which Kant claims that "particular laws [...] cannot be completely derived from the categories," as making the generic claim that laws are not *in general* grounded in the forms of experience; it is compatible with this that some laws, e.g., the metaphysical principles of matter, are formally necessary, though I will argue that they are not.

What is nomic necessity, if it is not formal necessity? First of all, I think it should be clear that nomic necessity is not logical necessity. Causal laws are not logically necessary; the negation of some law that actually obtains does not entail a contradiction. Nomic necessity, whatever it is, is a kind of real necessity.[8] Nor can nomic necessity be empirical-causal necessity, because empirical-causal necessity applies only to alterations in empirical substances, not to the laws that govern those alterations.[9] Nomic necessity, obviously, cannot be noumenal-causal necessity, or any other kind of necessity grounded in noumena, because we cannot know what is noumenally-causally necessary and what is noumenally-causally contingent, but we

[6] E.g., *MFNS* (Ak. 4: 468–9), *CJ* (Ak. 5: 183), and *Refl.*, 5414 (Ak. 28: 176). See also A91–2/B123–4, A159/B198, and A200/B246. Contra Guyer (1990*b*), 234, laws are already claimed to be necessary in the first *Critique*; on this point, see Walker's reply to Guyer in the same volume (Walker 1990, 252).

[7] See the texts quoted previously.

[8] See section 7 for a more complete argument that nomic necessity is a kind of real necessity.

[9] Guyer (1990*b*), 234 also makes the point that (in my terminology) neither formal necessity nor empirical necessity are nomic necessity.

can know of many truths (e.g., the truth that a particular event occurs) that they are nomically contingent.[10]

Several commentators have attempted to account for the necessity of laws in terms of systematicity.[11] One prominent representative is Paul Guyer. On Guyer's reading, laws are necessary in virtue of being parts of hierarchically ordered systems in which generic higher-order laws subsume and explain specific lower-level laws. [12] However, systematicity alone cannot explain nomic necessity, for two reasons. First of all, one of Kant's main points against Hume is that regularity does not suffice for necessity.[13] There is no reason why Hume cannot acknowledge systems of regularities hierarchically ordered from more general to more specific. If we accept Guyer's proposal then Kant is left without a response to the Humean who claims to have accounted for the necessity of natural laws *just as well* as Kant has. Secondly, Guyer's proposal leaves us without an account of the necessity of the highest-level laws, for instance, the law of universal gravitation. There is a sense in which a particular regularity in nature (e.g., the true orbits of the planets) is understood as non-accidental, and in that sense necessary, when it is brought under a higher-level law (e.g., the law of universal gravitation). However, the problem we are concerned with is the necessity of precisely those higher-level laws that, on Guyer's picture, 'inject' the necessity into the whole structure. Guyer might reply that it is the *a priori*, formally necessary principles of experience that 'inject' this necessity. But the texts quoted in section 2 show that even the highest-level and most general natural laws (e.g., the inverse-square law of gravitation) are not necessary in the same way those principles are, so Guyer's story leaves us without an explanation of how these highest-level laws are in some sense necessary but not formally necessary like the principles of experience. Guyer needs some necessity to attach to the highest-level natural laws, but does not tell us what that necessity is.

Finally, we might try to understand nomic necessity through the logical function of assertoric judgment. Kant derives the table of categories from the table of the logical functions of judgment.[14] The modal functions of judgment are problematic, assertoric, and apodictic, which correspond, respectively, to the modal categories: possibility, actuality, and necessity. One might, therefore, think that the way to understand the necessity of judgments (including judgments that state laws) is by understanding the logical function of apodictic judgment. In the following passage, Kant appears to equate the logical functions of judgment with their logical-deductive role and connects this in turn to the modal categories:

[10] Which does not mean that we generally know natural laws. See Kreines (2009) for a discussion of the limits of our knowledge of natural laws.

[11] E.g., Kitcher (1986), 209; Buchdahl (1969), 342; and Brittan (1978), 185. Kreines (2009) incisively critiques the 'best system' reading of Kant's theory of laws.

[12] Guyer (1990*a*), 287–8; see also Guyer (1990*b*), 238.

[13] *Prol.* (Ak. 4: 258). [14] See section 6.7.

The assertoric proposition speaks of logical actuality or truth, as, say, in a hypothetical syllogism the antecedent in the major premise is problematic, but that in the minor premise is assertoric and indicates that the proposition is already bound to the understanding according to its laws; the apodictic proposition thinks of the assertoric one as determined through these laws of the understanding itself, and as thus asserting *a priori*, and in this way expresses logical necessity. (A76/B101)

As I understand this passage, Kant is ascribing logical functions to judgments in virtue of their role in a syllogistic form of reasoning, e.g.,

(1) $p \supset q$
(2) p
(3) $\therefore q$

The middle premise, (2), Kant claims, is assertoric and therefore represents its content as 'actual.' The very same content, p, is the antecedent in the hypothetical first premise, and thus is represented as merely possible; (1) represents p as possible and states the consequence of its actuality, q. Kant describes (3) as apodictic because, in the context of the syllogism of which it is the conclusion, it represents q not merely as true but as true in virtue of a rule, (1), thus, non-accidentally true. Consequently, Kant claims, q is represented as necessary.[15]

But this should caution us in trying to model nomic necessity on apodictic judgment and Kant's association of that form of judgment with necessity. Natural laws correspond, not to the conclusions of syllogisms, but to their major premises, the rule (1) from which the conclusion (3) is deduced by subsuming (2) under the condition (antecedent) of (1). Put in less Kantian terms, by associating apodictic judgment with the category of necessity, Kant is giving an account of the necessity of judgments in terms of their roles in logical-deductive structures, but it is an account on which necessity consists in being the conclusion of a syllogism. This is completely unsuited to accounting for nomic necessity, because the laws in question are precisely the rules from which particular states of affairs (e.g., the true motions of bodies) are derived as conclusions. The necessity of laws is not the necessity of conclusions, but the necessity of the premises from which these conclusions are derived. In a similar spirit, Michael Friedman offers a sophisticated and intricate reading of Kant's use of the modal categories in the Phenomenology section of *MFNS*.[16] For reasons of brevity, though, I defer my critical engagement with that part of Friedman's reading to an Appendix that can be found on my website (see the Preface); I discuss Friedman's reading of the 'constructive' procedure of *MFNS* in section 5 below.

[15] See Leech (2012) for a nuanced analysis of Kant's views on the modal functions of judgment; Rosenkoetter (2013) argues for a quite different reading of the modal functions.

[16] Friedman (1992*a*), 159–64 and Friedman (2013), 531–61.

8.3. Essences: Real and Logical

In the next section I will argue for my 'essentialist' interpretation of nomic necessity. But first we must understand what *essences* are for Kant. That is the task of this section.

Throughout his lectures, Kant distinguishes between (i) the logical essence of a *concept* and the real essence of the *object* of that concept and between (ii) nominal definitions and real definitions. The former distinction typically appears in his lectures on metaphysics;[17] the latter, in his lectures on logic.[18] However, these distinctions are correlative: a nominal definition reveals a logical essence and a real definition reveals a real essence.

The best place to start understanding the distinction between logical and real essence is this long passage from the L_2 lectures on metaphysics:

> The concept of essence belongs properly in logic. Essence is either a logical essence or a real essence. A logical essence is the first ground of all logical predicates of a thing; a real essence is the first ground of all determinations of a thing. For an essence is either logical or real. We posit a logical essence through the analysis of the concept. The first ground of all predicates thus lies in a concept; but that is not yet a real essence. E.g., that bodies attract belongs to the essence of things, although it does not lie in the concept of the body. Accordingly, the logical essence is the first inner ground of all that which is contained in the concept. But a real essence is the first inner ground of all that which belongs to the thing itself [*was der Sache selbst zukommt*]—If I have the logical essence, I still do not yet have the real essence. In metaphysics, essence should never be understood as logical essence, for this belongs in logic. [...] Predicates belonging to the essence, but only as a consequence, are called attributes; what on the other hand belongs to essence as a ground is called an essential property. [...] The real essence is not the essence of the concept, but rather of the thing [*sondern der Sache*]. E.g., the predicate of impenetrability belongs to the existence of body. Now I observe through experience much that belongs to its existence; e.g., extension in space, resistance against other bodies, etc. Now the inner ground of all this is the nature of the thing. We can infer the inner principle only from the properties known to us; therefore the real essence of things is inscrutable to us, although we cognize many essential aspects. We become acquainted with the powers of things bit by bit in experience. (*ML*$_2$, Ak. 28: 553)

Kant makes a series of claims about logical and real essences in this passage, which I will unpack and analyze.

First, he distinguishes between the logical essence of a *concept* and the real essence of a *thing* ("a real essence is the first inner ground of all that which belongs to the thing itself"). He identifies the logical essence of a concept as the complete set of

[17] In the metaphysics lectures, *MH* (Ak. 28: 49), *MV* (Ak. 28: 411), *MvS* (Ak. 28: 492), *ML*$_2$ (Ak. 28: 553), and *MD* (28: 629).

[18] In the logic lectures, *LB* (Ak. 24: 113–18, 268–73), *WL* (Ak. 24: 838–40, 913–25), *LDW* (Ak. 28: 727–9, 756–60), *JL* (Ak. 9: 61, 140–5), *LP* (Ak. 24: 408–9, 456–9), *LBu* (Ak. 24: 634–5, 656–60), and *LPö* (Ak. 24: 535, 573–5).

marks contained in the concept. The logical essence of a concept is specified in its nominal definition; e.g., the nominal definition of <*gold*> might be: <*yellow*>, <*malleable*>, and <*metal*>.[19] Kant thinks that nominal definitions are relatively trivial for empirical scientific purposes. It may be important to make sure we all mean the same thing by our words—e.g., that we are all using 'gold' to denote yellow, malleable metal—but this does not by itself tell us anything substantive about *gold*, the stuff our concept refers to.[20]

A real essence, by contrast, is the essence of the *thing* picked out by a concept. The real essence of a thing is given by its real definition, and real definitions are definitions of *things* [*Sach-Erklärungen*],[21] not of concepts. Real definitions are not of individual things, however, but the kinds that these concepts pick out; for instance, the real definition of gold specifies the real essence of the kind *gold* rather than <*gold*> (the concept) or an individual sample of gold. Kant denies that we can give real definitions of individual things.[22] Kant also denies that we can know the complete real definition, hence the complete real essence, of kinds of empirical object. At this point, though, we are interested in what real essences are, according to Kant, not whether we know them.

Kant's definition of real essence in the L_2 text ("the first inner ground of all that which belongs to the thing itself") is idiosyncratic relative to other texts, because it does not make explicit the modal dimension of real essence. More typical is his definition in the Mrongovius lectures: "the first inner ground of all that belongs to the *possibility* of a thing is its [real] essence" (Ak. 29: 820).[23] But even this more typical formulation poses an interpretive challenge. It suggests that the real essence of *x* is the ground of the real possibility that there is *x*, but, as we have seen, the grounds of the real possibility of *x* are going to be something *other than x itself* (e.g., the forms of experience, the laws and the past, etc.). In a pre-Critical context we might even expect, on the basis of this definition alone, that the real essence of any finite being is God! However, Kant thinks of the real essence of *x* as the 'inner character' of *x* that explains its manifest character and its relation to other things, so the real essence is not some entity distinct from *x* (e.g., God). For instance, in the example from the L_2 lectures above, Kant describes the real essence of matter as including attractive force. So Kant thinks of the 'first inner ground of all that belongs to *x* itself' and 'the first inner ground of all that belongs to the possibility of *x*' as somehow being properties or characteristics of *x*, the inner characteristics of *x* that make it the possible being it is.[24] My interpretation of this is: where K is a kind (the appropriate

19 *Prol.* (Ak. 4: 267).

20 *LB* (Ak. 24: 116, 271), *LDW* (Ak. 24: 757), and *WL* (Ak. 24: 839, 918).

21 *JL* §106 (Ak. 9: 143); cf. *WL* (Ak. 24: 839) and *LDW* (Ak. 24: 757).

22 *LB* (Ak. 24: 118, 268).

23 See *LDW* (Ak. 24: 760), *LP* (Ak. 24: 456), *JL*, §106 (Ak. 9: 143, 144), *WL* (Ak. 24: 919, 920), *Refl.* 2916 (Ak. 16: 574–5), and *Refl.* 2995 (Ak. 16: 607).

24 In some texts Kant defines a real definition of a thing as providing marks sufficient to distinguish it from all other possible things; in other words, the real essence (what is specified in a real definition) makes

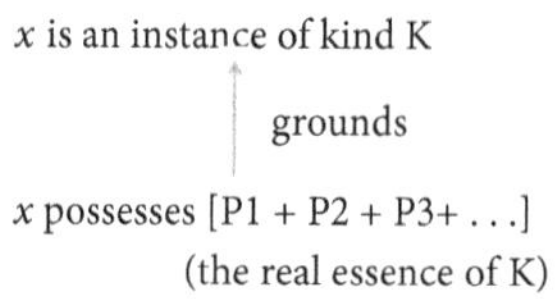

Fig. 8.1. The real essence of K

target of a real definition) and *x* is a possible instance of that kind, the real essence of K is the complex of properties possessed by *x* that ground *x*'s being an instance of kind K. In other words, the real essence of water is the complex of properties that make it the case, of any possible sample of water, that it is a sample of water (see Fig. 8.1).

What does 'grounding' mean here? Clearly, possessing the essential properties of matter does not *cause* an object to be matter. Nor is it logical grounding. If possession of the real essence of kind K logically entails being an instance of K there is no distinction between logical and real essence; the point of Kant's distinction between logical and real essence is that possession of the marks sufficient to logically entail membership in a kind (the logical essence) is not the same as its real essence. To return to a theme from the previous chapter (7.3), I think this is a non-logical non-causal asymmetric real grounding relation between mutually necessarily entailing propositions (necessarily, something is matter if and only if it has the essence of matter); to use Kant's terminology, the possession by *x* of the properties constituting its real essence is the *ratio essendi* of its being an instance of kind K. This is only to put a name and a structure on this relation; it is not yet even an *attempt* at a reductive analysis. Over the course of the chapter I will attempt to see what we can say about this real grounding relation and what connections it might have to the real grounding relations from Chapter 7.

Another important feature of Kant's theory of real essence is his distinction between the properties that compose the essence (essential properties) and the properties that are grounded in the essence but not contained in it (attributes).[25] As Kant says in the quoted passage: "predicates belonging to the essence, but only as a consequence, are called attributes; what on the other hand belongs to essence as a ground is called an essential property." The essential property/attribute distinction is important because not all laws will be 'parts' of essences. In general, that a law holds

a possible thing the possible thing it is; see A241–2 n., *LB* (Ak. 24: 658) and *Refl.* 2992 (Ak. 16: 606). However, there are other passages in which Kant appears to equate that 'differentiation' model of real essence with the 'ground of possibility' model (from the main text), or introduces one within the space of a page of the other, which strongly suggests he thinks of them as equivalent; see e.g., *LDW* (Ak. 4: 760), *LP* (Ak. 24: 473), and *WL* (Ak. 24: 919, 920). I take this to mean that 'the first inner ground of all that belongs to the possibility of *x*' is the set of marks that make *x* the possible being it is.

[25] The terminology comes from Baumgarten, who distinguishes essential predicates (*essentialia*), predicates entailed by essential predicates (*attributa*) and accidents (*Meta.* §§39, 41, 50); Kant takes over these distinctions almost verbatim in his metaphysics lectures: *MV* (Ak. 28: 411), *MvS* (Ak. 28: 492–3), *MP* (Ak. 28: 552–3), *MD* (Ak. 28: 629), and *MM* (Ak. 29: 820).

of a certain kind will be an attribute of that kind, not part of the essence. For instance, take some very complicated mathematical consequence of the inverse-square law of gravitation for systems of *n* bodies (a theorem of Newtonian dynamics); it is implausible that this law is literally a part of the essence of matter, but more plausible that it is grounded in that essence. This raises, of course, the question of the grounding relation between an essence and its attributes; I return to this issue in section 7.

Some empirical concepts are what I will call *empirical kind concepts*: they apply to objects in virtue of those objects belonging to the same *kind* (e.g., gold, water, etc.). This is liable to provoke the objection that it is anachronistic to attribute to Kant the view that there are empirical kinds at all, or that the function of empirical concepts is to pick them out. Consider, however, the following passage from the logic lectures:

> Can concepts of experience be defined analytically? When the inquirer into nature defines water, e.g., as a fluid body without taste or color, one readily sees how precarious the definition is. He who is not already acquainted with water will not thereby become acquainted with it. It is simply not necessary to define concepts of experience *per analysin* [nominal definition—NS], however. For why do I need such a definition? For when I say the word *water*, others understand me [...] For in the concept *water* there lies so little that I immediately go outside the concept and have to collect new marks through experience, i.e. I have to define the concept through exposition synthetically [partial real definition—NS], and not analytically. (*WL*, Ak. 24: 918–21)

Kant is claiming here that the nominal definition of empirical concepts like <*water*> is relatively insignificant for scientific purposes. The concept contains only some commonly observable marks, so that we can verify that we are all talking about the same thing (e.g., "the clear drinkable stuff in lakes and rivers"). The goal of scientific inquiry is to determine what all samples of water as such have in common, what makes them water, that is, the real essence of water. This requires that we use the concept <*water*> not to pick out everything that has the marks contained in its logical essence (because the nominal definition is dispensable, as Kant here claims) but to pick out objects in virtue of belonging to the kind *water*, that is, everything that has the marks contained in its real essence of water. Not all empirical concepts are like this. It is implausible, for instance, that weeds share a real essence; all there is to being a weed is our nominal definition of the concept <*weed*> because this is a concept arbitrarily invented by us. Concepts like <*water*>, that are used to pick out a kind in virtue of a shared real essence, I will call 'empirical kind concepts'; Kant calls them 'given' empirical concepts.[26] Kant does not think we can ever discover the complete real essence of any empirical kind, so we can never give the complete real definition of an empirical natural kind. But he does think that in natural science we can

[26] *LB* (Ak. 24: 272), *WL* (Ak. 24: 838–40), *LDW* (Ak. 24: 727–9), and *JL* (Ak. 9: 142).

determine some of the essential properties and attributes, as he makes clear in the L_2 passage quoted earlier: "therefore the real essence of things is inscrutable to us, although we cognize many essential aspects."

8.4. Essences, Natures, and Laws

If we want to understand Kant's views about essences and laws, one natural place to turn is *MFNS*. Kant begins the Preface to that work by distinguishing the 'formal' and 'material' meanings of the word 'nature':

> If the word 'nature' is taken simply in its formal meaning, where it means the first inner principle of all that belongs to the existence of a thing*, then there can be as many different natural sciences as there are specifically different things, each of which must contains its own proper [*eigenthümliches*] inner principle of the determinations belonging to its existence. But 'nature' is also taken otherwise in its material meaning, not as a constitution, but as the sum total of all things, insofar as they can be objects of our senses, and thus of experience. [...] (*MFNS*, Ak. 4: 467)

The formal sense of 'nature,' I take it, is the sense involved in claims of the form 'it is in the nature of ϕ to ψ' (e.g., it is in the nature of gold to melt at 1,064° C), while, in the material sense, 'Nature' refers collectively to all objects of sensible intuition (both inner and outer).[27] For ease of reference, I will capitalize 'Nature' when it refers to nature in the material sense. The asterisk in this passage indicates a footnote, where Kant clarifies his meaning:

> Essence is the first inner principle of all that belongs to the possibility of a thing. Therefore, one can attribute only an essence to geometrical figures, but not a nature (since in their concept nothing is thought that would express an existence). (*MFNS*, Ak. 4: 468 n.)[28]

'Existence' here means being causally efficacious, either as a substance or the accidents inhering in a substance due to its interactions with other substances. This is the sense of 'existence' in which the Analogies of Experience express conditions on the possibility of representing *existing* objects.[29] As Kant claims here, there are objects that do not exist in this technical sense of 'existing,' which we can think of as equivalent to *being concrete* in the contemporary sense. I will refer to this notion of existence as existence*. In the final chapter of this study I will explain Kant's Critical theory of existence, existence*, and actuality, but for now I just want to note that, by admitting that existence* (concreteness), is a 'real predicate' (one that applies to a proper subset of the objects there are), Kant is not opening space for an ontological

[27] Cf. the discussion of the formal sense of nature in the footnote to A418/B446, *Prol.* §14–17 (Ak. 4: 294–7), and *GMM* (Ak. 4: 437). The formal meaning of 'nature' is clearly a descendant of Aristotle's definition of nature; see *Physics* 2.1, 192b20–23.

[28] Cf. the discussion of nature and essence in the metaphysics lectures: *MH* (Ak. 28: 49), *MV* (Ak. 28: 411), *MvS* (Ak. 28: 492), *ML_2* (Ak. 28: 553), and *MD* (Ak. 28: 629).

[29] See section 6.6, and A160/B190, A176, B219, A182, B225, B233, B257, and A215–16/B262–3.

argument; he can freely admit that *God exists** is an analytic judgment (if there is a God, he is concrete), while denying that *there is a God* (that there is an object instantiating the concept <*existent* God*>) is analytic. The point of the quoted passage is that only existing* objects have what Kant calls 'natures.'

In the first sentence of the Preface, quoted above, Kant claims there can be as many different sciences as there are "specifically different things" that have a nature; I take this to refer to generic *kinds* rather than individual things, kinds constituted by things that share a common nature. Since only existing things have a nature, we might interpret this notion of (formal) nature as follows:

> (*Nature**) If ϕ is a kind of existing* thing, then property ψ is part of the nature of ϕs if and only if necessarily all ϕs have ψ.

But this is inadequate, because Kant asserts an explanatory connection between the nature and "what belongs to" the existence of the kind: the nature is "the first inner principle of all that belongs to the existence of a thing." The nature is a "first principle" because it explains the properties that things of that kind necessarily have. I propose instead:

> (*Nature*) If ϕ is a kind of existing* thing, the nature of ϕ is the complex of properties the possession of which by any possible instance of ϕ grounds its being ϕ and existing*.

The nature of a kind is the set of properties such that instances of the kind are instances of the kind in virtue of having those properties. Since natures are defined for *existing** things, possessing the properties that compose a nature grounds not only membership in the kind, but existence (causal effectiveness) as well. The nature of a thing is "the first inner principle of all that belongs to the existence* of a thing" because its nature is the set of properties in virtue of which it exists* as the kind of thing it is.

Kant begins his footnote about formal nature by distinguishing it from (real) essence—non-existing* mathematical objects have essences, but not natures—so we might wonder about the relation between these notions. It is clear that anything that is part of the nature of ϕ (if ϕ is a kind of existing thing) is also part of its real essence: if a set of properties grounds being ϕ *and* existing*, then (trivially) it grounds being ϕ. What is more, the essence of ϕ *is identical to* the nature of ϕ. If ϕ is a kind of existing* thing (concrete thing) then it is essentially a kind of existing* thing; existing* is either part of the essence of ϕ (essential property) or it is grounded in that essence (attribute). If some sample possesses the properties that constitute the real essence of ϕ then it possesses properties sufficient to ground its existence*; so the real essence of ϕ grounds not only its being ϕ but its existing* as well.[30] Consequently, the real essence of ϕ is its nature.

[30] Note, though, that this is the narrow conception of existence (=concreteness), not the 'absolute positing' (quantificational, non-nuclear) conception of existence, so to say that ϕ essentially exists* does not entail that there is an ontological argument for there being any ϕ (or any of its instances).

(*Nature=essence*) If ϕ has a nature (is a kind of existing* thing) then the nature of ϕ = the real essence of ϕ.

For instance, take ϕ to be a kind of physical substance, like molybdenum. Clearly, molybdenum is a kind of thing that exists*. The properties in virtue of which a sample of molybdenum is a sample of molybdenum, are also properties that ground its being causally efficacious (existing*). So the essence of molybdenum is the nature of molybdenum.[31] This interpretation is confirmed by Kant's metaphysics lectures, where he repeatedly equates nature and real essence (for existing* things) e.g., "this real essence is nature" (*MM*, Ak. 29: 821).[32] In section 3 I defined empirical kind concepts as concepts that apply to objects in virtue of belonging to a kind with a common essence; since essences for existing* things are natures, we can also call these 'empirical natural kind concepts': concepts that pick out objects in virtue of sharing a (formal) nature.[33] Since I will almost exclusively be discussing existing* things I will not always make this explicit. Likewise, I will often refer to real essence as essence *simpliciter* because logical essences will not play much of a role in my argument; when I mean *logical* essence I will make this explicit.

I have analyzed the notions of essence and nature in such detail because Kant's definitions of 'law' in *MFNS* depend on the formal concept of nature:

> Since the word Nature [in the material sense—NS] already carries with it the concept of laws, and the latter carries with it the concept of the necessity of all determinations of a thing belonging to its existence [...] (*MFNS*, Ak. 4: 468)
>
> For laws, that is, principles of the necessity of that which belongs to the existence of a thing [...] (*MFNS*, Ak.4: 469)

Kant's definitions of law in these passages (especially in the second) is so close to his definition of (formal) nature that we might be tempted to identify the nature of a thing with the laws concerning it, or even to claim that the nature of a thing is grounded in laws: what it is to be molybdenum, for instance, is to obey such-and-such laws (e.g., to have a melting point of 2,623° C). However, this would leave us without an analysis of what it is to be a law, and there is a more faithful reading of the texts available. There is a subtle difference between Kant's definition of formal nature in the very first sentence of *MFNS* ("the first inner principle of all that belongs to the

[31] Failure to recognize this point, I believe, is what leads Plaass to claim, implausibly, that 'nature in the formal sense' as defined in *MFNS* is distinct from 'nature in the formal sense' as defined in other texts (cited by Plaass himself) where Kant identifies nature with real essence (Plaass (1965), 217 n. 2).

[32] *MH* (Ak. 28: 49), *MV* (Ak. 28: 411), *MvS* (Ak. 28: 492), ML_2 (Ak. 28: 553), and *MD* (Ak. 28: 629). See also the May 12, 1789 letter to Reinhold (Ak. 11: 37).

[33] The qualification *empirical* natural kind concept is important, because it may be that there are *a priori* natural kind concepts: substances might form a kind in virtue of the common real essence of substance. But I want to remain neutral on that.

existence of a thing") and these definitions of law: the nature of a thing is the *first* principle of what belongs to its existence while these two passages define laws as necessary conditions on the existence of a thing.[34] I take this to mean that the nature of a thing grounds or explains the laws that are true of it, i.e.

(Law_{NAT}) If it is a law that ϕ has ψ then (a) it is true that all samples of ϕ have ψ, (b) ϕ is a natural kind ($<\phi>$ is a natural kind concept), and (c) every sample of ϕ has ψ in virtue of the nature of ϕ.

Consequently, there can be laws for each kind, and the laws of the more generic kinds (grounded in the generic nature of the species) will be generalizations that subsume the laws of the more specific kinds (grounded in the specific difference in nature that makes that species the species it is). For each such kind and its associated laws there can be in principle be a science of that kind.[35] This is the meaning of Kant's claim that "there can be as many different natural sciences as there are specifically different things, each of which must contain its own peculiar inner principle of the determinations belonging to its existence" (Ak. 4: 467). Given the identity of essences and natures for existing* things, it follows that:

(Law_{ESS}) If it is a law that ϕ has ψ then (a) it is true that all samples of ϕ have ψ, (b) ϕ is a natural kind ($<\phi>$ is a natural kind concept), and (c) every sample of ϕ has ψ in virtue of the essence of ϕ.[36]

Every law states a truth grounded in the essence of a natural kind. If this view is correct, there is a natural sense in which all laws as such are necessary: they obtain in virtue of the essences of natural kinds. Every law is nomically necessary if we define nomic necessity as follows:

(*Nomic necessity*) It is *nomically necessary* that p if and only if p is grounded in the essences of actual empirical natural kinds.

[34] Cf. his remark about nature in the material sense: "this signifies a derivation of the manifold belonging to the existence of things from their inner principle" (*MFNS*, Ak. 4: 468).

[35] For any nature, there are facts grounded in that nature, e.g., that objects with that nature have the properties contained in that nature. So, trivially, there will be laws: the law that all objects with that nature have those properties. However, as Allison (1994), 298 points out, this leaves open whether the laws are 'one off'—for there may be only one object with a given nature. In other words, this leaves open whether there is a different law for every empirical object, and whether there is any systematic unity among the laws. On my reading, it is the role of reflecting judgment and the regulative use of theoretical reason to justify us in holding that nature constitutes a system, in particular, that the natural kinds constitute a system and this 'one-off' situation does not obtain. However, interpreting Kant's idea of nature as a system lies outside the scope of this book.

[36] I have stated these as *necessary* conditions on lawhood, rather than necessary and sufficient conditions (or even definitions of lawhood) to make room for the possibility that there may be more to the Kant's conception of lawhood, such as the organization of laws into a system of Nature (of natures). Cf. Appendix to the Transcendental Dialectic, A643/B671–A668/B696.

My interpretation offers a unified account of Kant's conception of essence, nature, law, and nomic necessity.[37]

Some readers might suspect that this Aristotelian metaphysics of essence and nature attributes to Kant more metaphysics than he *needs* in *MFNS* and more metaphysics than he is *allowed*, given his epistemology. On the contrary, the metaphysics of essences and natures is crucial to his project in *MFNS*: to provide a foundation for the rational science of physics (of outer Nature). As he writes in the Preface,

> Any whole of cognition that is systematic can, for this reason, already be called *science*, and if the connection of cognition in this system is an interconnection of grounds and consequences, even *rational* science. If, however, the grounds or principles themselves are still in the end merely empirical, as in chemistry, for example, and the laws from which the given facts are explained through reason are mere laws of experience, then they carry with them no consciousness of their necessity (they are not apodictically certain), and thus the whole of cognition does not deserve the name of a science in the strict sense [...] (*MFNS*, Ak. 4: 468)

A science is a body of cognitions organized according to levels of increasing generality: a hierarchy of particular truths subsumed under more general laws, which are themselves subsumed under more general laws, up to and including one or more fundamental laws that are not subsumed under anything more general. In such a system, either the lower-level laws ground the higher-level laws, or vice versa.[38] In a rational science, the higher-level laws and principles (partly) ground the lower-level ones and the lower-level laws (partly) ground the facts at the bottom-level.[39] In a non-rational science, which does not properly deserve the title 'science' at all but is merely an "historical doctrine of nature" (Ak. 4: 468), the higher-level principles are simply generalizations about the bottom-level facts and, as such, are grounded in those facts. For instance, what Kant calls 'natural description' of objects is not a rational science because its highest-level principles are simply classifications grounded in similarities

[37] According to Watkins (1997), it was commonly held in the eighteenth-century German rationalist tradition that the laws of motion are grounded in the essence of matter; I am arguing that Kant held this view for laws in general and the natural kinds they concern.

[38] This is my gloss on Kant's claim that the principles that "order" a science are "either principles of *empirical* or *rational* connection of cognitions into a whole" (*MFNS*, Ak. 4: 468). Kant does not say that the principles *of* the science are either empirical or rational; he says the principles according to which the science is organized are either empirical or rational. A science organized according to empirical principles, I think, is a science in which general principles are *a posteriori* grounds of principles of lower generality; the truths of lower generality *ground* the truth of the general principles, but the general principles are a ground of *knowing* the more particular, less general truths. For instance, if all doves are white, this is made true by all of the individual white doves; but the fact that all doves are white is a ground of knowing, of each individual dove, that it is white. By contrast, in a rational science of ornithology the claimed ground of the whiteness of doves would be the nature of doves.

[39] The qualification 'partly' is crucial here because the nature of the genus (e.g., matter) does not wholly ground the laws of its species (e.g., water), for each species has a specific difference in its nature that grounds specific laws for that kind, which do not apply to matter as such (e.g., the law that water boils at 100° Celsius).

among individual objects. Because, by definition, Nature is a domain of objects governed by necessary laws, any science of Nature as such must be a rational science; a mere 'natural description' of Nature would not be a science of Nature *as* Nature (as a law-governed totality). Kant further requires of a rational science that its most fundamental, highest-level laws are known to be necessarily true. A putative science whose most fundamental laws are known merely through empirical generalization (as, Kant claims, is the case with chemistry as it currently stands)[40] is not really a rational science, for we cannot know through experience whether those generalizations are necessary.[41] Consequently, Kant's characterization of laws as "principles of the necessity of that which belongs to the existence of a thing" is crucial to his understanding of a rational science of Nature.[42]

Furthermore, Kant regards these essentialist claims as *scientifically* necessary. Consider his only explicit criticism of Newton in *MFNS*:

> It is commonly supposed that Newton did not find it necessary for his system to assume an immediate attraction of matter, but with the most rigorous abstinence of pure mathematics allowed the physicists full freedom to explain the possibility of attraction as they might see fit [...] But how could he ground the proposition that the universal attraction of bodies [...] is proportional to the quantity of their matter, if he did not assume that all matter as matter, therefore, and through its essential property, exerts this moving force? [...] [H]e could by no means say that the attractive force of two planets, those of Jupiter and Saturn, for example, manifested at equal distance of their satellites (whose mass is unknown), are proportional to the quantity of matter of these heavenly bodies, if he did not assume that they attracted other matter merely as matter, and thus according to a universal property of matter. (*MFNS*, Ak. 4: 514)

He refers here to the derivation of the relative masses of Jupiter, Saturn, and the Earth in the Corollaries to *Principia*, Proposition VII, which relies on the assumption that every body exerts an attractive force on *every other body* (not just on its satellites).[43] Kant claims that Newton would not be justified in claiming that all bodies *as such* attract all other bodies *without* the assumption that such a force is an essential property of matter. But why is that? Why can't Newton assume that it is a *fact*, perhaps even a *law*, that all bodies attract one another without assuming that this is *essential* to matter? My account delivers a straightforward answer: if this fact is not grounded in the essence of matter then it is merely a generalization about matter and hence is not a law and hence not part of a rational science of matter that *explains* the motions of bodies (rather than merely recording true generalizations about them).

[40] *MFNS*, Ak. 4: 468. Kant also calls chemistry a "systematic *art*" (Ak. 4: 469; cf. 470 f.), by which I take him to mean that chemistry tells us how to produce various items (e.g., salts) but does not explain *why* these procedures succeed (it is not *a priori*$_G$).

[41] "They carry with them no consciousness of their *necessity* (they are not apodictically certain), and thus the whole of cognition does not deserve the name of a science in the strict sense" (*MFNS*, Ak. 4: 468).

[42] On this point, I differ from Konstantin Pollok who downplays the importance of the formal concept of nature for the project of *MFNS*; see Pollok (2001), 46.

[43] I draw here on the discussion of these issues in Friedman (1992*a*), 153–9.

For Newton's theory to be genuinely explanatory and hence *scientific* (in the strict sense) it needs to assert that the force it attributes to bodies is essential to them. Consequently, Kant's account of how that science is possible in *MFNS* must be able to rationally ground that essentialist assertion.

8.5. Essentialism and Constructivism in *Metaphysical Foundations*

Up to this point I have discussed 'laws' in general, but Kant's system includes several different kinds of laws corresponding to different sciences. Figure 8.2 summarizes the relations among the most important Kantian sciences and their respective laws.

Transcendental philosophy is the science of the principles that govern any possible object of cognition; the critique of our cognitive faculty that prepares the way for the complete system of transcendental philosophy is the *Critique of Pure Reason* itself.[44] Objects of cognition divide into two kinds, objects of inner sense and objects of outer sense, corresponding to two different regions of Nature and two different sciences: the science of inner Nature (psychology) and the science of outer Nature (physics). Kant argues in the Preface to *MFNS* that there can be no science of psychology, so the only possible science of Nature is physics.[45] The metaphysical principles of physics (outer nature) are more determinate than transcendental principles restricted to outer objects, because metaphysics concerns an empirical concept, albeit the most general empirical concept of outer objects whatsoever: <*matter*>.[46] All outer objects fall under the concept <*matter*> and the metaphysical principles of physics are the *a priori* principles that make possible experience of matter. For instance, the transcendental principle that all alterations have a cause has a more determinate metaphysical correlate: all alterations in matter must have an external cause (matter cannot move itself).[47] Another important metaphysical principle of matter, derived by Kant in the Dynamics chapter of *MFNS*, is that all matter has attractive and repulsive forces. What I am calling 'rational physics' contains the *a priori* consequences of these metaphysical principles, for instance, the principle that there is some point at which the attractive and repulsive forces of a body cancel one another. More determinate than these principles of rational physics are the laws of empirical physics, the *a posteriori* laws that apply to all matter as such. For instance, if the inverse-square law of gravitation is *a posteriori*, then it belongs here, because it is a more determinate form of the *a priori* metaphysical principle of universal attraction and does not follow

[44] Axxi, Bxxii. [45] *MFNS*, Ak. 4: 470.

[46] *MFNS*, Ak. 4: 469–70. Cf. *CJ*, Ak. 5: 181 on the distinction between metaphysical and transcendental principles.

[47] *MFNS*, Ak. 4: 543; this is precisely Kant's example of a metaphysical specification of a transcendental principle in the *CJ*, Ak. 5: 181. For critical discussion, see Westphal (2004), 205–27.

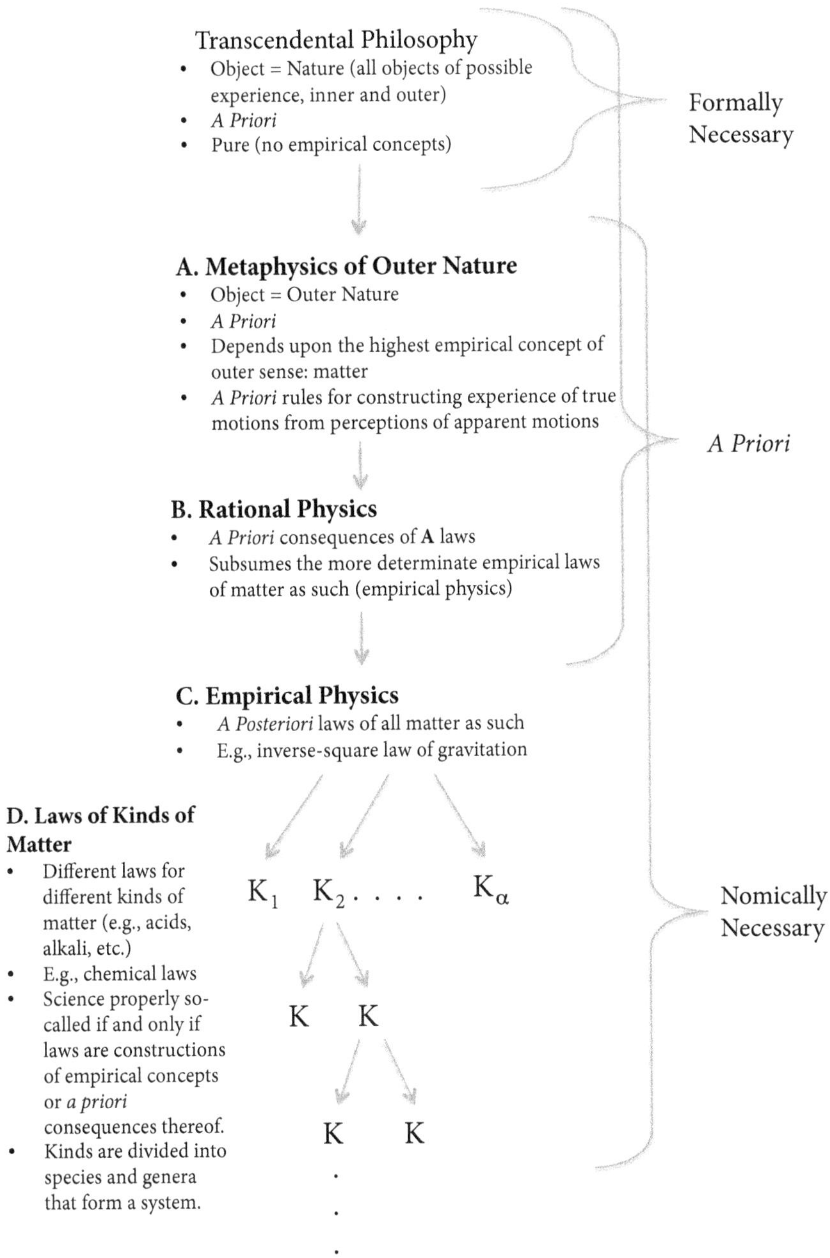

Fig. 8.2. The Kantian system of sciences

a priori from the principles of rational physics (by assumption).[48] Finally, below empirical physics we have various sciences of more determinate kinds of matter (since all outer objects are matter, all kinds of outer objects are kinds of matter); if chemistry can be a science (in the strict sense) and not a mere 'natural description,' it would belong here.[49] Kant refers to the principles of each of these sciences as *laws*; in some contexts, he even refers to the principles of transcendental philosophy (e.g., the causal principle) as *laws*. My focus in this chapter has been on A-D laws, the laws that are more determinate than the formally necessary principles of transcendental philosophy. It is a commitment of my interpretation that the laws of each of these sciences is grounded in the nature/essence of their respective domains: A-C laws are grounded in the essence of matter, while D-laws are grounded in the essences of the relevant empirical natural kind (e.g., various chemical kinds like acid or salt).

However, this division of Kantian sciences poses a potential problem for my essentialist interpretation: there is a tension between (what I claim are) Kant's essentialist conclusions and the methodology he uses to prove them in *MFNS*. Put roughly, Kant's method in *MFNS* is to describe the *a priori* rules we apply in constructing experience of the true motions of bodies from perceptions of their apparent motions and thus derive the A-laws and some B-laws (though not all of rational physics). I will call this the 'constructivist' method of *MFNS*. But how can Kant, merely by elucidating the rules of a constructive procedure, determine anything about the real essence of matter, much less determine that real essence *a priori*?

This poses, in an acute way, the relation of my interpretation to that of Michael Friedman, who develops a compelling interpretation of Kant's *a priori* method for establishing A-laws and (some) B-laws in his (1992*a*). What is more, Friedman's reading delivers a unified view about the modal status of A- and B-laws; in my terminology, they are not formally necessary *sensu stricto*, but (as I explain below) the consequents of formally necessary conditionals. However, I think I can retain within my larger essentialist view these merits of Friedman's interpretation. In particular, I will argue that my view delivers (i) a unified account of the necessity of *all* laws, which (ii) explains why some laws are *a priori* knowable (A- and B-laws), and (iii) delivers a better understanding of the relation between formal necessity and nomic necessity.[50] First, I will explain what I mean by the 'constructivist' method of *MFNS* (on which I largely agree with Friedman 1992*a*), and then I will explain how this can be embedded within my essentialist interpretation of laws.

[48] The epistemic status of the inverse-square law is problematic. In several texts, Kant appears to flirt with the idea of an *a priori* proof of it (e.g., *Prol.* §38 and *MFNS*, Ak. 4: 518–21), but backs off from endorsing such a proof. See Friedman (1992*a*), ch. 4 for a detailed and, to my mind, convincing case that Kant is not endorsing a purely *a priori* derivation of the inverse-square law in these passages.

[49] Kant holds open the possibility that chemistry might become a science (in the strict sense) if chemical concepts can be "constructed" (Ak. 4: 471); I discuss this further in §8.

[50] In the Appendix on Friedman (see my website) I argue that Friedman does not deliver (i).

In the Preface to *MFNS* Kant describes the method of proof he will deploy in the rest of the book:

> But in order to make possible the application of mathematics to the doctrine of body, which only through this can become natural science, principles for the construction of the concepts that belong to the possibility of matter in general must first be introduced. (*MFNS*, Ak. 4: 472)

The key term here is 'construction,' but Kant uses that crucial term in two incompatible ways in *MFNS*. In some passages, he claims that no empirical concept (e.g., <*matter*>), nor any concept that depends upon an empirical concept (e.g., the concept <*matter in motion*>), can be constructed, because construction only occurs in pure intuition and pure intuition is independent of any experience;[51] but in others he talks about the construction of empirical concepts like <*matter*> and <*motion*>.[52] I think the solution rests in recognizing that in *MFNS* 'construction' can have two senses, which I will call 'mathematical construction' and 'metaphysical construction.'[53]

Mathematical construction. Mathematical construction is construction in pure intuition. To construct a concept is to prove that an object that instantiates that concept can be presented in intuition. Because mathematical construction is pure and *a priori*, it does not depend upon the sensory content of experience; whatever can be mathematically constructed, can be mathematically constructed regardless of how we are affected. Consequently, no empirical concept can be mathematically constructed, for our possession of empirical concepts depends upon the matter of experience. Most importantly, in the context of *MFNS*, the concept <*matter*> cannot be mathematically constructed. Nor can any concept of a force be constructed.[54] In general, no concept of an existing* thing is mathematically constructible, because our ability to represent existing* things depends upon the sensory content of experience.[55]

Metaphysical construction. Metaphysical construction, unlike mathematical construction, is always construction of an empirical concept. To construct an empirical concept is to elicit the *a priori* rules we employ in transforming perception of the object of that concept into experience of that object. To recall the discussion of

[51] Cf. *MFNS*, Ak. 4: 469, and 487–8.

[52] Cf. *MFNS*, Ak. 4: 470, 472, 517, 518, 534, and 549.

[53] Kant himself suggests this distinction when he writes: "for this purpose I have considered it necessary [to isolate] the former [empirical natural science] from the pure part of natural science (*physica generalis*), where metaphysical and mathematical constructions are customarily run together" (*MFNS*, Ak. 4: 473). It is extensively developed in Plaass (1965), 272–81; my analysis owes much to his. Buchdahl (1992), 231–5, also distinguishes two kinds of constructions. However, I would not rest my case for the distinction on this passage, but on the sense it allows us to make of *MFNS* as a whole.

[54] See Westphal (1995), §II. I have also learned a great deal from Westphal (2004)'s meticulous analysis of the arguments of *MFNS* and their relation to the *CPR*.

[55] "For laws, that is, principles of the necessity of that which belongs to the *existence* of a thing, are concerned with a concept that cannot be constructed, since existence cannot be presented *a priori* in any intuition" (*MFNS*, Ak. 4: 469). "Existence" here refers, I take it, to existence*.

Chapter 7.3, the content of a perception records merely how the object appears to the subject having the perception.[56] Experience of an object, by contrast, is objectively valid. The problem of the (metaphysical) construction of matter is the problem of determining how we transform our merely subjectively valid perceptions of matter into objectively valid experience of matter.[57] I introduced the topic of construction in *MFNS* in order to understand Kant's claim in the Preface that "principles for the construction of the concepts that belong to the possibility of matter in general must first be introduced." This is a case of metaphysical construction: Kant means that principles for the construction of the experience of bodies from perceptions of bodies must be introduced, in order for there to be a rational science of physics.[58]

My explication of the notion of metaphysical construction rests on the distinction between *perception* and *experience*. Experience is a whole of perceptions, synthesized by *a priori* rules.[59] The perceptions that constitute experience also include the perceptions of other subjects. Consider the important distinction drawn in the *Prolegomena* between judgments of perception and judgments of experience:

> All of our judgments are at first mere judgments of perception; they hold only for us, i.e., for our subject, and only afterwards do we give them a new relation, namely, to an object, and intend that the judgment should also be valid at all times for us and for everyone else; for if a judgment agrees with an object, then all judgments of the same object must also agree with one another, and hence the objective validity of a judgment of experience signifies nothing other than its necessary universal validity. (*Prol.*, §18, Ak. 4: 288)

I interpret this as a distinction between judgments of how things seem to me (judgments of perception), which may not be valid for other subjects, and intersubjectively valid judgments about how empirical objects actually are (judgments of experience). Kant defines judgments of experience here in terms of their "necessary universal validity," which I take to mean: intersubjective validity. When I make a judgment of experience, I am making a valid claim to the agreement of all other

[56] However, the perception/experience distinction functions differently in *MFNS* than it does in the *CPR*. In the *CPR* it is the distinction between empirical cognition of mere sensory manifolds (perception) and combining them into representations of publicly available existing* objects (experience—see A176, B219, A182, B225, B233, B257, and A215–16/B262–3). The objects of *MFNS* (matter and its motions) are already publicly available existing* objects, so the *MFNS* distinction is a distinction within experience in the *CPR* sense: experience of relative motion (perception, in *MFNS*) and experience of true motion (experience, in *MFNS*).

[57] Here I hew closely to the interpretation of Friedman (1992*a*).

[58] Cf. *MFNS* (Ak. 4: 510) where Kant says that the impenetrability of matter makes it possible for us to "transform" our perceptions of outer sense into the concept of an experience (*Erfahrungsbegriff*) of "a matter as an object in general" [*eine Materie als Gegenstand überhaupt*]. I take this to mean: (representation of) repulsive force makes it possible to transform our *perceptions* of matter into *experience* of determinate material objects with determinate boundaries, bodies. See later discussion for the distinction between *Materie* and *Körper* in *MFNS*. Thanks to an anonymous referee for OUP for pressing me to clarify my interpretation on this point.

[59] A110, B161, A176/B218–19, A183/B226, B218–19, A179/B222–3, A221/B269, A225/B272–3, A227/B279, A230–1/B282–4, A495/B523, and A764/B792.

cognitive subjects like me: those with discursive spatiotemporal intellects. Kant identifies this with making a judgment 'with objective validity,' which I take to mean: a judgment about how the object is represented in the unified experience composed of perceptions of all subjects. A judgment of experience makes a claim to intersubjective validity because it represents its content as part of the unified experience constituted by the lawful coherence of all subjects' perceptions.

To bring this discussion back to *MFNS*, Kant there distinguishes the *appearance* of motion from the *experience* of motion (Ak. 4: 554). The appearance of motion is merely the motions that bodies are perceived to have relative to some other body we designate as fixed. A judgment of perception of motion records the appearance of motion, or, equivalently, the content of a perception of motion. The possible *true* (as opposed to apparent) motions of bodies are the motions it is possible to *experience*; as Kant states in the Phenomenology section of *MFNS*, "matter is the movable insofar as it, as such a thing, can be an object of experience" (Ak. 4: 554). To make a judgment of experience of motion is to make a judgment about the true motions of bodies, or, equivalently, to make a judgment about how their motions are experienced rather than perceived.

Constructing the concept of the motion of bodies, therefore, is determining the *a priori* rules that make it possible for us to transform perceptions of motion into experience of motion.[60] It is equivalent to answering the question, how is it (formally) possible for us to make judgments of experience about the motions of bodies? In what sense are these rules necessary for constructing the experience of motion? They are at least *partly* grounded in our forms of experience. For example, because we are only ever given *empirical spaces*, sets of bodies we take to be fixed and relative to which we define the apparent motions of bodies, we can never experience the motion of bodies relative to *absolute space*. For any pair of 'empirical spaces' (in contemporary terms, reference frames) in uniform rectilinear motion relative to one another, it is impossible for us to experience one of them as *stationary* with respect to absolute space. It is impossible for us to make a judgment of experience that the one empirical space is absolutely at rest while the other is absolutely in motion because the alternate judgment (or the judgment that both are in uniform rectilinear motion relative to absolute space) is equally supported by the empirical evidence and our *a priori* constructive rules.[61] This familiar point about the 'equivalence of hypotheses', when embedded in Kant's epistemology, entails that absolute space is not an object of possible experience at all, hence not even an object at all (absolute space cannot be a noumenon, because "space represents no property at all of things in themselves, nor any relation of them to each other"—A26/B42), but a mere idea of reason.[62]

But these *a priori* laws are not themselves, strictly speaking, formally necessary. For example, take the laws that Kant derives in the Dynamics chapter of *MFNS*: all

[60] This means that *MFNS* is a 'transcendental' argument. Cf. Watkins (1998*a*).

[61] *MFNS*, Ak. 4: 555. [62] *MFNS*, Ak. 4: 559.

matter exerts an attractive and a repulsive force on all other matter. This law cannot be formally necessary, for if it were, it would be grounded *wholly* in our forms of experience, which apply to all outer objects as such. If the laws are grounded in our forms of experience and apply to all objects of experience as such, this means, in terms of the distinction drawn earlier, that they are *transcendental* principles not *metaphysical* principles. This would mean that they could be derived from the forms of experience without using any empirical concept, including the most general empirical concept in physics, <*matter*>. This is clearly not Kant's view for it would collapse the distinction between transcendental philosophy (*CPR*) and metaphysics of nature (*MFNS*). The *a priori* metaphysical principles of matter must be at last partly grounded in the essence of matter.

But the same argument will apply to *any* law about matter, or any empirical natural kind, for every empirical natural kind is a kind of matter. No law about any empirical natural kind is formally necessary. What are formally necessary are various conditional principles like:

(1) If matter is an object of possible experience, it has its motions relative to a family of inertial reference frames (it is not absolutely at rest or in motion).
(2) If matter is an object of possible experience, it has attractive and repulsive forces.

The A- and B-laws (the metaphysical principles of matter and the laws of rational physics) are the consequents of such conditionals, that is:

(1c) Matter has its motions relative to a family of inertial reference frames (it is not absolutely at rest or in motion).
(2c) Matter has attractive and repulsive forces.

The tension between Kant's constructive method and his essentialist conclusion comes down to this: what warrants Kant in claiming that the consequents of these conditionals ((1c) and (2c)) are essential to matter?

To take a concrete instance, Kant argues that:

(FC) It is formally necessary that if matter is an object of possible experience it has attractive and repulsive forces. [Formal necessitation of conditional (2).]

Kant's argument, in brief, is that we can only experience matter if we experience it as extended, and this requires experiencing matter that has a determinate location. This in turn requires experiencing matter that occupies space by repelling other matter that attempts to move into its space; the other ways that matter could occupy [*einnehmen*] space (through absolute hardness, or by mere presence in space) would violate our forms of experience (absolute hardness)[63] or make it impossible

[63] *MFNS*, Ak. 4: 502.

for us to experience determinate location (mere presence in space).[64] If matter has only repulsive force it will become arbitrarily dispersed as time elapses (its quantity in a bounded space will become arbitrarily close to zero after a sufficient duration), and for any finite amount of time, more than that amount of time has dispersed (there is no first point in time, though there is no 'completed' infinity of past times);[65] so if matter had only repulsive forces, for any finite volume, it would be dispersed to a density arbitrarily close to zero within that volume, making it impossible for us to perceive and hence to experience matter (for to perceive matter we need to perceive it within some finite volume).[66]

But Kant does not merely draw the conclusion that it is formally necessary that if we can experience matter then it has attractive and repulsive forces; he concludes that these forces are *essential* to matter:

> A property on which the inner possibility of a thing rests, as a condition, is an essential element thereof. Hence repulsive force belongs to the essence of matter just as much as attractive force, and neither can be separated from the other in the concept of matter. (*MFNS*, Ak. 4: 511)

In case there was any doubt that he has 'real essence' in mind here, the attractive force of matter is precisely his example, in the L_2 metaphysics lectures, of a really essential property of a thing not contained in the logical essence of its concept: "the first ground of all predicates thus lies in a concept; but that is not yet a real essence. E.g., that bodies attract belongs to the essence of things, although it does not lie in the concept of body" (*ML_2*, Ak. 28: 553).[67] Let us just grant Kant that his constructive procedure can prove (FC). What warrants him in concluding that:

(EC) Matter essentially possesses attractive and repulsive forces.

[64] What I have in mind by 'mere presence' in space is the view that there is no constitutive connection between matter's location (the space it occupies) and a force to repel the motion of other matter into that location; matter is just primitively located. Kant dismisses such a view because it would render it impossible for us to experience the determinate locations of matter: if matter can be located in a space without exerting force in that space then any body could be located anywhere in space, for all we know by experience.

[65] Cf. Ch. 7.4 for a brief discussion of Kant's view that there is no such object as the complete past before a given time.

[66] This is my (highly condensed) reconstruction of the proofs of Propositions 1, 2, and 5 in the Dynamics chapter of *MFNS* (Ak. 4: 497–500, 508–9).

[67] My interpretation differs sharply from that of Hanna (2006), 140–90. Hanna does not distinguish Kant's different notions of real modality. He assumes that *if* a scientific essentialist claim like 'gold is the element with atomic number 78' is necessarily true, it is (in my terminology) formally necessary. However, on Kant's view, this claim is formally contingent but nomically necessary: if gold is in fact the element with atomic number 78, then this is part of its essence, and, although we cannot experience gold that has a different atomic number, this fact is grounded in the essence of gold, not in our forms of experience. Hanna's argument in the last section of that chapter—that there are experienceable worlds that involve gold but have different laws, in which gold might have a different atomic number—is beside the point. That shows the formal contingency of the fact that gold has atomic number 78; it does not address its modal status with respect to nomic necessity. Given that, as I have argued, laws, for Kant, are grounded in real essences of empirical natural kinds, in a world with different laws different empirical natural kinds are instantiated.

The relation between (FC) and (EC) is this: (FC) is a formally necessitated conditional; (EC) is the claim that the consequent of that conditional is essential to matter. That consequent—matter has attractive and repulsive forces—is the A-law in question. So my question can be restated: how do we go from the formal necessitation of the conditional principle to the claim that the A-law (the consequent of that conditional principle) is essential to matter?

8.6. From Forms to Essences

At the end of the last section I raised a problem whose general form is this: suppose we grant Kant that his constructive procedure can prove that:

> (FC*) It is formally necessary that (if matter is an object of possible experience, matter is ψ).

What warrants him in concluding that:

> (EC*) It is essential to matter to be ψ.

I argued in section 3 that the real essence of matter is the complex of properties in virtue of which, for anything that is an instance of matter, that thing is matter (see Fig. 8.3).

Let us assume, for the sake of argument, that being an object of a possible outer experience (an outer phenomenon) is part of the real essence of matter (matter is essentially the object of a possible experience), as shown in Figure 8.4.

In Chapter 7.3 I discussed the crucial claim of Kant's transcendental idealism: "the *a priori* conditions of a possible experience in general are at the same time conditions of the possibility of the objects of experience" (A111). I argued there that this means, where x is any possible object of experience, that conformity to the forms of experience is part of what makes x a possible object of experience; in the terminology of this chapter, conforming to the forms of experience is part of the *real essence* of being an object of possible experience. This is expressed in Figure 8.5.

x is matter

↑ grounds

x possesses [P1 + P2 + P3+ . . .]
(the real essence of matter)

Fig. 8.3. Real essence of matter

x is matter

↑ grounds

x possesses [possible outer object + . . .]

Fig. 8.4. Real essence of matter

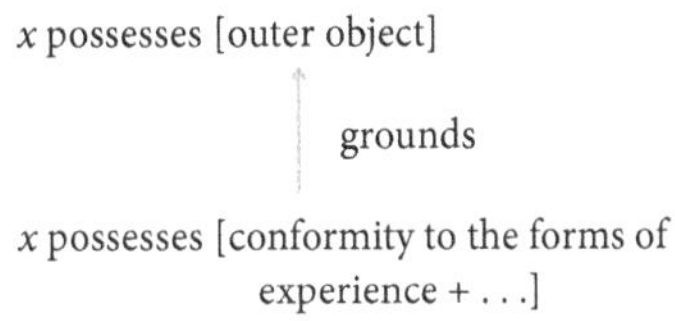

Fig. 8.5. Conformity to forms of experience

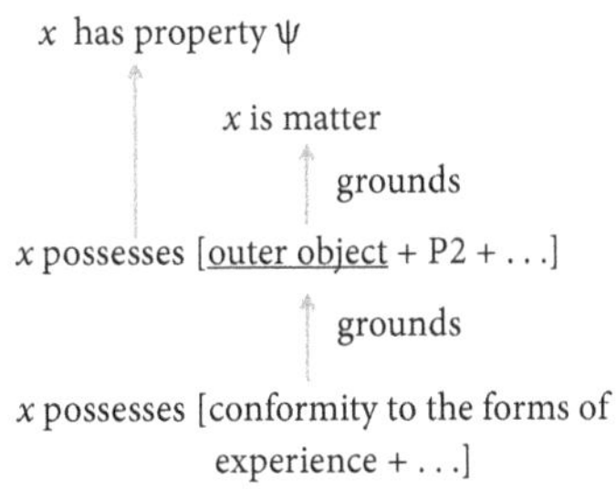

Fig. 8.6. Conformity to forms of experience

What does 'conformity to the forms of experience' mean? To return to a theme from Chapter 7.3, the forms of experience ground a certain content in experience, and this content grounds the properties of objects of possible experience; that x conforms to the forms of experience means that x has certain formally necessary properties, and it has them in virtue of the forms of experience. The possession by x of those properties is grounded in the forms of experience.

If we put these pieces together, x is matter partly in virtue of being a possible object of outer experience (Fig. 8.4), which it is partly in virtue of conforming to the forms of experience (Fig. 8.5). This is shown in Figure 8.6.

Now the conclusion of Kant's constructive argument is (FC*): in virtue of the forms of experience, if matter is the object of a possible experience it has property ψ. But I have argued that being an object of possible experience, and hence conforming to the forms of experience, is part of the essence of matter, part of what makes x matter. So the essential properties that ground x being matter also suffice to ground it being ψ, in other words, being ψ is grounded in the essence of matter, or, more simply, matter is essentially ψ.

To make this slightly more concrete, consider any possible object that is matter (any possible body).[68] It is matter partly in virtue of being a possible object of outer experience, and it is a possible object of outer experience in virtue of conforming to the forms of experience. So it is essentially a possible object of experience conforming to the forms of experience (though that is not its whole essence). Now we know from

[68] The relationship between the concept <*body*> and the concept <*matter*> in *MFNS* is not entirely clear; Kant writes that "a body, in the physical sense, is a matter between determinate boundaries" (Ak. 4: 524). I take this to mean that *body* is a count noun, while *matter* is a mass term. All bodies instantiate the concept <*matter*> because they are made of matter, but <*body*> adds more determinate content than is contained in the concept <*matter*>, namely, that bodies have determinate boundaries.

(FC*) that, in virtue of the forms of experience, any matter we can experience has property ψ (e.g., has attractive and repulsive forces). So it has property ψ (e.g., has attractive and repulsive forces) in virtue of properties that are essential to it. So it is essential to matter to have property ψ.[69]

To return to and, finally, answer our initial question: how can we know *a priori* that it is nomically necessary that matter has attractive and repulsive forces? Conformity to the forms of experience is part of the essence of matter (it is part of the essence of any kind of outer object), and that matter has these forces is a formally necessary consequence of our being able to experience it, and both of these claims are *a priori* knowable: the first is a consequence of the *a priori* principle that all kinds of outer objects are essentially kinds of objects of possible experience, and the second follows from the *a priori* argument sketched above.[70] The apparent tension between Kant's constructive procedure in *MFNS* and his essentialist conclusions is thereby resolved.[71]

8.7. Nomic Possibility as Real Possibility

In section 4 I argued for a definition of nomic necessity:

> (*Nomic necessity*) It is *nomically necessary* that p if and only if p is grounded in the essences of actual empirical natural kinds.

Although I did not pause to unpack this notion there, it is important to note that the nomic necessities are the truths grounded in the totality of facts about the essences of empirical natural kinds. For instance, if it is not essential to kind ϕ to have property ψ, but it is essential to kind ϕ^* that it cause samples of kind ϕ to have property ψ, we may not want to say that it is a law that ϕ has ψ (because this fact is not grounded in the essence of ϕ itself) but we will want to say that it is nomically necessary, that it 'holds with the necessity of law' (given the empirical natural kinds there actually are). We will also want to include among the nomic necessities, the logical consequences of nomic necessities. For instance, if it is nomically necessary that p and nomically necessary that q, it is, intuitively, nomically necessary that $p\&q$ even if we would not want to count it as a law per se; this holds with the necessity of law, for if it were false, at least one law would be violated. I propose, therefore, the following emendation to our definition of nomic necessity from earlier:

[69] Does this argument entail that it is formally necessary that matter is ψ? No, because it is not grounded in the forms of experience alone that matter is an object of possible experience; that is partly grounded in the essence of matter, which is not exhausted by 'possible outer object' (see Fig. 8.4).

[70] Note that I am not claiming it is part of the essence of x itself to be matter; although I do think that Kant accepts *de re* necessities like this, I am remaining neutral on that issue here.

[71] Buchdahl (1992), 233, notes that *MFNS* takes its orientation from the construction of the concept of matter rather than from claims about its real essence, but fails to note that Kant takes the former to justify the latter.

(*Nomic necessity*) It is *nomically necessary* that p if and only if p is a member of the set S, where S contains all of the facts about the essences of the actual empirical natural kinds there are, and S is closed under logical entailment (i.e. if p is in S, and p logically entails q, then q is in S) and under grounding (i.e. if p is in S, and p grounds q, then q is in S).

Given the interdefinability of necessity and possibility, this corresponds to the following definition of nomic *possibility*:

(*Nomic possibility**) It is *nomically possible* that p if and only if $\neg p$ is not a member of the set S, where S is as above.

This is not exactly right, though, for this definition will entail that, if there *are* no empirical natural kinds, then everything is nomically possible. But that, intuitively, is wrong; if there are no natural kinds then there is *no such thing* as nomic possibility. So I will define nomic possibility instead as:

(*Nomic possibility**) It is *nomically possible* that p if and only if $\neg p$ is not a member of the set S, where S is as above and S is non-empty.

For the most part, though, I will ignore these complications and regard nomic possibilities as propositions whose negations are not grounded in the essences of actual empirical natural kinds.

It should be clear that nomic possibility, so defined, satisfies the two conditions on real possibility from Chapter 7.2:

(*Real Possibility*) Nomic possibility $\Diamond_{\text{No}}p$ (and its associated kind of necessity $\Box_{\text{No}}p$, where $\Box_{\text{No}}p \leftrightarrow \neg\Diamond_{\text{No}}\neg p$) is a kind of *real possibility* (and $\Box_{\text{No}}p$ is a kind of *real necessity*) only if
(i) *Non-logicality*: it is not a conceptual truth that $\Diamond_{\text{L}}p \supset \Diamond_{\text{No}}p$ (equivalently, it is not a conceptual truth that $\Box_{\text{No}}p \supset \Box_{\text{L}}p$), and
(ii) *Groundedness*: if $\Diamond_{\text{No}}p$ then the fact that $\Diamond_{\text{No}}p$ has a real ground in some actual object or principle.

For obvious reasons, it is not a conceptual truth that every nomic necessity is logically necessary; as Kant points out, attractive force is not contained in the concept <*matter*> but it is essential to matter. It is likewise clear that the grounding relation that holds between the essences of empirical natural kinds and laws is not, in general, a *logical* one. In section 6 I argued that the synthetic, logically contingent *a priori* formal principles of experience, provide, in some cases, the relation between the real essence of matter and its nomically necessary properties (e.g., attractive force); the real essence of matter grounds the fact that matter has attractive force, but it does not logically entail it. Nor is the relation a causal one. The essence of matter does not *cause* these laws to obtain; the essence of matter grounds the fact that matter has attractive and repulsive forces with certain properties, and the operations of these forces cause motions. Kant needs this grounding relation to draw the distinction between essential properties *sensu stricto* (properties that are parts of real essences) and attributes (properties grounded in real essences).[72] As I pointed out earlier, it is not plausible that *all* laws state properties *contained* in an essence; it not plausible, for instance, that every complex mathematical theorem of Newtonian dynamics is contained in the essence of matter.

One constraint this grounding relation must meet is that the attributes be 'just as necessary' as the essential properties in which they are grounded. Readers might well wonder what it could mean for something to be 'just as necessary as', or more or less necessary, than something else.[73] Intuitively, the idea is this: we do not want it to be the case that the attributes are grounded in the essence but *might not have been*. This is not a problem if the grounding relation is logical grounding (logical relations hold with logical necessity, and logical necessities are true under any counterfactual supposition), but it would be if the grounding relation here were *causal* grounding (the essence might not have its actual effect). Since the grounding relation involved, I have argued, is non-logical non-causal grounding, we need an account of why this problem does not arise. The natural answer is that the grounding relation is itself *essential* to the natural kind in question: if it is essential to ϕ to stand in a grounding relation to ψ then there is no worry that, in some counterfactual situation (some other possible world), ϕ might lack ψ.

This would be the case if the constructive procedure described above were true of all laws as such. Assume that all laws have the form ϕ is ψ, where ϕ is an empirical natural kind and ψ is a formally necessary condition on experiencing objects of kind ϕ. Since the forms of experience are essential to ϕ (ϕ is a kind of matter, and the forms of experience are essential to matter), it follows, by the argument of the previous paragraph, that the grounding relation between ϕ and ψ is essential to matter. What is more, this would allow us to give a unified account of the grounding relation involved in nomic necessity. In Chapter 7.3 I explored the relation between two basic grounding relations in Kant's theory of experience: the content-content grounding (between the forms of experience and the content of experience) and content-property grounding (between the content of experience and the properties of phenomena). We might be able to analyze the grounding relation involved in nomic necessity as the product of these two grounding relations. If, as we have assumed, the forms of experience make it the case that if we experience ϕ we experience it as having ψ, then there is a content-content grounding relation between these contents: experiencing the content ϕ grounds experience of the content ψ. Likewise, there is a content-property grounding relation: since the objects in question are phenomena, our experience of them as ψ grounds their possessing that property (e.g., having their

[72] See the lectures on metaphysics: *MV* (Ak. 28: 411), *MvS* (Ak. 28: 492–3), *ML*$_2$ (Ak. 28: 552–3), *MD* (Ak. 28: 629), and *MM* (Ak. 29: 820).

[73] For a sophisticated account of 'degrees' of necessity, see Lange (2009).

motions relative to a family of equivalent inertial frames). Combining these two grounding relations, experiencing objects as ϕ grounds their possessing ψ; since ϕs are phenomena, they are essentially objects of possible experience, so their possession of ψ is grounded in the essence of ϕ. This would bring a satisfying degree of unity to Kant's theory of real grounds, but it would not constitute a reductive analysis of this real grounding relation.

I have been investigating the possibility that the real grounding relation in the definition of nomic necessity could, in general, be modeled on the real grounding relations involved in formal necessity, effectively generalizing this constructive procedure to all laws as such. There is reason, though, to be skeptical that this is Kant's view. First of all, in *MFNS* Kant adopts a skeptical attitude to the possibility that chemistry could ever be a rational science, which means: the possibility that the constructive procedure could ever be applied to chemical kinds.[74] He does not express skepticism that there are chemical *laws*, only that they could be discovered *a priori* through a constructive-mathematical operation like the one by which we discover the *a priori* laws of motion. This suggests that Kant thinks that the laws of nature may outstrip, in principle, all possible *a priori* constructions. At a more general level, Kant is committed to the idea that Nature forms a system of higher- and lower-level empirical (kind) concepts and laws[75] and this commitment appears independent of his commitment to the *a priori* constructive procedure for discovering A- and B-laws in *MFNS*. For reasons of space, though, I cannot discuss the real grounding relation in any more detail here.[76]

8.8. Back to the Beginning

I began this chapter by quoting certain texts in which Kant claims that it is formally necessary that there are laws, but not formally necessary which laws there are. But we are now in a position to understand some additional claims Kant makes in these passages, which might otherwise have been obscure. In the very same passages in the third *Critique* in which Kant claims that it is formally necessary that there are laws, he typically inserts a note of caution. For instance:

> [Laws] must be regarded as necessary on a principle of the unity of the manifold, even if that principle is unknown to us. (*CJ*, Ak. 5: 180)
>
> [...] given the constitution and the limits of our faculties of cognition we have no insight at all into [*gar nicht einsehen*] this necessity [of laws]. (*CJ*, Ak. 5: 183)
>
> [The understanding] does not and never can cognize [laws'] necessity. (*CJ*, Ak. 5: 184)

[74] *MFNS*, Ak. 4: 471.

[75] See the Appendix to the Transcendental Dialectic (A642/B670–A668/B696) and the two introductions to the *CJ*.

[76] Readers interested in further discussion can see the article "Further issues in grounding and nomic necessity" on my website (see the Notes on the Text).

Now we can understand Kant's claims in these passages: even if we cognize a law we never have *complete* insight into why it obtains because we never know the *complete* real essence in virtue of which it obtains.[77] As Kant puts it in the L_2 lectures: "the real essence of things is inscrutable to us, although we cognize many essential aspects" (*ML*$_2$, Ak. 28: 553).[78] This will be especially true of C- and D-laws, which are known *a posteriori* if known at all (assuming the constructive procedure does not generalize): even if it is a law that *p* and we know that *p*, we will typically not know what it is about the relevant natural kind in virtue of which *p* obtains. We will lack insight into the necessity of this law.

I have focused mainly on A- and B-laws in this chapter. But in conclusion I want to make a few remarks about C- and D-laws and how they fit into my interpretation. C-laws are defined as laws that are not *a priori* consequences of A- or B-laws. If there are such laws, they are not *a priori* because they are not grounded in the *a priori* cognizable components of the essence of matter; if the *a priori* constructive procedure generalizes to all laws (all laws can be proved from the constructive procedure necessary to experience objects of the relevant kind), then there *are* no C-laws.[79] The clearest candidate for a C-law is the law that the attractive force of matter varies inversely with the square of the distance; Kant's attitude towards the inverse-square law is somewhat complicated, but if he regards it as *a posteriori* then it is a C-law. If the inverse-square law is *a posteriori* then it is a C-law grounded in the real essence of matter that cannot be derived *a priori* from A- and B-laws.

Kant's views about D-laws are somewhat more complicated, because they are tied up with a complex set of doctrines about the systematicity of Nature, the regulative use of reason, and reflecting judgment; a complete articulation of his views on these topics would require extensive discussion of the Appendix to the Transcendental Dialectic in the *CPR* and (at least) the two Introductions to the *CJ*. All I can do here is indicate how my essentialist interpretation of nomic necessity fits into those issues. D-laws, remember, are laws about empirical natural kinds more specific than matter. I am unsure whether it is formally necessary that there be such laws; the formally necessary principle that all alterations in substances be governed by universal laws may be satisfied if all the laws applied to matter as such (A-, B-, and C-laws). However, in scientific inquiry we are bound by the rational demand to find systematic unity in nature, guided by the idea of a *system* of nature in which the highest

[77] Gibson (2011) claims we have no 'insight' into the natures of empirical objects (15). If 'insight' [*Einsicht*] is taken in its strict technical sense of *a priori*$_G$ knowledge, this is correct, for *insight* into essences would require knowing the whole essence and how it grounds essential properties. However, this is compatible with our knowing many of the essential properties of generic objects (e.g., matter), a point unappreciated by Gibson. I also think Gibson is wrong to claim that for Kant we can only know *de dicto* necessities of the form *necessarily all phenomena are F* but no *de re* necessities, e.g., *if x is a phenomenon, then x is necessarily F*; however, I do not have the space here to argue for that claim.

[78] Cf. *MM*, Ak. 29: 821 ("we can cognize much that belongs [to the real essence], but not all").

[79] There might still be D-laws: laws grounded in the formally necessary conditions for experiencing specific kinds of matter (e.g., chemical laws).

empirical concept, matter, is divided into *species*, which I take to be lower-level natural kinds, which themselves have further species, etc. It follows from my essentialist interpretation of nomic necessity that each of these species will have a specific difference from its genus, something in its essence that distinguishes it from the essence of the genus (e.g., metals will essentially be a specific kind of matter), and these specific differences will ground lower-level laws that apply to those more specific natural kinds. So the idea of a system of species (natural kinds) is equivalent to the idea of a system of laws. However, our forms of experience do not determine which such laws will obtain. In fact, our forms of experience leave it open whether the system of laws is so vast and complicated that we could never comprehend even a single one of them; it further leaves open whether the manifest similarities between objects correspond to the underlying system of natural kinds in any way discoverable by us. Consequently, empirical inquiry, whose goal is to infer the underlying system of natural kinds and laws from their manifest properties, is only possible if we reflect upon nature as the product of an intelligent cause that has ordered its kinds and laws in such a fashion as to be comprehensible to us. This does not give us the right to claim that there is such an intelligent cause, but rationality requires and enables us to investigate Nature as though there is one.

Since one of the themes of this book is the continuity of Kant's theory of modality from *Beweisgrund* through the *CPR*, I want to mention briefly, before concluding this chapter, that this essentialist view of laws, which I have argued Kant holds in the Critical period, was also his view in *Beweisgrund*:

> On this basis it can be seen that the laws of the motion of matter are absolutely necessary. That is to say, if the possibility of matter is presupposed it would be self-contradictory to suppose it operating in accordance with other laws. This is a logical necessity of the highest kind. (*OPG*, Ak. 2: 100)[80]

We should not let his pre-Critical claim that the possibility (essence) of matter logically entails its laws obscure the more important point. Kant in both the pre-Critical writings and the Critical period takes the essence of matter to ground the laws of its motions, although in the Critical period he takes this to be a real, rather than a logical, grounding relation.

[80] See also *OPG*, Ak. 2: 96, 99, and 131. See Insole (2011) for critical discussion.

9

The Unity of Kant's Modal Metaphysics

9.1. Introduction

Up to this point I have explored Kant's rejection of the logicist metaphysics of Leibniz, Wolff, and Baumgarten, his own pre-Critical theory of the grounding of real possibility in God, his Critical reasons for denying that any such transcendentally realist metaphysical theory could constitute knowledge, as well as his own distinction among various kinds of modality in the *Critique of Pure Reason* (*CPR*) and other works written in the 1780s (most importantly, *Metaphysical Foundations of Natural Science*). This may give the inaccurate impression that Kant's Critical modal metaphysics is a mere aggregate of different modal notions, and, furthermore, that it lacks any substantive connection to his pre-Critical theory, other than his consistent rejection of logicism. In this chapter I argue, on the contrary, that Kant's modal metaphysics is tightly unified, both historically and systematically. First, it is systematically unified because all of these different kinds of real possibility exhibit a common scheme in virtue of which they are kinds of real possibility. Some of this was sketched earlier, in Chapter 7; in this chapter I go into more detail about Kant's unified conception of real possibility in the Critical system. Second, it is historically unified because the content of Kant's pre-Critical theory of God as the ground of real possibility is retained within the Critical philosophy, while its epistemic status is changed; what was once the conclusion of a philosophical demonstration is now an article of rationally justified belief [*Glaube*] that falls short of cognition [*Erkenntnis*] or knowledge [*Wissen*].

The jumping-off point for my argument in this chapter is this passage in the Pölitz transcripts of Kant's rational theology lectures from the mid-1780s:[1]

> On this point rests the only possible ground of proof for my demonstration of God's existence, which was discussed in detail in a work I published some years ago. Here it was shown that of all possible proofs, the one which affords us the most satisfaction is the argument that if we remove an original being, we at the same time remove the substratum of the possibility of all

[1] Kant lectured on rational theology during the winter semesters 1783–4 and 1785–6, and it is unclear which series of lectures is the basis of the Pölitz transcripts. See Kant (1996*b*), 337–8 for more.

things.—But even this proof is not apodictically certain; for it cannot establish the objective necessity of an original being, but establishes only the subjective necessity of assuming [*annehmen*] such a being. But this proof can in no way be refuted, because it has its ground in the nature of human reason. For my reason makes it absolutely necessary for me to assume a being which is the ground of everything possible, because otherwise I would be unable to know what in general the possibility of something consists in [*worin etwas möglich sey*]. (*Pöl.RT*, Ak. 28: 1034)[2]

The "work published some years ago" is, of course, *Beweisgrund*. In this passage, Kant appears to be saying that the 1763 argument does not constitute a *proof* or even a *ground of proof*[3] of the existence of a first ground of real possibility. I take this to mean that by following the steps of the 1763 argument and concluding on that basis that there is a first ground of real possibility we do not thereby come to *know* that there is such a being. Nonetheless, it is rationally necessary for us to postulate the existence of such a being that grounds all real possibility and to postulate it for the very reasons given in *Beweisgrund*. Whereas in 1763 Kant argued that, since there are real possibilities, there must exist a ground of all real possibility, he now thinks that the fact that something is really possible rationally requires us to seek an explanation of that fact, and the only rationally satisfying explanation is that there is a ground of all real possibilities, so we are rationally necessitated to accept—I want to remain neutral for the moment on exactly what 'pro' attitude we are supposed to adopt towards this proposition so I am going to use the non-Kantian term 'accept'[4]—that there is such a being.

This passage raises a number of questions, the most important of which is: what is the nature and source of the rational necessity that we postulate such an original being? Before we go any further, though, I want to point out that this is not the only necessity invoked in this passage. Kant very quickly adds another necessity, the necessity of the being thereby posited:

For in addition to the logical concept of the necessity of a thing (where something is said to be absolutely necessary if its non-being [*Nichtseyn*] would be a contradiction, and consequently impossible), we have yet another rational concept of real necessity. This is where a thing is *eo ipso* necessary if its non-being [*Nichtseyn*] would remove all possibility. Of course in the logical sense possibility always precedes actuality [*Wirklichkeit*], and here I can think the possibility of a thing without actuality [*Wirklichkeit*]. Yet we have no concept of real possibility except through existence [*Existenz*], and in the case of every possibility which we think *realiter* we always presuppose some existence [*Daseyn*]; if not the actuality of the thing itself, then at least an actuality in general [*Wirklichkeit überhaupt*] which contains the data for everything possible. Hence every possibility presupposes something actually given [*wirklich gegebenes*], since if everything were merely possible, then the possible itself would have no ground; so this

[2] Cf. *Volk.RT* (Ak. 28: 1176) and *Danz.RT* (Ak. 28: 1259). The translation of the Pölitz text is from Kant (1996*b*), with minor modifications.

[3] See Chapter 5.3 for the difference between *Beweis* and *Beweisgrund*.

[4] In order to distinguish between 'accepting' (my neutral term) and Kant's technical term *Fürwahrhalten*, I translate the latter as 'holding to be true.' I discuss *Fürwahrhalten* and its different grades in §5.

ground of possibility cannot itself be merely possibly, but must be actually given [*wirklich gegeben*]. But it must be noted that only the subjective necessity of such a being is thereby established, i.e. that our speculative reason sees itself necessitated to presuppose this being if it wants to have insight at all into *why* something is possible [*wenn sie überhaupt einsehen will, warum etwas möglich sey*], but the objective necessity of such a thing can by no means be demonstrated in this matter. (*Pöl.RT*, Ak. 28: 1036)[5]

Kant attributes necessity not just to the postulation of an 'original being' but to the content of that postulate: it is rationally necessary that we postulate an absolutely necessary being. He says we have a logical concept of necessary existence—a being exists necessarily if and only if the proposition that it does not exist entails a contradiction—which, of course, is the conception of necessary existence that formed the basis of the ontological argument in Leibniz, Wolff, and Baumgarten. Kant rejects this conception of necessary existence because it requires the (according to him) false assumption that existence is a determination or 'real predicate.'[6]

I want to focus on what Kant calls the 'other' concept of absolutely necessary existence: a being exists absolutely necessarily just in case that being exists and, were it not to exist, nothing would be really possible. This is the very same definition of absolutely necessary existence Kant gave in *Beweisgrund* and which I analyzed in some detail in Chapter 5. He also introduces here a concept of 'real' possibility and a principle governing real possibility—if it is really possible that p then the fact that it is really possible that p is grounded in something that actually exists—that are familiar from that pre-Critical work and our earlier discussion of it.

Together, these passages from the Pölitz theology lectures raise a number of interesting questions:

(1) What does Kant mean when he says "in the case of every possibility which we think *realiter* we always presuppose some existence"?
(2) What is the concept of necessary existence that Kant claims we have (in the second passage)?
(3) Why is it rationally necessary that we postulate such a necessary being?
(4) What is the epistemic status of this postulate? How and why is this epistemic status different than the epistemic status of the conclusion of the 1763 argument?

In sections 2–4 I answer the first and second question by showing the unity of Kant's Critical theory of modality and how this concept of necessity fits into it. In section 4 I begin to answer the second two questions by examining Kant's discussion of the modes of 'holding to be true' [*Fürwahrhalten*] in the 'Canon of pure reason' section of the *CPR*. I argue for a category of 'holding to be true' that I call 'necessary

[5] This crucial text is also cited and discussed in Adams (1994), 182, n. 9; and Chignell (2007*b*), 349.
[6] See Ch. 1.6.

theoretical beliefs.' A necessary theoretical belief is a belief in a postulate that is necessary to satisfy theoretical reason's need for explanatory unity, but which cannot be cognized by us. The postulation of an absolutely necessary being that grounds real possibility is just such a necessary theoretical belief. It constitutes the only rationally satisfactory explanation of why anything at all is really possible, and thus is the only way to answer theoretical reason's demand for an explanation of this fact. I conclude the chapter by returning to the Pölitz passage and explaining in detail what it means.

9.2. Categories and Real Definitions

The first thing to note about the Pölitz passage quoted above is that these modal concepts—absolutely necessary existence and real possibility—are being applied beyond the range of experience—we are rationally necessitated to postulate a supersensible object that is absolutely necessary.[7] This involves using the modal category <*necessity*> in thinking of a supersensible object, a noumenon. But what is the content of a modal category when applied to noumena? This is an instance of a larger question in Kant's philosophy: what (if anything) is the content of a category when it is applied to noumena?[8] In this section I try to answer that question.

Although we can (and must) think of all objects under the categories,[9] there is an important difference in the epistemic status of these thoughts for objects we can intuit (phenomena) and those we cannot intuit (noumena).[10] We can know *a priori* that it is really possible for phenomena to fall under the categories; this, I argued in Chapter 6, is what it means that phenomena can be *cognized a priori* using the categories, or, equivalently, that the categories have *a priori objective reality* with respect to phenomena. Cognition [*Erkenntnis*] is distinct from, but necessary for, knowledge [*Wissen*]: *a priori* knowledge about objects in some domain (e.g., noumena) using a set of concepts (e.g., the categories) requires that we can come to know *a priori* the real possibility that objects in that domain instantiate those concepts (that those concepts are cognitions for us with respect to that domain). Since we cannot think about any objects without using the categories,[11] if the

[7] Some readers might object that the necessary being hypothesized at Ak. 28: 1034–6 is not supersensible; this being is merely the 'totality of experience' that grounds all possibility according to Kant at A581–2/B609–10. However, that claim occurs in the context of Kant's discussion of the idea that God is necessary as the ground of the complete determination of all objects. The *Beweisgrund* argument does not concern complete determination, so Kant's discussion in the Pölitz lectures, which refers back to that text, is not about the idea of complete determination. See Stang (2012) for a discussion of the complete determination principle in the *CPR*.

[8] I am using 'content' here in a very minimal sense (roughly, the meaning or intension of the category), not in Kant's technical sense: its relation to a (knowably really possible) object. In Kant's technical sense, categories do not have content (*Inhalt*) for noumena. See Ch. 6 and Tolley (2011) for further discussion.

[9] A253, *CPrR* (Ak. 5: 54, 103).

[10] On the 'one object' interpretation, empirical and non-empirical standpoints on objects.

[11] At B110 Kant distinguishes between the mathematical categories (of quantity and quality) and the dynamical categories (of relation and modality), a distinction to which he returns at A160/B199. The

categories lack objective validity for some domain of objects, then we cannot cognize those objects, which entails that we cannot know anything about them, aside from logical truths and perhaps some negative facts about them (e.g., that they are not in space and time).[12] Categories *as applied to phenomena* have objective validity because they can be *schematized*, i.e. provided with *a priori* rules for application to objects given in experience. I will refer to them as 'schematized categories.' Categories, as concepts of objects in general (phenomena and noumena), cannot be schematized; I will sometimes refer to them as *unschematized* categories. The question from earlier can be rephrased: what are we doing when we think of an object under the unschematized categories <*necessary*> or <*possible*>? What is the content of such thoughts?

Before going any further in trying to answer *that* question, I want to investigate further what is lacking or defective about unschematized categories. In the 'Phenomena and noumena' section, Kant makes the familiar point that unschematized categories do not have objective validity for noumena because we cannot prove the real possibility of noumena falling under them.[13] However, he goes on in the next paragraph to suggest that unschematized categories—the categories as applied to *all* objects in general, both phenomena and noumena—are defective for cognitive use by us for a slightly different reason:

> That this is also the case with all categories [that we cannot use them to cognize anything about objects in general] however, and the principles spun out from them, is also obvious from this: that we cannot even give a <u>real</u> definition of a single one of them, <u>i.e. make intelligible the concept of their object</u>, without immediately descending to conditions of sensibility, thus to the form of the appearances, to which, as their sole objects, they must consequently be limited, since, if one removes this condition, all significance, i.e. relation to the object, disappears, and one cannot grasp through an example what sort of thing was really intended by concepts of that sort. (A241/B300 – underlined material added in B edition)

distinction seems to be that in thinking about objects of mere intuition (mere appearances in the terminology developed above, e.g., objects constructed in pure intuition) we must use the mathematical categories but not the dynamical categories, but we must represent objects of full-blooded experience using both sets of categories. So it is not the case that we must use *all* moments of the table of categories in thinking about every object. However, Kant's reasons for thinking that we must use dynamical categories in thinking about certain objects is that those categories "concern the existence of the objects of a possible empirical intuition" (A160/B199). I take this to mean that we must use dynamical categories in thinking of these objects because they exist independently of any particular intuition of them (unlike objects constructed in pure intuition). However, in thinking about noumena we are also thinking about objects that exist independently of our thinking about them, so we must also use dynamical categories. So the point in the main text holds.

[12] Kant also makes at least one positive claim about noumena: that they are the causal ground of the matter of experience (see Ch. 7). However, I will not attempt in this study to reconcile that doctrine with Kant's doctrine of noumenal ignorance.

[13] A230/B298. The full title of this section is "On the ground of the distinction of all objects in general [*überhaupt*] into phenomena and noumena."

Kant here identifies the lack of objective validity of the categories in general with our inability to give real definitions of them for all objects in general (phenomena and noumena). He identifies a real definition of a category as making 'intelligible' [*verständlich*] to ourselves how it is really possible that objects instantiate it; this is most naturally read as meaning that a real definition of a concept reveals not merely *that* it is really possible for objects to instantiate it, but reveals or makes comprehensible to us *why* this is really possible. In his own copy of the A edition, Kant wrote in the margin of this passage: "we cannot explain their [the categories'] possibility." Giving a real definition of a concept is proving that it is really possible that an object instantiates that concept by eliciting the *grounds* of that real possibility and thereby explaining that real possibility. When we really define a concept we come to know what it is *in virtue of which* objects of that concept are really possible.

In Chapter 8 I discussed the distinction between logical and real definition (and the correlative distinction between logical and real essence) in the case of empirical concepts, but this distinction also applies to *a priori* concepts. A nominal definition of an *a priori* concept reveals what we 'think' in the concept, the marks contained in it, or what Kant sometimes calls the 'logical essence of the concept.' A real definition, by contrast, reveals the essential inner marks of the object that make it possible and that explain its other properties. A real definition would reveal not only *that* the object of the concept is possible but *why*. For instance, a real definition of a mathematical concept displays the grounds of the real possibility of such an object, and paradigmatically takes the form of a construction of the object in pure intuition. As with empirical concepts, Kant is quite skeptical about our ability to give real definitions of 'given' *a priori* concepts, concepts given by the nature of the mind itself, like the categories, as opposed to *made a priori* concepts like mathematical concepts, which we arbitrarily define. Kant allows that we may be able to analyze given *a priori* concepts up to a point and determine some of the marks contained in them, but, in general, we will not be able to analyze them fully and give them complete *nominal* definitions.

By denying that we can give real definitions of the categories for all objects *überhaupt*, Kant is denying that we can know the grounds of the real possibility of objects in general falling under the categories. In the A edition, Kant expands upon the lack of a real definition of the categories as concepts of objects in general:

> Above, in the presentation of the table of the categories, we spared ourselves the definitions of each of them, on the ground that our aim, which pertains solely to their synthetic use, does not make that necessary, and one must not make oneself responsible for unnecessary undertakings that one can spare oneself. [...] But now it turns out that the ground of this precaution lies even deeper, namely, that we could not define them even if we wanted to [...]. (A241)

The reason Kant did not define the categories earlier in the *CPR* is not, as he earlier claimed, that putative definitions "would distract us from the chief point of the investigation by arousing doubts and objections that can well be referred to another occasion" (A83/B109), but because *real definitions* of the categories in general are impossible for us; we can only give real definitions of categories restricted to phenomena (schemata).

Kant admits that we can give definitions *of a sort* for the categories as concepts of objects *überhaupt*, both phenomena and noumena, and gives a series of examples of such quasi-definitions:

> If I leave out persistence (which is existence at all times), then nothing is left in my concept of substance except the logical representation of the subject, which I try to realize by representing to myself something that can occur solely as subject (without being a predicate of anything). (A242/B300)

> From the concept of a cause as a pure category (if I leave out the time in which something follows something else in accordance with a rule), I will not find out anything more than that it is something that allows an inference to the existence of something else. (A243/B301)

> No one has ever been able to define possibility, existence, and necessity except through obvious tautologies if he wanted to draw their definition solely from the pure understanding. For the deception of substituting the logical possibility of a **concept** (since it does not contradict itself) for the transcendental possibility of **things** (where an object corresponds to the concept) can deceive and satisfy only the inexperienced. (A244/B302)[14]

The 'obvious tautologies' to which Kant alludes in the third passage are principles like: the necessary is that whose non-being is not possible.[15] The third passage also tells us what is incomplete in these putative definitions—it is not that they are false, but they do not show *that* it is really possible for an object to fall under the concept, much less *why*. At most they show that the concept is logically consistent. Kant does not deny, for instance, that a cause *in general* is something from which the existence of something else follows, but this principle does not show us what makes it really possible for there to be a cause. Similarly, it is surely the case that a necessary being is one whose non-existence is not possible. But this does not tell us what would make such a being necessary. Given that Kant has just invoked the notion of a real definition—and thus implicitly contrasted the real definition of the categories with mere nominal definitions—it is natural to think of these as partial nominal definitions. They unpack conceptual connections between categories (e.g., the interdefinability of the necessary, the possible, and the contingent) but they make no progress in making intelligible to us the real possibility that objects fall under these concepts.

[14] Cf. A459/B487 and esp. A593/B621, where Kant explicitly describes as a nominal definition the principle that a necessary being is one whose non-existence is not possible but criticizes this for not making "*verständlich*" to me "whether through the concept of an unconditionally necessary being I am still thinking something or perhaps nothing at all."

[15] Cf. *MV* (Ak. 28: 418), *MvS* (Ak. 28: 498), *ML*$_2$ (Ak. 28: 556–7), and *MD* (Ak. 28: 633).

9.3. Unschematized Modal Categories

I have argued that by not being able to give real definitions of the categories as concepts of objects in general we are not able to identify what, in general, grounds the real possibility that objects fall under these categories, e.g., we are not able to answer perfectly generally what makes it really possible that something is a cause, or a substance. But what does 'real possibility' mean here? In Chapters 7 and 8 I distinguished various different kinds of real possibility in Kant's Critical system (formal possibility, empirical-causal possibility, noumenal-causal possibility, nomic possibility), but these are defined only for propositions about phenomena. When Kant claims that we cannot give a real definition of a single one of the categories as concepts of objects in general, he presumably does not have these notions of real possibility in mind because they do not apply to all objects *überhaupt*.

Before I address that question directly, I want to point out that the modal categories should be treated slightly differently than the other categories. The problem of the real definition of <*cause-effect*>, for instance, is the problem of specifying the ground of the real possibility of objects in general standing in cause–effect relations, i.e. one object being such as to posit the existence of an object distinct from itself. But applying the same scheme to the problem of the real definition of <*possibility*> we get: the problem is to determine the ground of the real possibility of objects in general being possible. In other words, we get an iterated modality: the ground of the real possibility of possibility. This is very indirect and artificial. It is much more natural to think that what we want from a real definition of the category <*possibility*> is to determine the grounds of the real possibility of objects in general; no modal iteration is necessary. After all, possibility is already included in the idea of a real definition of a concept, so it makes sense that when we come to the question of the real definition of <*possibility*> itself we do not ask about the grounds of the real possibility of possibility, we ask about the grounds of the real possibility *period*. To take a related example, Kant's complaint about what he calls the 'nominal definition' of necessary existence (A593/B621)[16]—a being exists necessarily just in case its non-existence is impossible—is not that this fails to tell us what makes a necessary being possible (iterated modality) but that it fails to tell us, if there is a necessary being, what grounds the necessary existence of that being, i.e. *why* it is not possible for that being not to exist. Likewise, what is missing in our grasp of the category <*possible*> is an account of what grounds the possibility of all objects in general—not an account of what grounds the real possibility of the possibility of all objects in general.

But this means that the question being raised about the non-modal categories—e.g., what is the ground of the real possibility of objects in general being causes?—is the same kind of question being raised about the modal category—what is the ground of the *possibility* of objects in general?—only if the notion of 'real possibility' involved

[16] Kant makes the same point in *Beweisgrund* at Ak. 2: 81; I discuss these passages in Ch. 1.

in the first question is the *same* as the notion of possibility involved in the *second* question, which is itself the modal category <*possibility*>. This means that the sense of 'real possibility' involved in a demand for a real definition of the non-modal categories is the same notion of possibility involved in a demand for a real definition of the modal category <*possibility*>. In other words, the unschematized category <*possibility*>, applied to all objects in general, is the sense of real possibility at issue when Kant denies we have insight into the grounds of real possibility of objects in general falling under the categories.

The categories, including the modal categories, are concepts we apply when we think about all objects in general (*überhaupt*). So if the unschematized category <*possibility*> (henceforth, <*possibility*>$_{UC}$) is a concept of real possibility, then it must be the highest or most general concept of real possibility. If there were a more general concept of real possibility than <*possibility*>$_{UC}$ then in thinking about that kind of real possibility we would not be using the unschematized category, and this is impossible; any thought about objects requires thinking about them under the categories.[17]

Conversely, if the most general concept of real possibility were less general than <*possibility*>$_{UC}$—if logical possibility were a species of the unschematized category—then Kant's claim that we cannot give a real definition of a single one of the categories as concepts of objects in general would be misplaced with respect to <*possibility*>$_{UC}$. For in this case, we could give a real definition of a species of <*possibility*>$_{UC}$ for all objects in general, logical possibility, for we know *quite generally* what logical possibility is (not containing mutually contradictory marks).[18] The categories are concepts at exactly the level where Kant thinks we cannot give real definitions for all objects in general. If there is a higher genus of the concept and we cannot give a real definition *of that genus* for all objects in general, then we have too narrowly specified the category. And if there is a species of the concept for which we can give a real definition for all objects in general, then we have identified the category too generically.

What does it mean that the unschematized category is the most general concept of real possibility? First of all, it means that the category of possibility is a concept of real possibility, rather than of logical possibility. What does this mean? In Chapter 6 I claimed this principle captures part of what it is to be a kind of real possibility in Kant's Critical system:

(*Real Possibility*) For any kind of possibility $\Diamond_x p$ (and its associated kind of necessity $\Box_x p$, where $\Box_x p \leftrightarrow \neg\Diamond_x \neg p$), $\Diamond_x$ is a kind of *real possibility* (and $\Box_x p$ is a kind of *real necessity*) only if

[17] Or, at least, modal thought requires modal categories, just as thought about the quantities of objects requires categories of quantity, and, as I argued above, thought about the *existence** of objects requires the relational categories.

[18] *Disc.*, Ak. 8: 195.

(i) *Non-logicality*: it is not a conceptual truth that $\Diamond_L p \supset \Diamond_x p$ (equivalently, it is not a conceptual truth that $\Box_x p \supset \Box_L p$), and
(ii) *Groundedness*: if $\Diamond_x p$ then the fact that $\Diamond_x p$ has a real ground in some actual object or principle.

When I introduced this principle I described it as an interpretive hypothesis that will be confirmed in the course of my interpretation. We can now see that the Pölitz passage is an important source of evidence for clause (ii), for it is precisely what Kant means when he says: "yet we have no concept of real possibility except through existence [*Dasein*], and in the case of every possibility which we think *realiter* we always presuppose some existence" (Ak. 28:1036). This is a claim about the unschematized category of (*real*) possibility in general. This is *why* our concept of real possibility carries with it the rational necessity to inquire into the ground(s) of real possibility: in thinking about noumena we have to think about them using the unschematized category of (real) possibility and this carries with it the thought that (real) possibilities for noumena—as for all objects in general—have grounds in actuality.

Some readers might note that in the passage quoted from the Pölitz lectures Kant says that a real possibility must be grounded in *existence*, while I claimed it is grounded in *actuality*. As we have seen, Kant sometimes uses these terms interchangeably, and sometimes he distinguishes them. In previous chapters I distinguished a broad notion of existence (on which it is identical to what is 'absolutely posited') and a narrow notion, which I designated existence*, which applies only to the subset of (absolutely posited) objects that are causally efficacious. Formulated in terms of this distinction, I think Kant's view in the Pölitz lectures cannot be "in the case of every possibility which we think *realiter* we always presuppose something *that exists**," for clearly there are kinds of real possibility that are not grounded in existing* objects, for instance, formal possibility; nor can it be "we presuppose something *there is* [absolute positing]," for formal possibility is grounded in the unity of apperception, and the unity of apperception is not the object of an absolute positing for it is not an object of intuition.[19] So the notion of *actuality* involved in clause (ii) above must be more general than absolute positing or existence*; I explore it in Chapter 10.7.[20]

If we cannot give a real definition of real possibility for all objects in general, and cannot know a metaphysical theory about all objects in general, then one might wonder about the status of Real Possibility, as formulated above. It seems to be a global claim about the nature of real possibility for all objects in general. Recall,

[19] See B Deduction, §25.

[20] I also think there is a distinction between existence in the broadest sense (what *there is*, the objects of absolute positing) and actuality, but we will not be in a position to understand this until we examine, in Chapter 10, Kant's doctrine that the intuitive intellect represents no difference between actuality and possibility.

though, that although he denies that we can give real definitions of categories for all objects in general, Kant does allow that we can give partial (partial nominal definitions) analyses of them. In section 2 I argued that these partial analyses of unschematized categories include such principles as *a substance has predicates but is not predicated of anything else* and *a cause is something from which the existence of something else follows according to a rule.*[21] But these analytic truths also include modal principles:

> No one has ever been able to define possibility, existence, and necessity except through obvious tautologies if he wanted to draw their definition solely from the pure understanding. For the deception of substituting the logical possibility of a **concept** (since it does not contradict itself) for the transcendental possibility of **things** (where an object corresponds to the concept) can deceive and satisfy only the inexperienced. (A242–4/B300–2)

These obvious tautologies, I take it, are partial nominal definitions of the relevant unschematized categories (e.g., a cause is something from which the existence of something else follows according to a rule). I have included the interdefinability of possibility and necessity in Real Possibility, which I take to be a partial nominal definition of our most general concept of real possibility, <*possibility*>$_{UC}$. The first clause of Real Possibility just unpacks what we mean by talking about 'real' possibility in the first place. By itself, it does not entail even that real possibility and logical possibility are really distinct (i.e. differ in extension). Finally, when Kant says in the Pölitz lectures that, "yet we have no concept of real possibility except through existence, and in the case of every possibility which we think *realiter* we always presuppose some existence," I take him to be asserting the second clause of Real Possibility as an analytic mark of our most general concept of real possibility.

We can now return to Kant's claim that we can only give real definitions of the categories for phenomena. The problem of the real definition of the modal categories is the problem of knowing what in general the real grounds of possibility are. So we can now see that Kant's claim that we can give real definition of the categories only for phenomena means, in the case of the modal categories, that we only know the grounds of real possibility for phenomena. If p is some proposition, and p is really possible, then we know the grounds of the real possibility that p only if p is a proposition concerning phenomena only. But phenomena have two kinds of grounds: immanent grounds and transcendent grounds.[22] We only know the 'immanent' grounds of real possibility of phenomena, so the only kinds of real possibility for which we can give real definitions are kinds of real possibility with immanent grounds. Three of the kinds of real possibility distinguished in Chapters 7 and 8 have immanent grounds: formal possibility (forms of experience), empirical-causal possibility (the past and laws), and nomic possibility (real essences of natural kinds). Although we cannot cognize the noumenal grounds of phenomena, we can *think*

[21] A240/B300. [22] See Ch. 7 for this distinction.

about their noumenal grounds, and whether they ground certain real possibilities for phenomena. For instance, do the causal powers of noumena ground the real possibility that I could be so affected I would have an experience with a completely different matter than the matter of my actual experience? Likewise, we know that, since I actually made a lying promise at time *t*, that it was not empirically-causally possible for me to omit that promise at *t*; but could my will, considered as a noumenon, have determined itself to act according to a different maxim, one that would not have produced this action? These are questions about what, in Chapter 7, I called 'noumenal-causal possibility':

> (*Noumenal-causal possibility*) It is *noumenally-causally possible* that *p* if and only if (i) it is formally contingent that *p*, and (ii) there is some noumenon with the causal power to make it the case that *p*, where *p* is a synthetic proposition about phenomena.

Noumenal-causal possibility is one species of what I have called 'noumenal$_1$' real possibility in Figure 9.1, real possibility for phenomena that is grounded in noumena. However, since we cannot cognize noumena, we cannot give a real definition of this kind of real possibility.

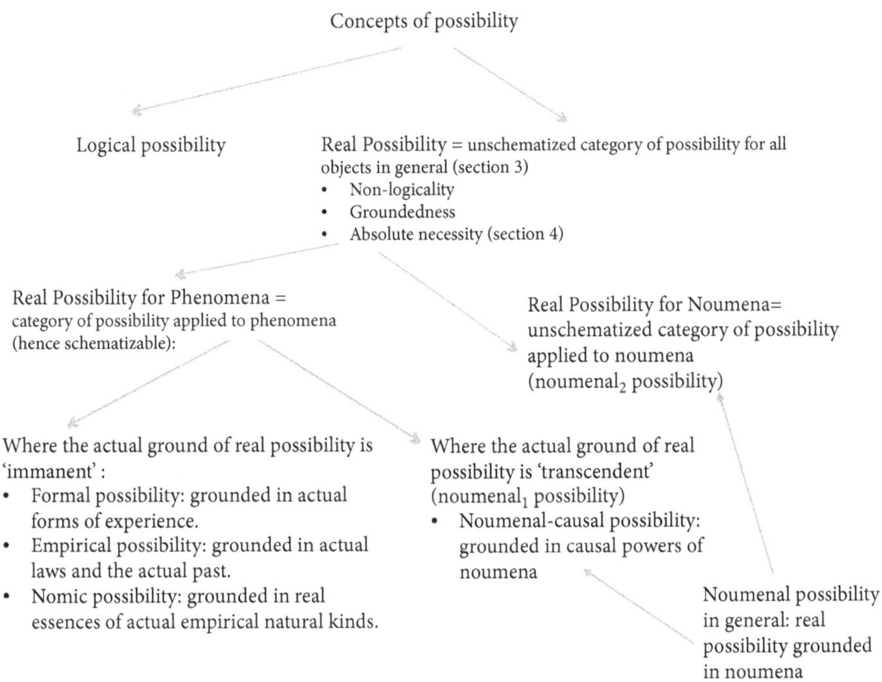

Fig. 9.1. Kinds of real possibility in Kant's Critical system

Finally, for propositions or states of affairs involving noumena *themselves* we might ask whether they are really possible, using the unschematized category of (real) possibility.[23] In doing so, we would be talking about a kind of real possibility whose grounds must be transcendent, that is, noumenal, because noumena ground phenomena, and not vice versa (so phenomena cannot be among the grounds of the real possibility of noumena or any state of affairs involving noumena). We cannot in principle know what the grounds of this kind of real possibility would be. This is the notion of real possibility that I have called 'noumenal$_2$' in Figure 9.1. But notice that 'noumenal$_1$' and 'noumenal$_2$' are really species of a single conception of real possibility, which I have dubbed simply 'noumenal possibility in general': the concept of a real possibility (whether for phenomena or for noumena themselves) that is grounded in noumena. We cannot give a real definition of this kind of possibility, for we do not know the noumenal grounds of real possibility. Figure 9.1 summarizes the distinctions between, and relations among, these different notions of possibility.

To return to the original passage, we can now answer one of the questions from the Introduction: what is the notion of 'real possibility' for which we are rationally necessitated to postulate an absolutely necessary ground? We can get a grip on that question by noting that Kant there regards this postulate as directed at the same object whose existence he took himself to have demonstrated in *Beweisgrund*: God, the absolutely necessary ground of real possibility. We need to ask ourselves, therefore, which of these various kinds of real possibility corresponds to the pre-Critical conception of real possibility? I think the answer is clear: noumenal real possibility. From the perspective of Kant's Critical theory, the God of *Beweisgrund* was a noumenon, a supersensible being, and he grounded real possibility for all beings, including other noumena. Consequently, the absolutely necessary being we are rationally necessitated to postulate is an absolutely *noumenally* necessary being that grounds all *noumenal* real possibility.

9.4. Absolutely Necessary Existence

Kant's primary concern in the Pölitz passage was the notion of 'absolutely necessary existence' and the claim that we must postulate an absolutely necessarily existing ground of real possibility, or, as we can now identify it, noumenal real possibility. In sections 5 and 6 I will explore why these considerations make it 'rationally necessary' to postulate such a being, but in this section I will focus on the content of this postulate.

[23] Kant seems to be applying modal concepts directly to noumena—rather than merely considering noumena as putative grounds of the possibility of empirical objects—at *Refl.* 5177 (Ak. 18: 109), 5184 (Ak. 18: 111), and 5723 (Ak. 18: 335).

Kant's notion of absolutely necessary existence in Pölitz is the same as his conception of absolutely necessary existence in *Beweisgrund*. Consider these two passages, the first of which is from Pölitz, the second of which is from *Beweisgrund*:

> In addition to the logical concept of the necessity of a thing (where something is said to be absolutely necessary if its non-existence [*Nichtseyn*] would be a contradiction, and consequently impossible), we have yet another rational concept of real necessity. This is where a thing is *eo ipso* necessary if its non-existence would cancel all possibility. (*Pöl.RT*, Ak. 28: 1036)

> Something may be absolutely necessary either when the formal element of all that can be thought is cancelled by means of its opposite [the 'logical concept of the necessity of a thing'—NS], that is to say, when it is self-contradictory; or, alternatively, when its non-existence [*Nichtsein*] eliminates the material element and all the data of all that can be thought. (*OPG*, Ak. 2: 82)

The "data of all that can be thought," in the context of *Beweisgrund*, is a reference to really possible predicates. In Chapter 5 I argued that absolute necessity should be interpreted as follows:

(1) It is absolutely really necessary that p if and only if $\neg p \Rightarrow$ (nothing is really possible).

Kant invokes the same notion of absolute necessity in the Pölitz lecture. Principle (1) defines absolute necessity in terms of a counterpossible conditional: if, *per impossibile*, God were not to exist, nothing would be really possible. As I argued in Chapter 5, this 'canceling' relation is a consequence of the *grounding* relation between God and real possibilities: because God is the ground of all real possibility, if, *per impossibile,* he were not to exist, none of those real possibilities would be possible.

Kant is claiming in the Pölitz lectures that we are rationally required to postulate the existence of an absolutely necessary ground of noumenal real possibility (God), where noumenal real possibility is one of several specific kinds of real possibility. If my interpretation is correct up to this point, then we should expect that, for each specific kind of real possibility, there will be (i) a corresponding notion of absolute necessity, and (ii) an absolutely necessary entity or principle that is the first ground of real possibility of that specific kind. After claiming at Ak. 28: 1034 that we are rationally necessitated to assume the kind of "absolute necessary being" whose existence Kant (he now thinks) mistakenly took himself to have proved in *Beweisgrund*, Kant claims at Ak. 28: 1036 that "we have no concept of real possibility except through existence [*Existenz*], and in the case of every possibility which we think *realiter* we always presuppose some existence [*Daseyn*]; if not the actuality of the thing itself, then at least an actuality in general [*Wirklichkeit überhaupt*] which contains the data for everything possible." Earlier I interpreted those as claims about real possibility as such; they are conceptual truths about our most general modal concept, the category of (real) possibility. If we are rationally required to form the idea of an absolutely necessary ground of real *noumenal* possibility, then it would stand to reason that we are rationally required to form the idea of an absolutely necessary ground for each different lower-level kind of real possibility. Thus, for each lower-level kind of possibility, we should expect to find some absolutely necessary principle *p* such that:

(1*) It is absolutely necessary$_x$ (absolute-$\Box_x$) that *p* if and only if $\neg p \Rightarrow$ (nothing is really possible$_x$).[24]

The subscript *x* refers to the different kinds of possibility, from above. Some care is required in deploying the formula, though, because the catch-all phrase 'nothing would be really possible' covers over differences in what the various kinds of real possibility apply to (experiences, alterations, etc.). In the case of the 'immanent' real modalities—modalities with immanent grounds—we can give more specific definitions:

(1_F) It is absolutely formally necessary (absolute-$\Box_F$) that *p* if and only if $\neg p \Rightarrow$ (there is no formally possible experience).
(1_E) It is absolutely empirically-causally necessary$_x$ (absolute-$\Box_E$) that *p* if and only if $\neg p \Rightarrow$ (there are no empirically-causally possible alterations).
(1_N) It is absolutely nomically necessary that *p* (absolute-$\Box_{No}$) if and only if $\neg p \Rightarrow$ (there are no nomically possible truths).

Precisely because we know the grounds of these kinds of real possibility, we should expect that we can know what is absolutely necessary in the case of each of them.

This expectation is most clearly satisfied in the case of formal possibility. In several texts Kant identifies space as the absolutely necessary ground of formal possibility for outer objects and explicitly compares space in this respect to God as the absolutely necessary ground of noumenal possibility:

All reality must be completely given, and so some actuality is prior to possibility, just as space is not merely something possible, but the ground of all possible figures. (*Refl.* 4119, Ak. 17: 424)

From this it follows only that the *ens realissimum* must be given prior to real concepts of possibility, just as space cannot be thought antecedently as possible, but as given. But not as an actual [*wirklich*] object in itself, rather as a merely sensible form, in which alone all objects can be intuited. (*Refl.* 6290, Ak. 18: 558)[25]

[24] 'Absolute-$\Box_x$' indicates that this is a definition of the absolute form of the relevant kind of necessity; the argument given above shows that this is not simply necessity of the relevant kind ($\Box_x$).

[25] Kant makes this connection between God and space even more explicit in the Transcendental Ideal section of the *CPR*, where he writes: "all manifoldness of things is only so many different ways of limiting the concept of the highest reality, which is their common substratum, just as all figures are possible only as different ways of limiting infinite space" (A578/B606). In this passage, Kant compares the role that the *ens realissimum*, God, plays in grounding possibility to the role that space plays in grounding the possibility of figures in space. However, the connection he makes here rests on the idea that God contains all realities, which reflects *neither* his pre-Critical understanding of how God is the ground of possibility *nor* his considered view in the Critical period. See A579/B607, where Kant makes clear that some realities are grounded in God without being limitations of infinite realities instantiated by him.

These texts suggest that something about space is absolutely formally necessary, in the sense specified by (1*) above (according to my regimentation, there is some proposition p about space that is absolutely formally necessary). It cannot be that space exists* because (as the second text reminds us) space is not causally efficacious (it is not *wirklich*). It is a pillar of Kant's Critical theory that without being given in space and time, no outer or inner object is (formally) possible; if objects were not given to us in space and time, there would be no experience.[26] The absolute formal necessity of space and time consists, then, not in the fact that space exists* absolutely formally necessarily, but in the fact that it is absolutely formally necessary that outer objects are given to us in space and inner objects are given to us in time. *Per impossibile*, if objects were not given to us in space and inner objects were not given to us in time, there would be no formally possible experience of outer or inner objects, respectively. Space is the absolutely necessary ground of the formal possibility of outer objects;[27] time is the absolutely necessary ground of the formal possibility of inner objects.

It is also a pillar of Kant's Critical theory of experience that experience has two kinds of grounds: sensible grounds (the givenness of objects in intuition) and intellectual grounds (thinking about objects). The unity of apperception, the source of the intellectual conditions of experience, is just as absolutely formally necessary as space and time, as Kant makes clear in this passage from the B Deduction:

> The supreme principle of the possibility of all intuition in relation to sensibility was, according to the Transcendental Aesthetic, that all the manifold of sensibility stand under the formal conditions of space and time. The supreme principle of all intuition in relation to the understanding is that all the manifold of intuition stand under conditions of the original synthetic unity of apperception. (B136)[28]

Kant's description of space and time as "the supreme principle of possibility [*der oberste Grundsatz der Möglichkeit*]" of objects of intuition is a reference, I take it, to the analogy between the role that space and time play in grounding the formal possibility of objects of intuition and the role that God plays in grounding (noumenal) real possibility. Kant here extends that analogy to the "original synthetic" unity of apperception: the unity of apperception is the ground of the real formal possibility of intellectual representation of sensibly given objects, that is, of objects of experience (phenomena) in the full sense. Formal possibility has three grounds, each of which is absolutely formally necessary: space and time (sensible grounds), and the unity of

[26] A23–25/B38–39 and A30–32/B46–47 are especially illuminating when read in light of the idea of absolute formal necessity.

[27] One would expect Kant to say similar things about time. However, from my overview of the relevant texts, it seems he more frequently draws the comparison between space and God as first grounds of (formal and noumenal, respectively) possibility.

[28] Cf. A107: "the numerical unity of this apperception grounds all concepts *a priori* just as the manifoldness of space and time grounds the intuitions of sensibility."

apperception (the intellectual ground). If any of these were 'canceled' there would be no formally possible experience of objects.[29]

Applying the definition of absolute necessity to empirical-causal possibility we get:

(1_E) It is absolutely empirically-causally necessary (absolute-$\Box_E$) that $p =_{def} \neg p \Rightarrow$ (no alterations occur that are compatible with the *actual* empirical laws and the actual past).

Since the actual alterations are all and only the empirically-causally possible alterations, we can replace the final consequent of (1_E) with: none of the actual alterations occur. Consider the following conditionals:

(i) (t)(An arbitrary non-zero interval of the past before t is different than it actually is $\Rightarrow$ no actual alterations occur at t).
(ii) (t)(An arbitrary non-zero interval of the future after time t is different than it actually is $\Rightarrow$ no actual alterations occur at t).[30]
(iii) (t)(The laws are different than the actual laws $\Rightarrow$ no actual alterations occur at t).

Conditional (i) is true only if the determinism that obtains at the empirical level excludes 'convergent' empirical series: empirical series, governed by the actual laws, that are distinct from the actual series before a time but then 'converge' to the actual series after that time. If the series before t were different from the actual series (for a non-zero interval), the laws would remain the same;[31] so whether any actual alterations would occur after t depends upon whether different possible world-series with deterministic laws can converge. However, Kant does not seem to think that the determinism of natural laws excludes convergence; to put it roughly, he holds a 'same cause, same effect' principle, but not a 'same effect, same cause' principle;[32] causation introduces temporal asymmetry, according to Kant, so the determinism that

[29] For each kind of possibility distinguished above, we can distinguish a *rigid* definition of that kind of possibility and a non-rigid definition. On the non-rigid definition, we eliminate 'actual' from the definition. This will produce an important difference when we embed these definitions of possibility within modal contexts, as we are doing with the definitions of absolute necessity above. In the main text I will focus on the *rigid* definitions, and thus on *rigid* absolute necessity. For the non-rigid notion, and its difference from the rigid notion, see the supplementary article "Rigid and non-rigid absolute necessity" on my website (see Notes on the Text).

[30] In Chapter 7.3 I pointed out that, for a given time t, "the past before t" and "the future after t" do not refer to anything, on Kant's view, because the complete totality of the past or future is not a possible object of experience. The same fix I proposed there also applies here: let "the past before t" refer to any non-zero interval before t, and let "the future after t" refer to any non-zero interval after t.

[31] I am assuming that laws are more 'counterfactually robust' than alterations; in the contemporary idiom of worlds and closeness, for any world in which the past is different and the laws are different, there is a world closer to the actual world where the past is different and the laws are the same. See Lange (2009) for a different way of formulating this idea.

[32] See his discussion, in the metaphysics lectures, of the principle that a cause posits a determinate (specific) effect, while an effect entails that there is some cause or other, but does not posit a determinate (specific) cause: *MV* (Ak. 28: 401–2, 408), *MvS* (Ak. 28: 487), and *MM* (Ak. 29: 808).

causation brings into the empirical series goes one way: the past determines a unique future, not the other way around.

Ironically, while the past is the *ground* of the empirical-causal possibility of alterations in the future, there are reasons to think that the *future* is absolutely empirically-causally necessary. Assume that events in the future (past time *t*) are different than they actually are during some non-zero interval. Holding fixed the natural laws, and given determinism, it follows that the empirical series before *t* is different than the actual series at every time. Given Kant's determinism, if the past were the same as the actual past for *any* non-zero interval, the future would be the same as the actual future, contra hypothesis. However, if the past is different than the actual past for *every* non-zero interval before *t*, it follows that if alteration E is *actually* empirically possible at time *t* (i.e. alteration E *actually* occurs), it does not occur.[33] So if we change the *future* after time *t* we cancel the empirical possibility of any actually empirically possible alterations at time *t*. There are good reasons, then, to think that (ii) is true and that the *future* is absolutely empirically-causally necessary.

Determining whether the actual laws are absolutely empirically-causally necessary (i.e. whether (iii)) is true, is somewhat harder. The laws (I argued in Chapter 8) are truths grounded in the essences of empirical natural kinds; therefore, for the laws to be non-actual is for there to be non-actual empirical natural kinds (because nothing could have a different essence than it actually has, and the relevant grounding relations are essential to the natural kinds as well). There are two ways for there to be non-actual empirical natural kinds: (a) there is a non-actual kind of matter, or (b) there is a non-actual kind that is not a kind of matter. Option (b) divides into two sub-options: (b1) there is matter in addition to the non-actual kind, or (b2) there is no matter. If (b2) obtains, then none of the alterations that occur are actual alterations because actual alterations are individuated by the substances of which they are alterations, and if there is no matter, then no actual substances exist (because they are all material substances).[34] So we are left with (a) and (b1). But intuitively, there could be a slight difference in laws from the actual world (either (a) because there is a non-actual kind of matter or (b) because there is a non-material kind) while some, if not most, actual alterations remain, if the difference were small enough (e.g., if the non-actual kind were very extremely rare). It is hard to evaluate the counterfactual

[33] See the discussion of continuity at A207-8/B253-4. There is a complication here. The argument in the main text shows that the 'complete state' of the world for any non-zero interval of time before *t* cannot be identical to the 'complete state' of the actual world during that interval (for, by determinism, this would entail that the series after t is identical to the actual series). But it may be that some actual alteration occurs, even though it is part of a different 'complete world-state' than it is in the actual world. This problem can be partly alleviated by individuating alterations very finely, to include the 'complete world-state' at the time of alteration. Another complication is that there is no such thing as 'the complete world-state' at a given time, just as there is no 'complete world-history' before a given time, because this is not a possible object of experience. Resolving these problems would require going into more detail on Kant's views on the infinite, ideas of reason, and the metaphysics of the natural world than I have space to do here.

[34] I am assuming that any material substance is *de re* necessarily a material substance.

conditional, *if there were a non-actual kind of matter no actual alterations would occur*, without knowing much more about Kant's views on causal determinism. It is clear, though, that if there were no *matter* (e.g., if the laws of matter did not obtain) then no actual alteration would occur; the existence of matter, the highest empirical natural kind, is absolutely empirically-causally necessary. Since the laws of matter are grounded in the real essence of matter, these are absolutely empirically-causally necessary as well.

When we apply this notion to nomic necessity we obtain:

(1_N) It is absolutely nomically necessary that p (absolute-$\Box_{No}$) $=_{def} \neg p \Rightarrow$ (nothing is compatible with the essences of the *actual* empirical natural kinds).[35]

The actual natural kinds form a system of species and genera: the highest natural kind is matter, and all of the other natural kinds are, ultimately, species of matter.[36] If there were no matter then there would be no actual natural kinds, and nothing would be compatible with essences of actual empirical natural kinds (because these would also be canceled). It is absolutely nomically necessary that there is matter. But no other actual empirical natural kind is absolutely nomically necessary; for any other such kind, if that kind did not exist, there would still be matter (matter is the highest kind), so some things would be compatible with actual empirical natural kinds (in particular, with matter). This is summarized in Table 9.1.

Now we are in a position to see that Kant's Critical modal metaphysics, far from being a mere list of different conceptions of modality, is highly systematically unified, and moreover, this unity is provided by the elements that are retained from his pre-Critical modal theory. Kant has a very general concept of real possibility, the unschematized category <*possibility*>$_{UC}$. Certain conceptual truths hold of this very

Table 9.1. Absolute necessities for immanent real possibility

Kind of possibility	*Absolute necessity*
Formal	Actual forms of intuition (space, time) and actual forms of understanding (unity of apperception, categories)
Empirical-causal	Actual future, actual highest natural kind (matter) and its laws
Nomic	Actual highest natural kind (matter) and its laws

[35] Some readers will object that in a world w where none of the actual natural kinds are instantiated, any p is compatible, given how I defined the compatibility relation: p is compatible with the essences of actual natural kinds if and only if $\neg(\neg p$ is ground in the essences of actual natural kinds). If there are no actual natural kinds in w, goes the objection, then, trivially, everything satisfies the right-hand of this biconditional. However, this objection does not hold for the more involved technical definition of nomic possiibility in Chapter 8.7.

[36] See Ch. 8.8.

generic conception of real possibility, truths which are retained from the pre-Critical theory of real possibility. In particular:

(*Real Possibility*) For any kind of possibility $\Diamond_x p$ (and its associated kind of necessity $\Box_x p$, where $\Box_x p \leftrightarrow \neg\Diamond_x \neg p$), $\Diamond_x$ is a kind of *real possibility* (and $\Box_x p$ is a kind of *real necessity*) only if

(i) *Non-logicality*: it is not a conceptual truth that $\Diamond_L p \supset \Diamond_x p$ (equivalently, it is not a conceptual truth that $\Box_x p \supset \Box_L p$), and

(ii) *Groundedness*: if $\Diamond_x p$ then the fact that $\Diamond_x p$ has a real ground in an actual object or principle.

(iii) *Absolute necessity*: there is an associated notion of absolute necessity (absolute-$\Box_x p$) defined as follows: absolute-$\Box_x p =_{def} \neg p \Rightarrow$ (for all q, $\neg\Diamond_x q$). There is some actual p about the grounds of all possibility$_x$ such that absolute-$\Box_x p$.[37]

Kant generalizes from his pre-Critical conception of metaphysical real possibility to a conception of real possibility *in general* by dropping the pre-Critical requirement that real possibility be grounded in some feature of the world independent of our minds. He retains the requirement that each kind of real possibility be interdefinable with a related concept of real necessity, be conceptually distinct from logical possibility, be grounded in some features of actuality, and that there be a corresponding conception of absolute necessity.

This conception of absolute necessity is most problematic, but also most interesting, in the case of 'noumenal' real possibility, my generic term for real possibility grounded in noumena. Kant claims in the Pölitz lectures that all noumenal real possibilities are grounded in an absolutely necessarily existing being. This was precisely the conception of God and his absolutely necessary existence that Kant defended in *Beweisgrund*, translated into the Critical terminology. In the Pölitz text he claims that it is "subjectively necessary" to posit such a being. In the next section I use Kant's theory of 'doctrinal belief' to explain what it could mean that it is "subjectively necessary" for us to posit the

[37] In Chapter 5.2 I discussed the 'reductive' reading that identifies absolute necessity and necessity *simpliciter*. Similarly, one might argue that, in the Critical system, 'absolute real necessity' of some kind (e.g., absolute formal necessity) is just real necessity *simpliciter*, interdefined with real possibility. However, this cannot be so, as this brief argument shows:

(1) It is really possible that p if and only if it is not absolutely necessary that $\neg p$. [Assumption]
(2) (It is absolutely necessary that p) if and only if ($\neg p$ $\Box\!\!\rightarrow$ nothing is really possible). [Definition of absolute necessity]
(3) ∴ (It is really possible that p) if and only if $\neg$ (p $\Box\!\!\rightarrow$ nothing is really possible). [From (1) and (2)]
(4) (There is a ground of the real possibility that p) if and only if (it is really possible that p). [Assumption]
(5) ∴ (There is no ground of the real possibility that p) if and only if ($\neg p$ $\Box\!\!\rightarrow$ nothing is really possible).

But this conclusion is absurd, for it entails that if there is no ground of the real possibility that I act otherwise than I actually do (perhaps because my noumenal will lacks the freedom to do so), the hypothesis that I act otherwise than I do cancels all real possibility. Absolute real necessity cannot be identified with real necessity *simpliciter* and thus interdefined with real possibility; premise (1) should be rejected.

existence of something, and why, in particular, this is the case with the absolutely necessary ground of all noumenal real possibility.

9.5. Postulates of Pure Theoretical Reason

Kant's theory of doctrinal belief is found in the Canon of Pure Reason in the *CPR*, in the section entitled "On opining [*meinen*], knowing [*wissen*], and believing [*glauben*]" (A820/B848–A831/B859).[38] Kant there distinguishes several different "modes" of "holding-to-be true" [*Fürwahrhalten*], different epistemic statuses that a judgment can have for an agent. Since the epistemic status of a judgment can vary from agent to agent (e.g., Friedrich merely believes that *p*, while Gottfried knows that *p*), Kant needs to distinguish between judgment-contents (or judgment-types individuated by their content, which I have called *propositions*) and the relation of that judgment-content to an agent who judges that it is true. Kant introduces *Fürwahrhalten* to refer to the second: an agent's act of judging that some content is true.[39] Consequently, one and the same judgment-content can be the target of different acts of *Fürwahrhalten* by different agents. I have used 'proposition' to refer to judgment-*contents* throughout this study; I will sometimes refer to *Fürwahrhalten* simply as *judging*.[40]

The different modes of *Fürwahrhalten* correspond to different epistemic statuses that an act of judging can have: the agent can have the mere opinion that *p*, believe that *p*, know that *p*, etc. Kant's classification of these different epistemic statuses is quite complex, but for our purposes the most important such status is belief [*Glaube*], which Kant defines as follows: "if taking something to be true is only subjectively sufficient and is at the same time held to be objectively insufficient, then it is called **believing**" (A822/B850).[41] This is only informative if we know what 'subjective sufficiency' and 'objective (in)sufficiency' are. Kant writes in the same passage that "subjective sufficiency is called **conviction** [*Überzeugung*] (for myself)," a remark he does not elaborate on until two pages later, where he writes that "the usual touchstone of whether what someone asserts is mere persuasion or at least subjective conviction, i.e. firm belief, is **betting**."[42] The 'subjective sufficiency' that defines belief that *p* as a mode of *Fürwahrhalten* is what would now be called 'credence' and is

[38] See also the parallel discussions of *Fürwahrhalten* in *JL* (Ak. 9: 65–73) and the logic lectures: *LB* (Ak. 24: 148–53), *LP* (Ak. 24: 436–41), *LPö* (Ak. 24: 541–5), *LBu* (Ak. 24: 637–9), *LDW* (Ak. 24: 732–5), and *WL* (Ak. 24: 850–7). My discussion of these issues is deeply indebted to Chignell (2007*a*) and (2007*b*). I will not bother to indicate minor differences in our interpretations, though, for I am interested primarily in Kant's theory of 'doctrinal belief'. See also Hanna (2006), 258–61.

[39] Cf. *JL*: "the judgment through which something is represented as true, the relation to an understanding and thus to a particular subject, is subjectively, holding-to-be-true" (Ak. 9: 65–6). See A74–5/B100–1.

[40] Kant's complete theory of judgment includes the modality of judgment, a complication I suppress in the body of the text. A subject can relate to a judgment-content in one of three ways: problematically (entertaining the content without asserting it), assertorically (asserting it), or apodictically (asserting it by deriving it from a rule). See A74–6/B100–1.

[41] Cf. *JL* (Ak. 9: 67), *LPö* (Ak. 24: 541), *LBu* (Ak. 24: 638), *LDW* (Ak. 24: 732), and *WL* (Ak. 24: 852).

[42] Cf. *JL* (Ak. 9: 73) and *WL* (Ak 24: 852–3, 855).

measured by what bets one is willing to take on the proposition that *p*: the more one is willing to wager, the greater the degree of 'subjective sufficiency' or 'subjective conviction' in one's judgment that *p*. Kant does not state a precise lower bound on 'subjective sufficiency' that differentiates belief from mere opinion, but the idea, I take it, is clear enough; opining that *p* is subjectively *in*sufficient insofar as one entertains the proposition that *p* but would not be willing to bet anything on it. Kant allows that we can believe things we could never (in principle) verify or falsify, but maintains that the 'usual touchstone' still applies: we are to imagine that there is some way of verifying or falsifying the proposition and ask 'what would one be willing to wager that *p*?' "If it were possible by any sort of experience to verify whether there are inhabitants of at least some of the planets that we see, I might well bet everything I have on it" (A825/B853). Kant is expressing his strong, albeit unverifiable (at least in 1781–2016), belief that some planets in the universe other than Earth are inhabited. Whether it makes sense to apply this test to propositions that could not in principle be verified because they concern non-sensible objects (e.g., that God exists, or that the soul is immortal) is a point Kant does not address. Even if it is incoherent to ask 'how much would you bet that there is an absolutely necessary ground of possibility?', Kant's idea still applies: belief requires subjective conviction, this subjective conviction comes in degrees (beliefs require more than mere opinion), and the 'betting' test ("the usual touchstone") is one intuitive way to measure this degree of subjective conviction. This means that belief is defined in terms of one's subjective degree of credence, not in terms of any positive epistemic status (like justification); there can be beliefs that lack any rational standing, and it is psychologically possible to have greater credence in what one believes than in what one knows to be true.[43]

Before we understand the positive epistemic status of 'rational belief,' though, we need to understand the 'objective sufficiency' that defines knowledge [*Wissen*] and distinguishes it from belief (so that we can understand why even rational beliefs are not knowledge). In the paragraphs leading up to his discussion of belief, Kant distinguishes mere persuasion [*Überredung*] from conviction: "the touchstone [*Probierstein*] of whether taking something to be true [*Fürwahrhalten*] is conviction or mere persuasion is therefore, externally, the possibility of communicating it and finding it to be valid for the reason of every human being to take it to be true [*Fürwahrhalten*]" (A820–1/B849). Kant here is talking about the intersubjective validity of a judging, that is, whether it is based on grounds (reasons) that are grounds (reasons) for other epistemic agents to concur in the judging that *p*. I interpret 'touchstone' [*Probierstein*] here as meaning a necessary but not sufficient condition on conviction [*Überzeugung*],[44] so that if some judging is an instance of

[43] See *JL* (Ak. 9: 70), *LB* (Ak. 24: 148–50), *LPö* (Ak. 24: 543), *LBu* (Ak. 24: 638), *LDW* (Ak. 24: 734), and *WL* (Ak. 24: 852, 855). I will not discuss here Kant's further claim (made in several of these texts) that it is *rational* to have a higher degree of credence in moral beliefs than in what one knows to be true.

[44] Not the *subjective* conviction discussed above, but *objective* conviction. Cf. Chignell (2007*a*), 40–4.

conviction, then it is based on intersubjectively valid grounds (but not vice versa). Since this notion of judging on intersubjectively valid grounds will be crucial for Kant's theory of rational belief, it is worth being precise about it:

> (*IS-valid*) S's judging that *p* is intersubjectively (IS) valid if and only if $\exists g$ such that (i) S bases her judging that *p* on *g*, and (ii) *g* is an epistemic ground (reason) for S and other rational agents to judge that *p*.

The range of rational agents for which *g* is a ground to judge that *p* can vary; this allows us to distinguish *degrees* of IS-validity. At one extreme will be judgments based on grounds that are valid for all finite rational agents as such; I will refer to this as *maximal* IS-validity. The quoted sentence continues with this remark: "for in that case [IS-validity] there is at least a presumption that the ground of the agreement of all judgments, regardless of the difference among the subjects, rests on the common ground, namely the object, with which they therefore all agree and through which the truth of the judgment is proved" (A821/B849). I interpret this to mean that, in the case of judging on IS-valid grounds, there is a presumption that these IS-valid grounds are objectively valid (in a sense to be explained below), and hence constitute a case of conviction (knowledge), but that there can be cases of judging on IS-valid grounds that are not conviction (knowledge).

This is Kant's point when he introduces opinion, belief, and knowledge: "taking something to be true, or the subjective validity of judgment, has the following three stages in relation to conviction (which at the same time is valid objectively): having an opinion, believing, and knowing" (A822/B850). When I opine that *p*, I do not even have subjective conviction that *p* because I would not bet that *p*. When I believe that *p*, I am willing to bet that *p*, and in some cases (which I will discuss further below), my belief that *p* is based on IS-valid grounds, but I am aware that my grounds are not objectively valid. Only in the case of knowledge do I judge that *p* with subjective sufficiency (I am willing to bet that *p*) on the basis of grounds that are IS-valid and objectively valid.[45] This is shown in Figure 9.2.[46]

Before continuing to the main topic of this section, Kant's theory of doctrinal belief, I want to discuss, briefly, the 'objective validity' of grounds that distinguishes knowledge from belief. It is crucial to realize that the grounds in question (the grounds for S's judging that *p*) are not *a priori*$_G$ grounds of the truth of *p*, either causal or non-causal grounds, but epistemic grounds. If they were *a priori*$_G$, then all knowledge (all knowledge from 'sufficient grounds') would be *a priori*$_G$ knowledge, but this is clearly not Kant's view; he allows that in both empirical and mathematical cases there is genuine knowledge from the consequences (*a posteriori*$_G$ knowledge).[47] So the

[45] I am persuaded that *p* when I have subjectively sufficient grounds for *p* (I am willing to bet that *p*) but I mistakenly take these grounds to be either IS-valid or objectively valid, or even both. Persuasion [*Überredung*] is *not* simply the complement of *conviction*. When one has rational belief that *p* one is neither persuaded nor convinced that *p*.

[46] Cf. Chignell (2007*b*), 358.

[47] *JL* (Ak. 9: 70–1), *LPö* (Ak. 24: 544).

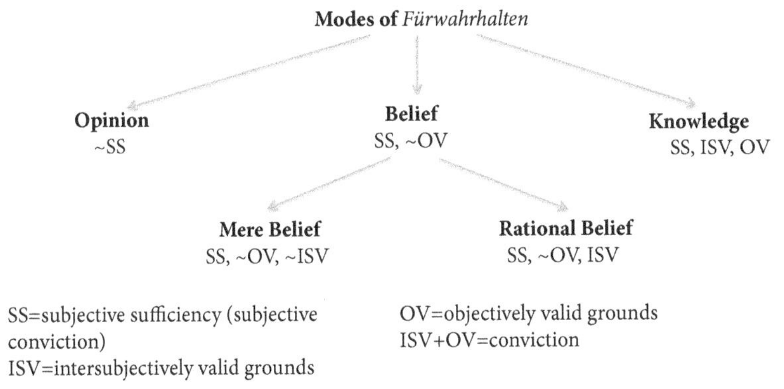

Fig. 9.2. Modes of validation

notion of grounding that defines knowledge is not the metaphysical notion of a ground that makes something possible or makes something exist, but the epistemic notion of a ground that confers warrant or justification on a judgment (although the epistemic ground *may also be* the ground of possibility or of existence, in the case of *a priori*$_G$ knowledge).

In the empirical case, Kant seems to regard *g* as an objectively sufficient ground for judging that *p* just in case *g* makes *p* sufficiently probable (where *p* is also true); this allows for empirical knowledge by testimony, for instance, knowledge of history from fallible sources.[48] In the *a priori* case, though, Kant requires "consciousness of necessity" (*JL*, Ak. 9:71).[49] This cannot simply mean that the proposition (judgment-content) *p* is itself necessary, for this will not distinguish between cases where one knows a necessary truth and where one merely thinks one knows a necessary truth. The 'necessity' characterizes not the proposition (judgment-content) itself, but the relation between one's ground and the proposition one judges: the ground must necessarily entail the proposition (in the *a priori*$_J$ case).[50] In other words:

> (*Obj. validity*) S's judging that *p* is objectively valid if and only if ∃*g* such that (i) S bases her judging that *p* on *g*, and (ii) *g* obtains, and either (iii) *g* makes *p* sufficiently probable (empirical knowledge) or (iv) necessarily, if *g* obtains then *p* is true (*a priori* knowledge).[51]

48 *JL*, Ak. 9:70–1. Cf. Chignell (2007*a*). 49 Cf. *LPö*, Ak. 24: 544.

50 Hence Kant's characterization of *Wissen* as apodictic judgment (*JL*, Ak. 9: 66); the apodictic mode of judging is defined at A76/B101 as judging that some content is a necessary consequence of a rule ("the apodictic judgment [*Satz*] thinks the assertoric one as determined through these laws of the understanding itself"). The example in that passage is an example in which the apodictic judgment is a logical consequence of the epistemic ground on which it is based (the premises in a hypothetical syllogism). But Kant must generalize the necessity here beyond logical necessity in order to make sense of, for instance, synthetic *a priori* mathematical knowledge.

51 Chignell rejects my clause (i), which requires that *S* be aware of the ground and base her judgment on that ground. On this issue, he opts for a counterfactual analysis: on reflection, *S* would cite *g* as her ground; see Chignell (2007*a*), 45. He cites the Blomberg logic (Ak. 24: 87–8), but I don't see how that passage

The obvious question about this principle, especially in the context of this study, is what notion of necessity is involved? My answer is that the notion of necessity involved depends upon the content of *p*: if *p* is an analytic judgment, knowing that *p* requires judging on the basis of a logically necessitating ground; if *p* is a mathematical claim, knowing that *p* requires judging on the basis of a mathematically (formally) necessitating ground; if *p* is about an alteration in a substance then knowing that *p* requires judging on the basis of an empirically-causally necessitating ground;[52] and if *p* is about laws, then knowing that *p* requires judging on the basis of a nomically necessitating ground. My definition also remains neutral on the ontological status of *grounds*; in some cases they might be other propositions, or objects, etc. The phrase '*g* obtains' is meant to be category-neutral: whatever kind of thing a given ground is, necessarily, if that ground is present (true, in the case of propositions; absolutely posited in the case of objects, etc.) then the proposition for which it is the objectively sufficient ground is sufficiently probable or true.

Since Kant defines belief as subjectively sufficient judging consciously based on objectively invalid grounds, we can also now state his definition of belief:

> (*Belief*) S believes that *p* if and only if *S* subjectively sufficiently judges that *p* and S is aware that her judging that *p* is not objectively valid.

Note that the definition of belief leaves open whether it is based on IS-valid grounds (rational belief) or whether it lacks rational grounds (mere belief).

The rest of the section "On opining, believing, and knowing" is devoted to rational belief, of which kant gives various examples, but never precisely defines what makes rational belief *rational*. The one feature all of his examples share, though, is that they are based on intersubjectively valid grounds. The degree of inter-subjective validity varies from example to example, and the nature of the grounds themselves vary (e.g., practical or theoretical grounds), but their in-principle shareability does not vary. I propose, consequently, that we define rational belief in terms of IS-validity, from above:

> (*Rat. Belief*) S rationally believes that *p* if and only if (i) S believes that *p*, and (ii) S's judging that *p* is IS-valid.

supports his reading. Clause (iv) is not intended as an analysis of *a priori*$_J$ knowledge, but merely to indicate that in the *a priori*$_J$ case one's ground must do more than merely make *p* probable. One might also want to insert a clause (v) to the effect that S must be aware of (ii)–(iv). However, adjudicating whether Kant maintains that knowledge that *p* requires being aware that one's grounds for judging that *p* are objectively sufficient (rather than merely requiring that they are objectively sufficient) would require more space than I have here.

[52] There is a complication here. Knowledge of alterations in substances is not *a priori*$_J$ so we might think this should fall under the empirical clause (mere probability is sufficient). However, I think there are two ways to know alterations in substances: purely empirically (in which case an objective ground that makes it sufficiently probable is enough) and knowledge *a priori secundum quid* (see Stang 2011 and Ch. 7.4), in which one derives the alteration from the antecedent state of its cause and thus cognizes its empirical-causal necessity.

Consequently, we can distinguish the range of intersubjective validity that rational beliefs have: some rational beliefs will be intersubjectively valid only for agents who share my contingent ends (what Kant calls 'pragmatic' beliefs, A824/B852), while others will be intersubjectively valid for all agents who share the moral end of bringing about the highest good (an end shared by those with a good moral disposition [*Gesinnung*], according to A829/B857).[53]

While the lion's share of scholarly attention has focused on the latter class of practically rational beliefs (what Kant calls 'moral beliefs'), my focus will be on what he calls *doctrinal* belief, a "theoretical analogue" (A825/B853) of practical beliefs. Kant's clearest articulation of what doctrinal belief is occurs in this long passage:

> Now we must concede that the thesis of the existence of God belongs to doctrinal belief. For although with regard to theoretical knowledge of the world I have nothing **at my command** that necessarily presupposes this thought as the condition of my explanations of the appearances of the world, but am rather obliged to make use of my reason as if everything were mere nature, purposive unity is still so important a condition of the application of the reason to nature that I cannot pass it by, especially since experience liberally supplies examples of it. But I know of no other condition for this unity that could serve me as a clue for the investigation of nature except insofar as I presuppose that highest intelligence has arranged everything in accordance with the wisest ends. Consequently, the presupposition of a wise author of the world is a condition of an aim which is, to be sure, contingent but yet not inconsiderable, namely that of having a guide for the investigation of nature. The outcome of my experiments also so often confirms the usefulness of this presupposition, and nothing can be decisively said against it, so that I would say too little if I called my taking it to be true merely having an opinion, but rather even in this theoretical relation it can be said that I firmly believe in God; but in this case this belief must not strictly be called practical, but must be called a doctrinal belief, which the **theology** of nature (physico-theology) must everywhere necessarily produce. (A826–7/B854–5)[54]

Kant's point here is that the judgment that there is an intelligent author of nature (the role in which God is being cast here) is not based on an objectively valid ground. My reason for judging that there is an intelligent author of nature is not that I directly perceive one or possess some proof of its existence (Kant undermined such proofs in the Transcendental Dialectic), but that postulating an intelligent author is a means to the end of finding natural explanations of natural phenomena.[55] For instance,

[53] That these ends are morally obligatory does not entail that they are actually shared by other rational agents; however, the fact that the moral obligation to set those ends extends to all finite rational agents entails, presumably, that all finite rational agents have sufficient *reason* to set those ends, and hence to share those beliefs. Cf. *CPrR* (Ak. 5: 143), *JL* (Ak. 9: 86 n.), and *WL* (Ak. 24: 851).

[54] Kant claims that belief is only possible on practical grounds (ends of action) at A823/B851, but he appears to take this back three pages later when he gives his theory of doctrinal beliefs. He is more skeptical about the possibility of theoretical rational belief in the logic lectures: *JL* (Ak. 9: 68 n., 69), *LBu* (Ak. 24: 638–9), and *WL* (Ak 24: 852, 855 f.).

[55] Some caution is required here. Kant denies that belief in an intelligent cause of nature or (by parity of reasoning) in an absolutely necessary ground of possibility is a "*Hypothese*" because it does not contain a

if I postulate that nature is the product of an intelligent author, if I were to find that there are two species of natural object that seem to have nothing to do with one another, to be entirely incommensurable and governed by entirely different laws, I will not stop at this stage of inquiry, but will press on until I find some more fundamental understanding of both species that explains how they are particular determinations of a more general kind. I will do so because I have postulated a wise author who would not create a system of nature that is so disunified, one that lacks 'purposive unity.'[56] The postulate will lead to greater and richer empirical knowledge. Consequently, postulating the existence of an intelligent author of Nature is a means to one of my theoretical ends—to have richer and deeper explanations of the natural world. What is more, Kant claims it is a *necessary* means to that end: "purposive unity is still so important a condition of the application of the reason to nature that I cannot pass it by." This fact is an intersubjectively valid ground for all other agents who share the end of understanding Nature to believe that there is an intelligent author, which makes this belief a case of *rational* belief. Since this ground does not necessarily entail (or even significantly raise the probability) that there is an intelligent author of Nature, this is a case of mere belief, not knowledge. Notice, though, that this is a case of *rational* belief only if it is based on an intersubjectively shareable ground. Kant is assuming, not implausibly, that our theoretical ends are in principle shareable by other epistemic agents; since our theoretical ends are given by the nature of reason itself (more about this in section 6), and reason has the same nature in all rational agents, theoretical ends of one agent are in principle shareable by others.

Kant goes on in the rest of the section to develop his theory of moral beliefs, cases where believing something (e.g., that the highest good is possible) is a rational requirement on setting an end (e.g., the achievement of the highest good) that is morally obligatory for all (finite) rational agents. Because these ends are morally obligatory, the relevant belief is based on grounds that are valid for all rational agents.[57] In his discussion of doctrinal belief, Kant claims that the theoretical end in question—systematicity in our explanations of nature—is merely a "contingent"

determinate enough concept of its object (A827/B855) to explain why nature exists or why anything is really possible. As we have already seen, though, there has been a problem in Kant's view at least since 1763 about representing the determinate content of the idea of God as the ground of real possibility (see Ch. 4); in the next chapter I will address this issue in the Critical philosophy, and thus address (in the terms of A827/B855) whether rational belief in the existence of an absolutely necessary being is a genuine 'hypothesis' or not. Cf. the discussion of hypotheses in *LP* (Ak. 24: 439–40). Thanks to Noam Hoffer for pressing me to clarify my views on this point.

[56] I take 'purposive unity' to be the unity a thing has if it is created by a wise and intelligent being. The notion of 'purposiveness' is discussed extensively in the *CJ*. This passage, on the necessary doctrinal belief in a wise nature, should be compared with the two Introductions to that work, as well as to the discussion of the regulative use of the idea of God in the *CPR* in the Appendix to the Transcendental Dialectic (A669–704/B697–732).

[57] A828/B856 and *CPrR*, Ak. 5: 119–46.

theoretical end of the agent in question.[58] This raises the intriguing possibility, though, of a theoretical analogue of moral belief: judgments that must be believed in order to satisfy a necessary end of theoretical reason.[59] Kant calls moral beliefs—beliefs mandated by the nature of practical reason itself—'postulates of pure practical reason.' We might therefore think of necessary theoretical beliefs as postulates of pure *theoretical* reason. To make the analogy complete, we will need to impose on theoretical belief a requirement that Kant imposes on moral belief: its object must be logically possible, and must not be knowably really impossible.[60] No matter how much it would serve the ends of reason for us to believe that *p*, if *p* is logically inconsistent or provably really impossible, then we cannot have any rational warrant for accepting it. I will call this the 'no impossibilities' rule.

Note, though, that if all rational agents as such have end E, and subjectively sufficiently judging that *p* is a necessary means to E, then all rational agents as such have grounds to judge that *p*. So, if there is some end E that is set by the nature of theoretical reason (that is, solely in virtue of having theoretical reason we set theoretical end E) and there is some *p* such that subjectively sufficiently judging that *p* (holding *p* to be true with subjective conviction) is a necessary means for satisfying *E*, then we can have rational belief that *p* (belief based on IS-valid reasons). Recall, however, what Kant says about the postulate of an absolutely necessary being: it is "subjectively necessary" and "grounded in the nature of human reason." This suggests that belief in an absolutely necessary being is a necessary theoretical belief, where the end in question is explaining why anything is (noumenally) really possible, that this end is given by the nature of theoretical reason, and that this belief is a necessary means for that end because it is the only means by which we can explain why anything is (noumenally) really possible.[61]

In the next section I argue that, on Kant's view, these conditions are satisfied. We can have the theoretical rational belief that there is an absolutely necessary being if we base our belief on the fact that, without this being, we cannot explain noumenal real possibility. What is more, this theoretical belief is rationally obligatory for us: it is the only way to satisfy a necessary theoretical end.[62] This is *belief* rather than *knowledge*

[58] Although he denies this in both the Appendix to the Transcendental Dialectic and in the *CJ*; see previous discussion.

[59] Kant encourages the thought when he describes doctrinal belief as "a merely theoretical [...] analogue of practical belief" (A825/B853).

[60] Cf. Chignell (2007*b*), 347–50.

[61] Chignell (2007*b*), 349 anticipates my point here, even discussing the Pölitz text with which I began and to which I return later in this chapter. Chignell also points to the following *Reflexion*: "the principle of the self-preservation of reason is the basis of rational belief, in which assent [*Fürwahrhalten*] has the same degree as knowledge, but is of another kind which comes not from the cognition of grounds in the object but rather from the true needs of the subject in respect to *theoretical* as well as practical application" (*Refl.* 2146, Ak. 16: 371–2).

[62] Robert Hanna develops an analogous explanation of Kant's belief that cognitive subjects are 'noumenally' affected by things in themselves: reason demands an explanation of our sensory states. See Hanna (2001), 117.

because it is based on a ground that is not objectively sufficient for the existence of God; the fact that this postulate allows us to satisfy a necessary end of theoretical reason (explain why anything is noumenally really possible) does not make more probable or necessarily entail that God exists.[63]

9.6. The Necessary Ends of Theoretical Reason

The conclusion of the previous section is that we can have a species of Kantian belief—which I characterized as 'necessary theoretical belief'—in the existence of an absolutely (noumenally) necessary being that grounds all (noumenal) real possibility if two conditions are met:

(a) It is a necessary end of theoretical reason to explain why anything is (noumenally) really possible at all.
(b) The necessary means to explain this is to posit the existence of an absolutely (noumenally) necessary being that grounds all (noumenal) real possibility.

In this section I explore Kant's theory of (theoretical) reason in the Transcendental Dialectic of the *CPR* and his Critical reconstruction of the *Beweisgrund* argument in the Transcendental Ideal. I argue that they entail (a) and (b). This is what Kant means when he says in the Pölitz lectures that "[the *Beweisgrund* argument] can in no way be refuted, because it has its ground in the nature of human reason" (*Pöl.RT*, Ak. 28: 1034).[64]

Claim (a) concerns the faculty of theoretical reason [*Vernunft*], which differs in crucial respects from the faculty of understanding [*Verstand*]. The understanding is the faculty by which we think about sensibly given objects using concepts; consequently, the understanding is the faculty of cognition: representations of knowably really possible objects (those that can be given in intuition).[65] The faculty of reason does not produce cognition but brings unity to the cognitions generated by the faculty of understanding. Reason has two uses: a logical use, and a 'pure' use (which I discuss below). The logical use of reason is to bring logical unity to our cognition. Our cognition has logical unity to the extent that individual cognitions (e.g., individual judgments) are explained by higher-level principles. Reason discharges its logical

[63] Kant does remark in "Opining, believing, and knowing" that "there is something unstable about doctrinal belief; one is often put off from it by difficulties that come up in speculation, although, to be sure, one inexorably returns to it again" (A828/B856). I take this to be a reference to the antinomial conflicts that afflict reason when it attempts to think about supersensible objects. Kant's resolution of those antinomial conflicts, however, clears the way for the possibility of a non-wavering doctrinal belief in an absolutely necessary being, once we have proved that it is the unique way to answer reason's demand for a ground of possibility and have understood that it is merely a belief, not a claim to know supersensible reality.

[64] In the rest of this section 'reason' means specifically *theoretical* reason, or reason in its theoretical use. I am not attempting to give an interpretation of Kant's theory of practical reason, much less a unified account of reason as such.

[65] See Ch. 6.6.

function by subsuming individual items of cognition under higher-level explanatory principles.[66]

The logical use of reason is governed by the rule: seek a condition [*Bedingung*] for every conditioned [*bedingt*] object. I will refer to this as the *logical principle of reason* (LPR for short). I take 'condition' in the Transcendental Dialectic to mean *ground* in the very general sense of 'ground' used in pre-Critical German rationalism (including by Kant himself): it includes logical grounds, real grounds, causal grounds, etc., though it excludes merely "consequentially determining grounds" (*ND*, Ak. 1: 392), grounds of mere *a posteriori*$_G$ knowledge. Reason seeks conditions, not consequences.[67] Kant uses this term in the Dialectic to indicate that it is *wider* than any of the grounding relations he has considered up to that point in the *CPR*. Consequently, the relation of 'a condition to what it conditions' is a very broad notion that embraces, among other things, the relation of a spatial region to its parts, the relation of a material part to a material whole, the relation of a cause to its effect, etc.

Kant writes: "we see very well that the proper principle of reason in general (in its logical use) is to find the unconditioned for conditioned cognitions of the understanding, with which its unity will be completed" (A307/B361). The LPR leads to the idea of the *complete* series of conditions for a given object; since the complete series contains all of the conditions of the object, it is not conditioned by anything outside it, so we can also think of it as the *unconditioned* series of conditions for that object. Since there are different relations of condition to conditioned (of ground to consequence), the concept of the unconditioned series of conditions is always the concept of the series of objects that stand in a specific grounding relation to the original conditioned object, which has no grounds of the relevant kind outside that series.[68] For instance, given an alteration in time, reason forms the idea of the *complete series of causes of that alteration*, which itself has no causal condition external to the series itself (though it might have *spatial* conditions, e.g., the larger region of space which bounds and conditions the region in which this causal series occurs). The concept of the unconditioned is an *idea* in Kant's technical sense: a (logically consistent) concept of an object that could never be given in experience because it requires a completeness (of conditions) that is never present in any object of experience.[69]

This "principle of reason in general (in its logical use)" expresses a necessary end of theoretical reason, an end theoretical reason sets us in virtue of being the faculy that it is.[70] It "is merely a subjective law of economy for the provision of our understanding" (A306/B362). That we should search for a condition for every conditioned object and form the idea of an unconditioned being does not entail, or presuppose, that this goal could ever be reached, or that there is an unconditioned being. It does not tell us whether there is an unconditioned series of conditions for a

[66] A305/B361. [67] A331/B388. [68] A323/B379. [69] A320/B377.

[70] See Stang (2014*a*) for more on the idea that Kantian faculties have certain ends in virtue of being the faculties they are.

given object. In particular, Kant warns us not to make a "pure use" of reason and assume that "when the condition is given then the whole series of conditions, subordinated to one another, which is itself unconditioned, is also given" (A308/B364). To avoid confusion, I will refer to this assumption as the *pure principle of reason* (PPR). Mistaking the LPR, a mere "subjective law of economy," for the PPR is what Kant calls "transcendental illusion," and the goal of the Transcendental Dialectic is to uncover transcendental illusion as the source of dogmatic metaphysics and, as much as possible, dispel it.[71] What does the 'givenness' of an unconditioned series mean here? Recall that intuitions 'give' us objects; in Chapter 5 I interpreted this to mean that intuitions allow us to have thoughts about *the objects there are* (q-objects). Applying this reading of 'givenness' to the PPR, I interpret it as the assumption that, necessarily, if there is a conditioned object, then *there is* an unconditioned series of its conditions. The two requirements on theoretical belief, (a) and (b) from above, can be reformulated as claims about the principle of reason:

(a*) The LPR requires us to think of an unconditioned series of conditions for the fact that anything is (noumenally) really possible (the conditioned).
(b*) The only conception of the unconditioned series of conditions of the (noumenal) real possibility that would satisfy the LPR is that it terminates in an absolutely necessary being, one whose non-existence would cancel all (noumenal) real possibility.

I will argue that both are the case.

The Transcendental Dialectic contains a discussion of reason's search for a ground of (noumenal) real possibility that mirrors Kant's *Beweisgrund* argument, "The transcendental ideal (*prototypon transcendentale*)" (A571–83/B579–611). However, the close connection between that text and *Beweisgrund* is partly obscured by Kant's framing of the discussion around the issue of complete determination. Readers will recall from Chapters 4 and 5 that the complete determinacy of existents or of possibilities played no significant role in the positive argument of *Beweisgrund*. The reason he gives them such prominence in the Ideal is that he assigns the Ideal an additional dialectical function: to diagnose the transcendental illusion involved in Baumgarten's complete determination theory of existence. Hence, my explication of the Transcendental Ideal, and Kant's Critical attitude towards the *Beweisgrund* argument, will begin with a brief discussion of complete determination. I will then attempt to bracket the issue of complete determination and read the Transcendental Ideal as Kant's reconstruction, in Critical terms, of the original *Beweisgrund* argument.

It is impossible for us to cognize objects *as* completely determinate. For any object of a possible experience, there are always (empirical) predicates of the object that we do not experience. To make things concrete, take a body, which exercises mutual

[71] A297/B353. However, Kant describes transcendental illusion as *inevitable* at A297/B354, A298/B355, and A339/B397. For a discussion of the 'inevitability' thesis, see Grier (2001), 126–30.

attractive force on all other bodies in space. For any possible experience of that object, there is some distant region of space R, such that the experience leaves open whether there are bodies in R on which the given object exerts a (very small) mutual attractive force, so our experience leaves open whether it has, or whether it lacks, the predicate *exercises mutual attractive force on bodies in region R* (call this predicate F_R). Now, for any such region R, we could have a more complete experience that includes experience of region R, which would represent the body as determinately having or lacking predicate F_R. But for that experience there will be *another* distant region R* such that the more complete experience leaves it indeterminate whether the body has predicate F_{R*}. The point is that *for any such* predicate, there is a possible experience that would settle whether the object determinately *has* or determinately *lacks* the predicate. But for any such experience, there is some predicate such that the experience does not settle whether the object determinately has or lacks it. Consequently, for any possible experience the set of predicates it represents its objects as having is incompletely determinate; there will always be a predicate F such that neither F nor its negation ~F is included in the set.

An experience of an object that does not represent it as having or lacking (being determinate with respect to) a given predicate is a limitation of (is *conditioned* by) some possible 'larger' experience that would be determinate with respect to that predicate. So the logical principle of reason (LPR) sets us a task: to have a more complete experience of the object that would represent a more determinate set of predicates. Reason thus forms the idea of an unconditioned series of such conditions: an unconditioned series of ever more inclusive experiences, corresponding to an ever more determinate set of experienced predicates. Reason also forms the idea of the complete set of predicates that *could* be represented in any such series of experiences and thus appear in any corresponding series of sets of experienced predicates. Call this the set of all possible predicates.[72]

The first question one might have about this story is: what does 'possible' mean here? Let us take this, for now, to be a discussion of *formal* possibility, so the idea of the complete space of possible predicates is the complete space of *formally* possible predicates. But notice that there are two places where we can (illegitimately) apply the PPR. First, from the fact that we set ourselves the theoretical end of more completely experiencing objects (experiencing a more complete set of their predicates) it does not follow that there *is* a completely determinate experience, or even an unconditioned series of ever more completely determinate experiences. We cannot

[72] This is my reconstruction of Kant's *highly* condensed discussion in the last paragraph of the Ideal, A582–3/B610–11. First, we illegitimately apply the PPR and assume that the unconditioned series of experiences is given, and hence that the complete space of possible predicates of phenomena is given. Then (as Kant explains in the previous paragraph, A581/B609) we represent every phenomenon as having its properties in virtue of how it is represented in this one sum total of all experience. *Then* we apply this principle (individual things have fully determinate sets of possible properties in virtue of their relation to some sum-total of all possible properties) to objects in general (phenomena and noumena).

assume that the unconditioned series of conditions is given. Consequently, it does not follow that the set of predicates of objects forms an unconditioned series. In other words we have no guarantee that 'the set of all possible predicates of objects' refers to any set. This constitutes Kant's explanation of the transcendental illusion (the illegitimate application of the PPR) that generates the assumption that objects are completely determinate and that there is a complete space of all possible predicates of objects. This in turn, Kant thinks, leads to Baumgarten's view that beings as such (not just objects of possible experience) are completely determinate.

But the Transcendental Ideal contains a thread that is separable from the idea of complete determinacy, the thread that represents Kant's original thinking in *Beweisgrund*: "if we consider all possible predicates not merely logically but transcendentally, i.e. as to their content which can be thought in them *a priori*, then we find that through some of them a being is represented, and through others a mere non-being [...] all concepts of negations are thus derivative, and the realities contain the data, the material, so to speak, or the transcendental content, for the possibility and the complete determination of all things" (A574–5/B602–3). Not only does this recapitulate exactly the line of thought from *Beweisgrund*—logically complex predicates are possible only if their atomic constituents, the *data* of all possibility and all thought, are themselves really possible—it remains a compelling line of thought for reason [*Vernunft*] even if we eliminate the issue of complete determination. We do not need to assume that there is a set that contains all possible predicates in order for reason to ask: what is the condition that explains why there are really possible predicates at all? Nor do we need to assume that there is a set of all possible predicates in order to ask whether there is a set of *fundamental* predicates (a subset of the realities Kant speaks of above at A575/B603) such that there is no really possible predicate (at no point in the ascending series of sets of possible predicates) that is not ultimately derivative of those fundamental predicates. To return to my concrete example, it is implausible that the predicate *exerts mutually attractive force on matter in region R* is a *fundamental* empirical predicate. So even if ever more complete experiences will reveal ever more predicates in a series of conditions *in indefinitum*, this does not mean that they will reveal further *fundamental* predicates.

To apply LPR and its quest for conditions to the possibility of properties we need to specify what concept of real possibility is involved here. We already have a complete account of the conditions of formal, empirical-causal, and nomic possibility. However, even if we can know that some putatively possible predicate is compatible with our forms of experience (it is formally possible that an object have it), with the laws (it is nomically possible that an object have it), and with the laws *given* the past (it is empirically-causally possible that an object have it), we can *still* ask whether it is noumenally possible, that is, whether it has a noumenal ground of its possibility.[73]

[73] Cf. the discussion of the possibility of a different "single, all-encompassing" experience (A231–2/B284) in Ch. 7.5.

Kant himself explicitly attributes this question to *reason* (not the understanding) in the Postulates of Empirical Thinking in General: "[these] are proper questions, and are, be sure, to be solved synthetically, though they fall under the jurisdiction of reason alone" (A230/B284). Consequently, Kant's discussion of possibility in the Transcendental Ideal (*prior* to his discussion of the complete determinacy of experience at A581/B609) is a discussion of noumenal possibility. We are to consider, he says, "the whole of possibility, as the sum total [*Inbegriff*] of all predicates of things in general (*Dinge überhaupt*)" (A572/B600). I take it that Kant is talking about 'things' here to remind us that we are not talking only about objects of experience, but about 'beings as such,' as Wolff or Baumgarten might put it, which includes phenomena and noumena; we are trying to reconstruct a certain kind of pre-Critical metaphysics (indeed, Kant's own) as the product of transcendental illusion. In other words, we are now to apply LPR to the noumenal real possibility of predicates of objects in general, phenomena and noumena.

Take some putatively noumenally really possible predicate P of a phenomenon or a noumenon. By applying the LPR we are set the theoretical task of thinking of the *grounds* (conditions) of the possibility of P. By definition, these are noumenal grounds. By applying the LPR again, we are set the task of thinking of the noumenal grounds (conditions) of the grounds (conditions) of the possibility of P, i.e. what predicates in noumena make possible P. We can separate out logically complex properties; the possibility of the logically complex properties is ultimately grounded in (conditioned by) the possibility of their constituents. If a predicate is a mere *absence* of positive determination, then it is a negation of a reality. The reality is its condition, so we are interested in the real possibility of *that* reality (the condition). Likewise for properties that are limitations of realities; since limitations are conditioned, we can inquire into the ground of the possibility of the unlimited reality of which they are limitations. As Kant writes in the Ideal, "all negations [. . .] are mere limitations of a greater and finally of the highest reality; hence they presuppose it, and as regards their content they are merely derived from it" (A578/B606). 'Content' here, as in *Beweisgrund*, means (noumenal) real possibility.

This means that by applying the LPR to the concept of noumenal real possibility we are led to consider the fundamental predicates of noumena that make all other preperties of noumena really possible, and which make noumenally possible (specifically, noumenally-causally possible) predicates of phenomena possible. We can distinguish predicates that are derivative from those that are fundamental, as we did above. We can ask, further, whether the fundamental predicates are instantiated by a plurality of different noumena, or whether there is a unique noumenon that possesses all fundamental predicates, and whose possession of those predicates makes all other noumenally possible predicates really possible. Note, though, that all of this is independent of the idea of the complete determinacy of existing or possible things, and from the assumption that there is a space of all possible predicates. For the application of the LPR requires us to think for every possible predicate of the ground of its possibility; this is compatible with it being the case that the space of all possible predicates is indefinitely extendable, as may be the case with the space of all possible predicates of objects of experience (see previous discussion).

This regressive search for the conditions of (noumenal) real possibility in the Transcendental Ideal is the project of *Beweisgrund*, but transposed from a dogmatic-metaphysical to a critical-subjective key: the project is not to give a metaphysical theory of reality but to satisfy our "subjective law of economy," the LPR. What are presented in 1763 as metaphysical arguments appear, from the perspective of 1781, as expressions of reason's search for conditions. For instance, in *Beweisgrund*, having established (in Critical terms) that noumenal real possibilities must have noumenal real grounds, Kant argues that these possibilities cannot be 'parceled' out among a plurality of grounds (the 'pluralist' view). In Chapter 5.5 I argued that, even on the most charitable reconstruction, Kant's argument relies on assuming (i) that there cannot be an infinite ascending chain of grounds[74] and (ii) that there must be a single ground that explains why the various grounds of possibility postulated by the pluralist can interact with objects in the same world.[75] From a Critical standpoint we can see these assumptions as grounded in reason's regressive search for the unconditioned and for unity. From this perspective, (i) expresses reason's idea of a series of conditions that terminates in an unconditioned condition; (ii) expresses reason's demand for a condition of the plurality of grounds. Kant addresses precisely this point in the Ideal: "because one also cannot say that an original being consists in so many derivative beings, since each of the latter presupposes the former and so cannot constitute it, the ideal of the original being must also be thought of as simple" (A579/B607). If the fundamental predicates are 'parceled out' among a plurality of fundamental grounds (grounds of possibility, which are themselves ungrounded), any one of which grounds only a subset of them, this will violate reason's demand for unity: the series of conditions that explain why various predicates are really possible will terminate in multiple unconditioned beings. The precise point at which Kant's *Beweisgrund* argument was weak is the point at which Kant can make philosophical progress by transposing his claims from the register of metaphysical arguments to that of the subjective demands of reason: a plurality of fundamental grounds is rationally unsatisfying because it lacks explanatory unity.

This is why it is crucial in Kant's *Beweisgrund* conception of the *GARP* that its existence *and possibility* lack grounds; if they had grounds, then reason's regressive search for conditions could not stop there. The absolute necessity of *Beweisgrund*'s *GARP* is ultimately 'outward looking': the *GARP* is necessary because it grounds all real possibility, and, if it were canceled, nothing would be really possible. This is perfectly suited to Kant's conception of reason and the unconditioned in the Dialectic, because the *GARP* there plays the role of the

[74] See Fig. 5.2 and the discussion surrounding it.

[75] See Figs. 5.3 and 5.4 and the discussion surrounding them.

stopping point of reason: once reason reaches it, its regressive search for the conditions of possibility can cease.

What is more, the picture of the ground of all (noumenally) really possible predicates at which reason arrives in the Transcendental Ideal is precisely Kant's picture of the *GARP* in *Beweisgrund*:

The derivation of all other possibility from this original being, strictly speaking, also cannot be regarded as a limitation of its highest reality and as a **division**, as it were; for them the original being would be regarded as a mere aggregate of derivative beings, which, according to the above, is impossible, even though we represented it in such a way at the beginning in our first crude outline. Rather, the highest reality would ground the possibility of all thing as a **ground** and not as a **sum total**; and the manifoldness of the former rests not on the limitation of the original being itself, but on its complete consequences; to which our whole sensibility, including all reality in appearance, would then belong, which cannot belong to the idea of a highest being as an ingredient. (A579/B607)

This is just the point, which I discussed extensively in Chapter 4, that (contra Chignell 2009*a*) the *GARP* does not ground really possible predicates by instantiating *all* unlimited realities; some really possible predicates are "consequences" of its fundamental predicates but are not limitations of them. As Kant points out earlier in the Ideal, not all unlimited realities are equally fundamental, and some are inconsistent with one another: the idea of a *GARP* "excludes multiplicity of predicates, which, as derived from one another, are already given [in the fundamental predicates—NS] or cannot coexist with one another" (A573–4/B601–2). This was precisely Kant's anti-logicist point in *Beweisgrund*: some primitive (hence unlimited and unnegated) realities are inconsistent with one another, so no being can possess all of them.

Let us return to the passage with which I began this chapter, where Kant says that in *Beweisgrund* "it was shown that of all possible proofs, the one which affords us the most satisfaction is the argument that if we remove an original being, we at the same time remove the substratum of the possibility of all things" (*Pöl.RT* Ak. 28: 1034). The most rationally satisfactory explanation (the explanation that best satisfies LPR's regressive search for conditions) of why anything is (noumenally) really possible is that there is an absolutely necessary being that grounds these possibilities. But, Kant now claims, this should not have been mistaken for a proof that there is an absolutely necessary being: "this proof can in no way be refuted, because it has its ground in the nature of human reason. For my reason makes it absolutely necessary for me to assume a being which is the ground of everything possible, because otherwise I would be unable to know what in general the possibility of something consists in" (Ak. 28: 1034). I take this to mean that:

(a*) the LPR directs us to find an explanation (condition) for the (noumenal) real possibility of predicates (the conditioned).

We have the concept of noumenal real possibility, and noumenal real possibilities are subject to reason's search for conditions. Furthermore:

(b*) the only conception of the ground of (noumenal) real possibility that would satisfy the LPR is that of an absolutely necessary being, one whose non-existence would cancel all (noumenal) really possible predicates.

Kant asserts that his own pre-Critical conception of God as the ground of all real possibility is the only way to satisfy reason's demand for conditions of possibility ("otherwise I would be unable to know what in general the possibility of something consists in").

Kant's pre-Critical view of God as the absolutely necessary ground of all real possibilities satisfies two other important requirements on theoretical belief: the availability of intersubjectively valid reasons grounded in the nature of theoretical reason itself, and our inability to know such a being to be impossible. Since the putative grounds for theoretical belief in God are a necessary means to a necessary theoretical end of reason ((a*) and (b*)), they are grounds that are intersubjectively valid for all (finite) discursive rational intellects. Since this is a concept of a noumenon and it is logically consistent, we can never know (at least on theoretical grounds) that it is impossible. This entails that if a human epistemic agent (i) judges that [*hält für wahr*] there is such an absolutely necessary ground of (noumenal) real possibility, (ii) bases this judgment on (a*) and (b*), (iii) judges further that (a*) and (b*) do not (noumenally) necessarily entail that there is such a being, then that agent has theoretical belief in the metaphysical theory of modality first offered in *Beweisgrund*. Any human epistemic agent can come to have rationally grounded theoretical belief in an absolutely necessary being.

Earlier in the same passage Kant says of his *Beweisgrund* proof that "even this proof is not apodictically certain; for it cannot establish the objective necessity of an original being, but establishes only the subjective necessity of assuming such a being" (*Pöl.RT*, Ak. 28:1034). Read in the light of Kant's theory of the necessary ends of theoretical reason in the Transcendental Dialectic, this means that the original 1763 proof of a *GARP* was the product of transcendental illusion: it was illegitimately assumed that if the conditioned (a really possible predicate) is given then so is its unconditioned condition (*GARP*). Kant's Critical attitude towards the *Beweisgrund* proof is that it mistook the intention for the deed: although the postulate that God is the ground of all real possibilities is the most rationally satisfactory explanation of real possibility, and this is sufficient to warrant theoretical belief that there is such a being, this does not entail that there is such a being. Consequently, the epistemic ground on which we base our *belief* in the *GARP* does not necessitate its existence, so this is not a case of knowledge. Furthermore, because we cannot so much as cognize [*erkennen*] the real (noumenal) possibility of such a being, it cannot in principle be an object of knowledge [*Wissen*] for us.

10

The Antinomy of Kant's Modal Metaphysics

10.1. Introduction

The title of this chapter does not refer, as some readers might expect, to the fourth of the Antinomial Conflicts of Pure Reason, which concerns the conflict between a putative proof that there is a necessary being that grounds all contingent beings (Thesis) and a putative proof that there can be no such necessary being either in space and time or outside them (Antithesis). Like each of the Antinomial conflicts, the Fourth Antinomy is *resolved* in the Transcendental Dialectic: there can be no necessary being in space and time, but there may be (for all we know) a necessary being outside space and time, among the noumena. The title of this chapter refers instead to what I take to be an *unresolved* antinomy in Kant's modal theory, between, one the one hand, his theory of freedom, and, on the other hand, his famous claim in §§76 and 77 of the *Critique of Judgment* (*CJ*) that an intuitive intellect does not cognize modal properties in its objects.

As I discussed in Chapter 7, Kant's theory of freedom is intended to reconcile the fact that I am bound by the moral law with the fact that all of my actions are empirically-causally necessary (they are the consequences of deterministic laws, given the past). That I am bound by the moral law means that, although I actually commit an impermissible act, I *ought not* to have done so, which in turns entails that I *could have* omitted that act (*ought* implies *can*). But, considered merely as a phenomenon, it is not (empirically-causally) possible that I do otherwise. However, I am also a noumenon, which means that I have a nature independently of how I cognize myself (and how other discursive intellects cognize me) theoretically, a nature that appears to me (and to others) in space and time. This noumenal nature includes a will, and this will has the power to determine itself otherwise than it actually does (to adopt a maxim of action other than its actual one). For any act I commit, I possess the power, as a thing in itself, to have so determined my noumenal will that I *would not* have committed the action *if I had* so determined my will. As Kant writes in the second *Critique*:

> So considered [as a noumenon] a rational being can now rightly say of every unlawful action he performed that he could have omitted it even though as appearance it is sufficiently determined in the past, and so far, is inevitably necessary [...] (*CPrR*, Ak. 5: 98)

This is not a separable or minor commitment of Kant's theory of free will; Kant's proof of the compatibility of the empirical necessity of my actions with freedom and *thereby* with the bindingness of the moral law *rests* on attributing a modal property to my noumenal will: as a noumenon, I could have determined my will otherwise, so my actual acts are contingent.

However, this means that Kant's theory of freedom stands in tension with his claim, articulated most forcefully in §76 of the *CJ*, that an intuitive intellect would not represent objects as merely possible or as contingent:

> For an intellect [*Verstand*] to which this distinction [between intuiting and thinking] did not apply, all objects that I cognize would **be** (exist), and the possibility of some that did not exist, i.e. their contingency if they did exist, as well as the necessity that is to be distinguished from that, would not enter into the representation of such a being at all. (*CJ*, Ak. 5: 403)

An intuitive intellect is one whose faculty of intuition is spontaneous rather than receptive (like ours); consequently, the distinction between passive intuition and spontaneous thought, which holds for us, would not hold for such a mind. Kant is claiming in this passage that a being with an intuitive intellect would not represent some objects as merely possible (those it had conceived but not intuited), some as contingent (those it intuited but which it could conceive not being present) and some as necessary. In what follows, I will refer to this thesis as the thesis that an intuitive intellect does not cognize modal properties in its objects: it does cognize its objects as being possible, contingent, etc.

These two claims—that noumena have modal properties, and that an intuitive intellect does not cognize modal properties—stand in tension because God has an intuitive intellect and God is omniscient; God cognizes not merely the truth about noumena but the *whole* truth about them. As Kant says in the Pölitz lectures on rational theology, "God cognizes all things as they are *in themselves* immediately and *a priori* through an intuition of the understanding" (*Pöl.RT*, Ak. 28: 1052). He is therefore committed to an inconsistent set of claims:

(P0) If there is a God, he has an intuitive intellect.[1]
(P1) Some noumena have modal properties. [Consequence of the theory of freedom.]
(P2) No intuitive intellect cognizes its object(s) as having modal properties. [*CJ* §76.]
(P3) Noumena have modal properties if, and *only if*, if there is a God, God would cognize those properties.[2] [Omniscience of God, see previous discussion.]

(P1) is inconsistent with the 'noumenal amodalist' consequence of (P2) and (P3): noumena lack modal properties. Although I argued at length in Chapter 9 that Kant

[1] *Pöl.RT* (Ak. 28: 996, 1043, 1052), *Danz.RT* (Ak. 28 :1259, 1267), and *Volck.RT* (Ak. 28: 1165).

[2] My presentation of this problem is indebted to Kohl (2015). Whereas Kohl tries to use the intuitive intellect to refute what he calls the 'Leibnizian' reading of Kant (that noumena have categorial properties), I try to reconcile the intuitive intellect and the 'Leibnizian' reading.

is committed, not only for practical, but for *theoretical* reasons, to belief [*Glaube*] in the existence of God, I have formulated these premises as conditional on the existence of God in order to highlight that this problem is independent of Kant's own positive commitment to the existence of God; even if there is no God the problem remains. Nor can it be dismissed by pointing out (correctly) that God is merely an object of (theoretical and practical) *belief* rather than *knowledge*; regardless of the epistemic status of these inconsistent claims, as long as Kant is committed to *each* of them, there is an inconsistency in his view. Kant's doctrine that the intuitive intellect does not represent modal properties appears to be incompatible with what, according to Kant himself, is one of the two foundations of the Critical system: freedom of the will. The modal properties in question are modal properties of noumenal wills, e.g., *contingently* subordinating the moral law to self-interest, *possibly* doing otherwise, etc.

Although it does not take the canonical Kantian form of two incompatible claims (Thesis and Antithesis), this potential inconsistency deserves to be called the (or least *an*) antinomy of Kant's modal metaphysics, for, unlike the Fourth Antinomy, it is neither recognized nor resolved by Kant himself. What is more, while it may have been *recognized* by readers of Kant, to my knowledge no sustained attempt has been made to resolve it within Kant's system.[3] In the case of post-Kantian figures like Fichte, Schelling, and Hegel, who thought the intuitive intellect was one of Kant's most important ideas, this is because they were less interested in reconciling the various aspects of his system (which they tended to see as shot through with inconsistencies, at least if taken at face value) than with developing an adequate form of idealism on the basis of key Kantian insights. If they recognized this inconsistency at all, Schelling and (at least the early) Hegel would regard it as a sign that the intuitive intellect is incompatible with certain limitations in Kant's own system (e.g., the sharp distinction between inclination and duty, and hence between necessity and freedom) and embrace what they saw as the necessitarian (indeed, Spinozistic) consequences of *CJ* §76; [4] nor would Fichte, who understood intellectual intuition very differently than Schelling or Hegel, necessarily see this 'modal antinomy' as a problem in need of a solution.[5]

[3] There is an extensive literature on the Kantian notion of intuitive intellect and its fate in post-Kantian German idealism; see esp. Gram (1981); Tüschling (1990) and (1992); Allison (2000); Longuenesse (2000); Westphal (2000); Beiser (2002), 294–301, 580–2; Förster (2011); and Leech (2014).

[4] Cf. the discussion of intuitive intellect and *CJ*, §76 in *Faith and Knowledge* (in H.*Werke*, 2: 316–7, 325–7). Tüschling (1992) argues that *CJ* §76 is already Spinozistic and proto-Schellingian; Allison (2000) rebuts that interpretation. On Schelling's and the young Hegel's embrace of what they saw as the Spinozistic consequences of intellectual intuition, see Beiser (2002), 580–2. Longuenesse (2000) points out correctly, though, that *CJ* §76, taken at face value, does not entail noumenal necessitarianism, but noumenal amodalism: noumena have no modal properties whatsoever (Longuenesse (2000), 279 n. 19).

[5] For an analysis of Fichte's various notions of 'intellectual intuition' see Breazeale (1998) and Beiser (2002), 294–9.

In this chapter I explore the path not taken by these German Idealists, and attempt to resolve this apparent inconsistency in Kant's theory of the intuitive intellect (represented by (P2) and (P3)) and the modal commitments of his theory of freedom (represented by (P1)) using the resources of Kant's own system. I will take premise (P0) as a fixed point in what follows and consider the inconsistency among (P1), (P2), and (P3) *given* Kant's commitment to (P0). In section 3 I explore several ways one might resolve this inconsistency and argue that none of them succeed. In section 4 I present my own resolution of the problem; in sections 5 and 6 I expand upon that solution and explain how Kant's theory of freedom can be made consistent with 'noumenal amodalism.' In section 7 I use *CJ* §76 to clarify the relations among three notions that have been important throughout this book: existence, existence* (causal efficacy), and actuality. I conclude in section 8 by showing how my solution to the 'modal antinomy' helps clarify, retrospectively, central issues in Kant's modal metaphysics in *Beweisgrund*. First, though, in section 2, I briefly explain what an 'intuitive intellect' would be.

10.2. Intuitive Intellect, Intellectual Intuition

Although the concept of an *anschauende Verstand* (less frequently, '*intuitive Verstand*') is one of the most historically influential concepts in the whole Critical philosophy (due to its reception in post-Kantian German philosophy), it is only explicitly discussed in a few passages in the entirety of Kant's published writings.[6] Although I have elsewhere followed standard practice and translated '*Verstand*' as 'understanding', in this book I am going to translate '*intuitiver Verstand*' as 'intuitive intellect' to bring out the fact that it refers to a very different kind of mind than our discursive *understanding*.

Kant introduces the idea of intuitive intellect by contrasting it with our *discursive* form of intellect. Objects are given to a discursive intellect like ours passively through causal affection of our sensible faculty (hence, our intuition is said to be 'sensible'), and we then must *think* spontaneously about those objects using general (discursive) representations, concepts. The idea of an intuitive intellect is the idea of a kind of mind whose understanding (spontaneous faculty) would *give* it objects, that is, a mind that would spontaneously intuit objects.[7] A representation is spontaneous if the

[6] The main texts for the concept of an intuitive intellect are *CJ* §§76–7 (Ak. 5:401–10), various remarks in the B Deduction about "an understanding that would intuit" (B135, B138–9, B145, B149, B159); see also the first Introduction to *CJ* (Ak. 20: 227), *Prol.*, §34 (Ak. 4: 316–17), *Disc.* (Ak. 8: 216), Corr. (Ak. 10: 130), and ID, §10 (Ak. 2: 396–7); in the unpublished writings, see *Refl.* 1832 (Ak. 16: 131), 6048 (Ak. 18: 433), MK_1 (Ak. 28: 1256), ML_1 (Ak. 28: 241, 330), *LP* (Ak. 24: 361), *MM* (Ak. 29: 759), *MV* (Ak. 28: 372), MK_2 (Ak. 28: 782), and MK_3 (Ak. 29: 972), in addition to the theology lectures cited above. Förster (2011) provides a comprehensive survey of German philosophy from Kant to Hegel, with a focus on intuitive intellect; see subsequent discussion, though, for some reservations about his reading of *CJ* §76. See also Westphal (2000) for the reception of this Kantian concept in post-Kantian German philosophy.

[7] *CJ*, Ak. 5: 406; B135.

ground of that representation is contained within the mind whose representation it is, so a *spontaneously* intuiting intellect is one that is the ground of its own intuitions. We are the ground of the *possibility* of the objects we intuit because our minds contain the *forms* (space and time) in which any object can be given to us. An intuitive intellect is not merely the ground of the possibility of its objects but of its objects themselves (their existence, absolute positing); an intuitive intellect 'posits its objects absolutely' in intuiting them. This is why Kant characterizes an intuitive intellect as "an intellect through whose representation [*durch dessen Vorstellung*] the objects of this representation thereby [*zugleich*] exist [*existieren*]" (B138–9).[8] Since an intuition is a *singular* and *immediate* relation to an object, it may not be fully coherent to say that an intuitive intellect's intuition of an object *grounds* the existence of that object, since the object must exist (absolute positing) to be the object of an intuition in the first place.[9] Instead, I think it is more coherent to say that: the intuition of an object by an intuitive intellect *just is* its being absolutely posited. For the objects of an intuitive intellect, their existing (being absolutely posited) *just is* their being intuited by such an intellect. This is why Kant also describes an intuitive intellect as possessing 'intellectual intuition' [*intellektuelle Anschauung*]: it intuits its objects spontaneously (intellectually) rather than passively (sensibly).[10,11]

In this book I have focused on two technical meanings Kant associates with the term 'cognition' [*Erkenntnis*]: (1) concepts we can know to be really possibly instantiated, and (2) thinking a given object under a concept. But neither of these

[8] In this passage and elsewhere in the B Deduction Kant is mainly focusing on an intuitive intellect's representation of *itself*, but the same point applies, *mutatis mutandis*, to an intuitive intellect's representations of objects distinct from itself: its intuition of them is their existence. Cf. Corr. (Ak. 10: 30), ID §10 (Ak. 2: 397).

[9] The idea is this: *x* cannot exist in virtue of standing in relation *R* to some other object, for the fact that it stands in that relation is partly explained by the fact that *x* exists (otherwise, it would not be 'available' to stand in that relation). So it is more perspicuous to say: *x* existing just is *x* being an object of an intuitive intellect.

[10] In fact, '*intellektuelle Anschauung*' occurs more frequently in Kant's writings; the main discussion is in the 'Phenomena and noumena' section (A252–6, B307–313; cf. Kant's handwritten marginal notes at A248, E CXXX and CXXXI (Ak. 23: 36)). See also Bxl, B68, A279/B335, *Prol.* (Ak. 4: 375 n.), *CPrR* (Ak. 5: 31, 99, 123), *CJ* (Ak. 5: 409), *Disc.* (Ak. 8: 219, 220), *Rel.* (Ak. 6: 67), *Prog.* (Ak. 20: 267), *Tone* (Ak. 8: 389), Corr. (Ak. 10: 123, 11: 51), *Refl.* 4207 (Ak. 17: 456), *Refl.* 4677 (Ak. 17: 658), *Refl.* 5637 (Ak. 18: 274–5), *Refl.* 5653 (Ak. 18: 306), *Refl.* 6050 (Ak. 18: 434), *MK_I* (Ak. 28: 1520), *ML_I* (Ak. 28: 179, 206, 241), *MM* (Ak. 29: 800, 880), *MV* (Ak. 28: 371), *MD* (Ak. 28: 653), *MK_3* (Ak. 29: 972, 974), and *VE* (Ak. 29: 14–15). In two pre-Critical *Reflexionen*, *Refl.* 4228 (Ak. 17: 467) and 4336 (Ak. 17: 509–10), Kant claims that I know my freedom through intellectual intuition of myself.

[11] The thesis that 'intellectual intuition' and 'intuitive intellect' refer to well-defined concepts, much less the same thing, is controversial. Gram (1981) argued that Kant has several non-equivalent notions of intellectual intuition. Förster (2011), 150–60 argues that the concept of an intellectual intuition and the concept of an intuitive intellect are distinct. I concur with the argument of Leech (2014), though, that while these are different *concepts* (one is a concept of the kind of intuition a mind might have, the other is a concept of a kind of intellect) they are 'two sides of the coin,' or, more precisely, a being with intellectual intuition is an intuitive intellect, and an intuitive intellect has intellectual intuition. They are concepts, respectively, of a part of a kind of mind (intellectual intuition), and that kind of mind itself (intuitive intellect), which necessarily come as a package.

can be what Kant means when he talks of the cognitions of the intuitive intellect, because (1) the intuitive intellect does not represent its objects modally, and (2) it does not use general representations (concepts) to think about given objects.[12] So Kant must mean cognition in some even more general sense when discussing the cognition of the intuitive intellect, which I take to mean, roughly: representing an object correctly as the object it is.[13]

10.3. Some (Unsuccessful) Attempts to Resolve the Antinomy

In this section I argue that several tempting strategies for resolving the inconsistent triad of (P1)–(P3) do not succeed.

1. The first attempted solution to the problem arises from Kant's distinction between 'negative' and 'positive' noumena. In the A edition Kant distinguishes between phenomena and noumena and defines the latter as objects of a non-sensible or 'intellectual' intuition ("*coram intuitu intellectuali*"—A249). In the B edition he makes a further distinction between negative and positive noumena, defined, respectively, as objects that are *not* objects of a sensible intuition, and objects of a non-sensible intuition (noumena *simpliciter*, in the terminology of the A edition).[14] Notice, though, that <*negative noumenon*> is, in principle, a more general concept than <*positive noumenon*>: if something is an object of a non-sensible (intellectual) intuition then it is *not* the object of a sensible intuition, but the converse does not hold as a matter of definition alone. The concepts <*negative noumenon*> and <*positive noumenon*> are not logical contradictories, but stand in a relation of logical subordination. I will refer to negative noumena that are *not* positive noumena as *merely negative noumena* (see Fig. 10.1).

With these distinctions in place, we can see that 'noumena' in (P2) and (P3) means *positive* noumena (objects of God's non-sensible intuition) but the inconsistency arises only if (P1) is also true of positive noumena as well. Our grounds for holding (P1) is that our noumenal wills have modal properties. But why should we assume that our noumenal wills are positive noumena? If our noumenal wills are 'merely negative noumena' then the contradiction is resolved. However, the concept of God is the concept of an *omniscient* intuitive intellect, an intellect that cognizes all

[12] *CJ*, Ak. 5: 403, 406. Nor can the intuitive intellect have *Wissen* since *Wissen* is a mode of *Fürwahrhalten*, which is a subject's relation to judgment, and judgments are discursive representations of objects.

[13] It is, of course, hard to determine Kant's most general concept of cognition for he almost always discusses specifically *discursive* cognition; my hypothesis (cognition in general = objective representation of an object as the object it is) receives some indirect support from Bix-x, A58/B83, B137, B166, A258/B314, A320/B376, and *JL* (Ak. 9: 51). I am drawing here on some ideas of Smit (2000) and Schafer (forthcoming). For more, see the supplementary article "Intuitive Intellect and Cognition" on my website (see Notes on the Text).

[14] B307.

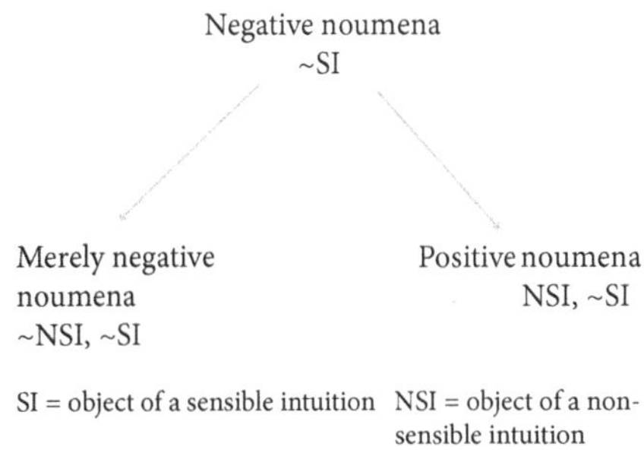

Fig. 10.1. Negative and positive noumena

noumena.[15] It follows that, if there is a God, then all noumena are positive noumena, and, thus, they cannot have any properties that positive noumena lack, including modal properties. The proposed resolution of the antinomy does not succeed, for it makes freedom incompatible with the existence of God, but we know that both are commitments of Kant's view.

In a similar vein, though, one might point out that the grounds for holding (P3) is that it is an instance of a more general principle:

(P3*) For any property F, noumena have property F, if, and *only if*, if there is a God, God would cognize them as having property F.

However, (P3*) has a counterexample. If there is a God then noumena have the property *being cognized by God* and God cognizes them as having that property (he is aware that he cognizes them). If we substitute *being cognized by God* for F in the right-hand side of (P3*), (P3*) entails *all by itself* that noumena are cognized by God, hence that there is a God. Intuitively, though, (P3*) should not entail that God exists! The point behind this objection is that the presence of God might alter the properties (at least the relational ones) of noumena and so we should restrict (P3*) to the 'God-independent' properties of noumena. However, having modal properties may be a God-dependent property: if there is a God then all noumena are positive noumena, as I argued in the last paragraph, and by (P2), it follows that none of them have modal properties. If (P3*) is restricted to God-dependent properties (properties noumena have whether or not they are positive noumena) then (P3) is not an instance of it, so the original contradiction is resolved. However, this objection to (P3) only succeeds *if there is no intuitive intellect*. But Kant thinks that we are rationally required by practical reason and (I argued in Chapter 9) by theoretical reason to believe (in Kant's technical sense of *Glaube*) that there is a God, and he thinks that if there is a God, he has an intuitive intellect (premise (P0)), so this way of resolving the antinomy is not available to him.[16]

[15] *Pöl.RT*, Ak. 28: 1052.

[16] A more ambitious response to this objection is possible. See the unpublished paper "Kant's Concept of an Object" on my website (see Notes on the Text).

2. Another way one might attempt to resolve the triad is pointing out that (P1) is cognized on practical grounds while (P2) (and perhaps (P3)) represents a mere postulate (rational belief) of theoretical reason. Consequently, the conflict is resolved: from a theoretical standpoint noumena lack modal properties, while from a practical one we have to attribute modal properties to them (e.g., the property *could have done otherwise*).

This solution involves relativizing (P1)–(P3) to different 'standpoints' or 'perspectives.' As such, it depends upon the 'epistemic' or 'methodological' interpretation of Kant's idealism pioneered by Graham Bird, Gerold Prauss, and Henry Allison.[17] In this book I have attempted to refrain from directly engaging with the debate about the meaning of Kant's transcendental idealism, but it may be necessary, at this point, to briefly discuss the relation between the 'modal antinomy,' this proposal for resolving it, and the epistemic interpretation of transcendental idealism. I will focus on Allison, whose version of the 'methodological' reading has been most extensively discussed by Anglo-American scholars.

The core idea of Allison's interpretation of transcendental idealism is to relativize various claims we might make about objects to a standpoint and to the conditions on objects appropriate to that standpoint. There is a *theoretical* standpoint, in which we consider objects (including ourselves) as subject to the conditions under which knowledge of them is possible, and a *practical* standpoint, in which we consider objects (most importantly, ourselves and our actions) as subject to the conditions under which judgments about morally right action is possible. My problem will not arise, on Allison's reading, because it is really a problem about the competition between standpoints—the moral standpoint (in which (P1) is made) and what we might call the standpoint of speculative theology (in which (P2) and (P3) are made)—and Kant, according to Allison, can dismiss such conflicts because there is no way things are independent of a standpoint, no 'meta-standpoint' from which the different standpoints can be judged. The Allisonian answer to my problem is to say that in considering ourselves as moral agents we must consider ourselves as possibly having done otherwise (P1) and when doing speculative theology we have to consider the objects of divine cognition as lacking modal properties ((P2) and (P3)), and there is no coherent question about which one is *correct*.[18]

I do not have the space here to argue against Allison's interpretation. I will simply point out that Allison's view is best equipped to resolve conflicts between the *practical* standpoint and the *empirical-theoretical* standpoint, the standpoint we adopt when we cognize empirically given objects using the faculty of understanding.[19] But (P1)–(P3) represent a conflict between the practical standpoint and what

[17] The *loci classici* of this view are Bird (1962), Prauss (1974), and Allison (1983) and (2004).

[18] See Allison (2004), 20–49.

[19] This comes out clearly in the essays on the freedom of the will collected in Allison (2012).

I referred to above as the 'speculative-theological' standpoint, the standpoint we adopt when we merely think about how God's intuitive intellect would cognize our noumenal wills. Allison has not explained how the latter standpoint functions, and how it is compatible with, but different from, the practical standpoint. It is not clear, then, that Allison's reading can be extended to make sense of (P2) and (P3).[20] What is more, even if the Allisonian reading can be so extended, the resulting view will inherit all of the problems of Allison's interpretation of transcendental idealism generally.[21] While none of this entails that Allison's view is false, or the Allisonian response to the modal antinomy cannot succeed, it provides significant motivation for a solution to the 'modal antinomy' that does not depend upon the epistemic reading of transcendental idealism.

3. A third alternative would be to draw a distinction, within the set of modal properties, between *theoretical* modal properties and *practical* modal properties. The idea would be that the modal categories as they are introduced in the *CPR* have a certain content or meaning and *that* content is what Kant denies applies to noumena in *CJ* §76, but the modal categories acquire a new *practical* content from their relation to the moral law and the will. Kant is *not* denying that content applies to noumena.

In the *Critique of Practical Reason* (*CPrR*) the modal categories do acquire a kind of practical content in the guise of what Kant calls 'categories of freedom': the category <*possibility-impossibility*> becomes "the permitted and the forbidden," the category <*existence-nonexistence*> becomes "duty and what is contrary to duty," and <*necessity-contingency*> becomes "perfect and imperfect duty."[22] The modal categories can now be used, not to express the relation of sensibly given objects to our cognitive faculties, but the relation of maxims of action to the moral law, which, for sensible beings like us, is expressed as duty (what we *ought* to do). The category <*possibility*>, which can be used to express the fact that an object is compatible with the sensible and intellectual forms of our experience, can also be used, in the guise of the concept <*permissible*> (a 'category of freedom'), to express the fact that a maxim is compatible with duty.[23]

[20] This is a point on which Allison's reformulation of his view in (2004) is especially vulnerable. Allison there identifies transcendental idealism as the rejection of the "theocentric" view that there is a "way things are," the way they would be disclosed to God's mind (if there is a God), and that human cognition is defective because it does not fully grasp the "way things are." But Kant endorses precisely such a theocentric model in *Pöl.RT* (see esp. Ak. 28: 1053–4). The conflict I am discussing (P1–P3) is really a conflict between the practical standpoint and that 'divine' standpoint. Cf. the discussion of the intuitive intellect in Allison (1983), 340 n. 2 and (2000), as well as the discussion of a related problem in Brewer and Watkins (2012).

[21] See Ameriks (1982*a*); Guyer (1987), 336–8; Robinson (1994); Langton (1998), 8–12; Van Cleve (1999), 4, 148–9; and Stang (2014*b*) and (forthcoming). See also Allison's response to his critics in (2004), 6–11, 42–9.

[22] Ak. 5: 65–6. For critical discussion see Beck (1960), 144–53 and Bader (2009).

[23] There is an important parallel between Kant's denial that an intuitive intellect would cognize modal properties in its objects and his denial that duty (and hence the modal categories of freedom, e.g., the permissible and the forbidden) applies to a holy will. The context in which Kant introduces (P2) in *CJ* §76 is

But the modal categories of freedom are *not* the modal properties attributed to noumena in (P1);[24] (P1) concerns *alethic* rather than *deontic* modality. To attribute a modal property to my noumenal will is not thereby to claim that any of my actions are permissible or impermissible; it is to make a claim about the *necessary condition* of that very permissibility or impermissibility: I could have done otherwise. In Kant's paradigm case, I do what is impermissible (I subordinate the moral law to my self-interest) but I could have done otherwise, where 'could have' is alethic *not* deontic. It is not that my doing otherwise was permissible (it was obligatory, in fact); it is that *because* I could have (alethic possibility) done otherwise that the moral law applies to me and makes it the case that I *ought to* (deontic necessity) have done otherwise. It is true that the modal categories, indeed *all* of the categories, acquire a new kind of practical content as "*modi* of the concept of causality of freedom" in the second *Critique*, but Kant's theory of freedom is committed to applying the modal categories with their original non-practical non-deontic *alethic* content to our noumenal wills and *that* appears inconsistent with *CJ* §76.[25]

4. Finally, it might be argued that this represents a *development* in Kant's view from the first edition of the *CPR* in 1781 (the first published work that contains the transcendental idealist theory of freedom)[26] to the publication of the third *Critique* in 1790. On this narrative, Kant originally wanted to attribute real modal properties to noumena and, in particular, to distinguish the (empirical-causal) necessitarianism obtaining at the phenomenal level from the contingency of the free acts of rational agents (noumenal wills) that obtains at the noumenal level, but was eventually pushed towards a more unified holistic and necessitarian conception of noumenal reality under the pressure of the idea of a divine intuitive intellect. Some might even see this as a *positive* development in Kant's thought, *away* from his earlier concern with eighteenth-century German debates about free will and determinism, and *towards* a more 'speculative' or 'absolute idealist' view on which the appearance of contingency is merely an artifact of our (discursive) cognitive architecture, which would fall away if we had intellectual intuition of noumenal reality.[27]

his claim that in three different cases we must recognize that our need to represent objects in a certain way is a subjective consequence of our kind of mind, which we express by thinking of an alternate kind of mind that would not so represent its objects: modality (intuitive intellect), ought (holy will), and purposiveness (intellectual intuition). I do not have space to explore this parallel further here.

[24] More precisely, (P1)'s denial that noumena have modal properties (fall under modal concepts) is not the denial that they fall under modal categories of freedom.

[25] *CPrR*, Ak. 5: 97–8.

[26] By which I mean the combination of phenomenal determinism and noumenal freedom. Another component of Kant's mature Critical view, that we can only know our freedom through awareness of ourselves as bound by the moral law, may not have been in place by 1781. Some scholars hold that Kant still entertained the possibility of a theoretical proof of freedom in 1781 and perhaps even in 1785 (*GMM*); see Ameriks (1981) and (1982*b*), 189–209.

[27] This is essentially the view of Tüschling (1990) and (1992). According to Beiser (2002), 580–2, it also represents the view of Schelling and Hegel at the turn of the nineteenth century.

As attractive as such a developmental narrative may be to some, it does not fit well with the texts, because Kant's doctrine that the intuitive intellect would cognize objects without modal properties is already present *in nuce* in the *CPR*, and based on reasons similar to those in the *CJ*. I will hold off on a detailed analysis of Kant's reasons for holding (P2) in the *CJ* until the next section, but very broadly, his reasons are these. Discursive intellects like ours are given objects and must think them under concepts. This gives rise to a distinction between what is actual (what is given in intuition) and what is possible (what agrees with the conceptual and intuitional form that the actually given objects fill out or instantiate). So, Kant reasons, an intellect that immediately intuited the whole nature of its object in virtue of being the ground of the existence of that object would not represent a distinction between what is actually the case and what might be the case.

The roots of this doctrine are already present in Kant's claim in the *CPR* that "the categories of modality have this special feature: they do not augment the concept to which they are ascribed in the least, as a determination of the object, but rather express only the relation to the faculty of cognition" (A219/B266). If the modal categories of the intuitive intellect (assuming it has them) express the relation of its objects to its cognitive faculty then there are grounds to suspect that these modal categories will at least be *very* different from ours, because the cognitive faculty to which they are related is quite different. Consider that, in our case, the category of actuality is closely connected to *intuition*[28] (which gives us the objects there *actually are*) while the category of possibility is tied to *concepts* (of objects that *can be* given in intuition).[29] From this alone, we should expect that an intellect in which there is no distinction between concepts and intuitions would represent no difference between what is actual and what is possible. The discussion of the intuitive intellect and modality in §§76 and 77 of the *CJ* draws out a consequence of doctrines already present in the *CPR*; it does not represent a radical new development in Kant's thinking.

10.4. Resolving the Antinomy

In this section I present my solution to the inconsistent triad of (P1)–(P3). Recall Kant's reasons for denying that the intuitive intellect cognizes modally:

> if our understanding were intuitive, it would have no objects except what is actual. Concepts (which pertain merely to the possibility of an object) and sensible intuitions (which merely give us something, without thereby allowing us to cognize it as an object) would both disappear. Now, however, all of our distinction between the merely possible and the actual rests on the fact that the former signifies only the positing of the representation of a thing with respect to

[28] A218/B265–6.

[29] A599/B627. Another way of seeing this connection: actuality is singular (there is only one actual world), while possibility is general (there are many ways the world could be).

our concept and, in general, our faculty for thinking, while the latter signifies the positing of the thing in itself (apart from this concept). Thus the distinction of possible from actual things is one that is merely subjectively valid for the human understanding, since we can always have something in our thoughts although it does not exist, or represent something as given even though we do not have any concept of it. (*CJ*, Ak. 5: 402)

Kant claims that we represent a modal distinction in objects because we have two different sources of cognition: intuitions, by which objects are given to us, and concepts, by which we think about objects. This allows us to represent a distinction between possibility and actuality: if a concept agrees with the conditions under which objects can be given to us, then it is the concept of a possible object, whether or not it has any instances; if, furthermore, the concept is instantiated by some object given to us in intuition, then the concept is the concept of an *actual* object. Since the representations of the intuitive intellect immediately present objects, and the 'absolute positing' of these objects is either identical with or grounded in their presentation to the intuitive intellect (see section 2), the intuitive intellect will not have uninstantiated general representations (concepts) of merely possible objects (objects it *could* spontaneously intuit). Since the representations of the intuitive intellect spontaneously generate their objects, the intuitive intellect will not represent a difference between the *actual* and the *merely possible*.[30]

The principle that the concept <*actual*> relates an object to our receptive faculty of intuition, while <*possible*> relates it to our spontaneous faculty of understanding, applies to these concepts as they are used by discursive intellects like ours. That the intuitive intellect would not represent a difference between objects consisting in their relation to receptivity and spontaneity might mean that the intuitive intellect would not represent the modal difference between objects that *we* do. It does not by itself entail that the intuitive intellect would not represent *any modal difference at all* between objects. When Kant claims that necessity and contingency "would not enter at all into the representations of such a being" (*CJ*, Ak. 5: 403), he might mean something more specific by possibility and necessity than necessity and contingency *überhaupt*; he might mean that the intuitive intellect would not represent the modal properties of objects in the specific way *we* do.

This idea of a less specific way of representing possibility and actuality should put us in mind of the threefold distinction among the unschematized categories (the

[30] While I agree with the critique of Gram and Förster in Leech (2014), I think that her account of why a discursive intellect does, and an intuitive intellect does not, represent its objects modally, conflates epistemic and 'metaphysical' (for lack of a better word) modality. According to Leech, we use modal categories because we face an epistemic problem: some of our concepts are not instantiated by objects. The intuitive intellect does not face this problem, because its objects are 'guaranteed' to exist (Leech 2014, 347). But this is incorrect about the role of modal concepts in Kant's theory; even if we had a 'guarantee' that our concepts are instantiated, we would still need the modal concepts to think about the *form* of our experience, hence to think about which objects and which of their properties are necessary in virtue of the form of our intellect.

categories as pure concepts of the understanding of all objects *überhaupt*), the categories as concepts of objects of a discursive intellect in general, and the (temporally) schematized categories (the categories insofar as they can be cognitively applied to objects of our specifically spatio*temporal* intuition).[31] The modal categories do not, as we have seen, add to the content of the concept of an object, since they are not, like the categories of the other three moments, rules for the synthesis of an intuition of an object in a manifold of representations.[32] Instead, schematized modal categories represent objects' relations to our specifically discursive and spatio*temporal* form of cognition:

> The schema of possibility is the agreement of the synthesis of various representations with the conditions of time in general (e.g., since opposites cannot exist in one thing at the same time, they can only exist one after another), thus the determination of the representation of a thing to some time.
> The schema of actuality is existence at a determinate time (A144/B184).

These are the schemata of the modal categories of possibility and actuality for our discursive spatiotemporal intellect; in my terminology, they are the real definitions of formal possibility and of actuality for specifically spatiotemporal objects. At one level of generality higher, there are principles that apply to modal categories for any discursive intellect whatsoever:

> Whatever agrees with the formal conditions of experience (in accordance with intuition and concepts) is possible.
>
> That which is connected with the material conditions on experience (of sensation) is actual. (A218/B266)

For any discursive intellect the actuality of an object is its relation to intuition and the possibility of an object is its relation to its intellectual and sensible forms; *these* are essentially the same as the principles for possibility and actuality Kant refers to in *CJ* § 76 above.

At the highest level of abstraction there are the unschematized modal categories, but of these we cannot say much more than a few 'obvious tautologies':

> No one has ever been able to define possibility, existence, and necessity except through obvious tautologies if he wanted to draw their definition solely from the pure understanding. (A244/B302)

These 'obvious tautologies,' I argued in Chapter 9, are partial nominal definitions of the unschematized category of (real) possibility. I argued further that the nominal definition of the unschematized modal categories includes various principles: Non-Logicality, Groundedness, Absolute Necessity, etc. But nothing in the nominal

[31] See B148–9. Kohl (2015) effectively collapses the first two ways of thinking of the categories.
[32] A219/B266.

definition of the unschematized category of (real) possibility says anything about a cognitive faculty for any mind and definitely nothing about our specifically discursive faculties. Consequently, to think of an object under the unschematized category of (real) possibility is not to think of its relation to our discursive intellect.

But surely Kant's point in the *CJ* is not merely that the intuitive intellect does not use the modal concepts a discursive intellect does, much less the specifically *temporally* schematized modal categories we use. That is a trivial consequence of the definition of the intuitive intellect and the idea of a schema of a pure concept of the understanding: a rule for combining *sensory* manifolds in an intuition of an object.[33] The substantive question is whether Kant means to be claiming that the intuitive intellect does not represent its objects even using the unschematized modal categories. Of course, the unschematized modal categories are *concepts*, and the intuitive intellect does not use (general, mediate) concepts, so if there is modal content in the intuitive intellect's cognition then it is not *conceptual* modal content. It might be that the intuitive intellect represents modal content but does so through non-conceptual means, that is, that there is some non-conceptual way of representing modal properties of objects, or something that corresponds to them.

In fact, this is the view Kant articulates in the Pölitz lectures, but with an unexpected twist. In those lectures Kant claims that the divine intellect (which he explicitly characterizes as an *intuitive* intellect)[34] *does* cognize the modal difference between the possible and the actual:

> We represent to ourselves that in cognizing his own essence (*simplex intelligentia*) God must also cognize everything possible, since he is the ground of all possibilities. Thus we derive the cognition of all possibilities from his nature and call it *cognitio simplicis intelligentiae.*—We think of *scientia libera* as God's cognition of the actual, insofar as he is simultaneously conscious of his free choice of things; for either all things are actual by the necessity of God's nature, which would be the principle of emanation; or else they exist through his will, which would be the system of creation. We think of *scientia libera* in God to the extent that in his cognition of everything possible, God is at the same time conscious in his free will of those possible things which he has made actual; hence his representation is grounded on the system of creation, according to which God is the author of all things through his will. (*Pöl.RT*, Ak. 28: 1054)

God is the ground of all possibilities, so in cognizing his own essence, God cognizes everything that is possible. God is the causal ground of everything that exists (other than himself) through his will, so in cognizing himself as the *author* of those objects, God cognizes what is actual. In both cases, God's cognition is *a priori*$_G$: he cognizes all possibilities and actualities through their ground (his essence and his will, respectively), and this cognition depends upon nothing external to him. In both cases God's cognition is a kind of self-cognition; awareness of his essence and

[33] A79/B105, B128, and A141–2/B180–1.

[34] *Pöl.RT* (Ak. 28: 996, 1043, 1052), *Danz.RT* (Ak. 28: 1267), and *Volck.RT* (Ak. 28: 1165).

everything that is made possible through it, and awareness of what he is actually, timelessly, creating. This means that God does cognize a difference among the objects of his intellect, their different manners of depending upon him: God cognizes some things as depending solely upon his essence (possible ways objects could be), and a subset of those ways (ways objects actually are) as depending upon his will (all of the things whose actuality depends upon God's will depend for their possibility upon his essence).[35, 36]

However, earlier in the same paragraph we find Kant saying, "in regard to God there is no distinction between the possible and the actual" (*Pöl.RT*, Ak. 28: 1053), which I take to mean: God does not cognize anything *as* actual or *as* possible. In fact, the sentence that immediately precedes the quoted passage might be thought to undermine my interpretation completely:

> The distinction between *scientia simplicis intelligentiae* (knowledge of simple intelligence) and *scientia libera* (free knowledge) is to be found only *in our* human representation of God's cognition, and not in this cognition itself. (*Pöl.RT*, Ak. 28: 1054)

This is why Kant emphasizes the relation to our thinking in the long passage quoted earlier: we think of God as differentially cognizing what depends upon his essence and what depends upon his will, but this does not constitute a modal distinction. However, the text on which Kant is lecturing at this point is *Metaphysica* §§874–5, where Baumgarten defines these kinds of cognitions:

> §874. God knows all determinations of all things, insofar as these are considered merely possible. This is *scientia simplicis intelligentiae.*
>
> §875. God knows all the determinations of the actual beings of this world, which is *scientia libera* (of vision).

Notice, though, that Baumgarten defines these kinds of cognitions in explicitly *modal* terms. In claiming that the distinction between "*scientia simplicis intelligentiae*" and "*scientia libera*" lies only in our human representation of God's cognition (Ak. 28: 1054), Kant is claiming that the representation of this distinction as a *modal* distinction is due to us and that modal representation is not part of God's cognition. He is not denying that there is a difference in how God cognizes his own essence and his actions.[37]

[35] As Kant remarks: "the actual is already included within the possible, since what is actual must also be possible, for otherwise it could not be actual. Thus if God is thinking of everything possible, he is already thinking of everything actual" (*Pöl.RT*, Ak. 28: 1053–4). But in cognizing actual objects as depending on his essence, God is not thereby cognizing them as *actual*, a point Kant recognizes, for he gives different accounts of God's *scientia libera* and *cognitio simplicis intelligentiae.*

[36] In §5 I will be more precise about what 'object' means here; for now I am painting with a broad brush.

[37] If the distinction between what depends upon God's essence and his will were only an artifact of our representations, then the distinction between the "system of emanation" (the existence of the world is a necessary consequence of God's essence) and the "system of creation" (the world is freely created by God) would be merely subjective. But this is not Kant's view. See *Pöl.RT* (Ak. 28: 1092–3), *Volck.RT* (Ak. 28: 1193–5), *Danz.RT* (Ak. 28: 1298–9), ML_1 (Ak. 28: 330), *MD* (Ak. 28: 701), and MK_2 (Ak. 28: 808).

On this reading, there is a real noumenal distinction in how objects relate to God and indirectly, to his intuitive intellect: some things depend upon his essence (nature), while others also depend upon his will. God is omniscient, so these objects and their different manner of dependence on him are within the scope of his intuitive intellect; God cognizes this 'dependence difference' among objects. In the Pölitz passage quoted above he claims that it is correct to describe this distinction in terms of the modal concepts of possibility and actuality *only from our standpoint*. Since we are using modal concepts to think about noumena, we are using the unschematized concepts of possibility and actuality (see Chapters 8 and 9). We use these concepts to think about a real noumenal distinction in objects that is not in itself (in its own nature, considered independently of its relation to our cognitive faculties) a *modal* distinction.

I am going to refer to these noumenal properties as noumenal 'correlates' of our modal concepts. The noumenal correlates of the unschematized categories of possibility and actuality are the properties, respectively, of depending upon God's nature or upon his will. Kant's view is that God differentially cognizes objects as depending either upon his will or merely upon his essence, and we represent that difference using the unschematized modal categories of possibility and actuality. This means that these noumenal correlates are in the content of God's intellectual intuition of objects, but the modal categories are not, *not even* the unschematized modal categories.

This allows us to see how Kant's denial of divine cognition of modality in the *CJ* is consistent with his view in the Pölitz lectures. In the *CJ* Kant emphasizes the point that the intuitive intellect would not represent modal content. In the Pölitz lectures he repeats that point, but expands upon it by claiming that the difference in our thoughts about his objects (between those we think as actual and those we think as merely possible) corresponds to a real distinction in the way objects depend upon God and in his awareness of them as so depending.

This means that (P1)–(P3) can be reformulated as follows:

(P1*) Noumena have noumenal correlates of unschematized modal categories (noumenal properties that *we* think under the unschematized modal categories).

(P2*) The intuitive intellect does not cognize its objects as having modal properties (though it does cognize the non-modal noumenal correlates of unschematized modal categories).

(P3*) Noumena have noumenal correlates of modal categories if, and *only if*, the intuitive intellect cognizes them as having these noumenal correlates (whether or not it cognizes them *as* correlates of modal categories).

In other words, for each unschematized modal category there is a corresponding non-modal property N such that God cognizes N in objects but does not cognize it modally (as being a modal property). These are consistent, so the original modal antinomy is resolved. In the next section I clarify this interpretation and extend it to the modal properties of noumenal wills.

10.5. Noumenal Freedom without Noumenal Modality

Up to this point I have referred somewhat loosely to the 'objects' of divine intuition and their 'modal properties,' but from now on I will be more precise. Since existence is not a real predicate, there are no merely possible objects (q-objects, objects of absolute positing); translated into the noumenal correlate of possibility, this means there are no noumenal *objects* that only depend upon God's essence without being created by his will. All the noumenal objects exist and are (represented by us as) actual. Our concept of noumenal (real) possibility does not have a set of noumenal *objects* as its correlate, but a set of *properties* of noumena, the properties we represent as being (really) possibly instantiated. These properties stand in relation R (to be specified further below) to God's essence. This relation is the noumenal correlate of our concept of the relation of a really possible property to a ground of its real possibility; the properties that stand in R to God's essence are the noumenal correlates of our concept <*possibly instantiated*>. The noumenal correlate of our concept of the *actual* properties of noumena is the set of properties possessed by noumenal objects; some of those properties are created by God in noumenal objects, while some are causally determined by those noumena themselves (see later discussion). All of the properties instantiated by noumena (which we represent as their *actual* properties) stand in R to God; in other words, all of the properties we represent as actual we also represent as possible. God cognizes the objects there are (and some of their properties) as depending for their existence (instantiation) on his will, and cognizes that all of these properties, as well as some uninstantiated ones, stand in R to his essence. In this fashion God cognizes the non-modal noumenal correlates of our unschematized modal concepts.[38]

In the Pölitz lectures Kant does not explain what relation to his essence God is cognizing in *scientia simplicis intelligentiae*, that is, what relation to God's essence is the noumenal correlate of possibility. Nonetheless, it is not hard to fill in this gap. Elsewhere he says this about God's cognition of actuality:

> We think of a *scientia libera* in God to the extent that in his cognition of everything possible, God is at the same time conscious in his free will of those possible things which he has made actual; hence this representation is grounded on the system of creation, according to which God is the author of all things through his will. (*Pöl.RT*, Ak. 28: 1054)

The possible is the possibly *actual*, so in representing a property as possible we are representing it as something possibly actualized by God. Since actuality is our modal representation of what depends upon God's will, it stands to reason that possibility is dependence upon God's potential will, his power. The properties God cognizes

[38] A problem arises about how God can be said to cognize properties since he does not cognize discursively (hence does not represent properties conceptually as predicates of possible judgments, as we do). I do not have space to address this problem here; see, however, the supplementary note "Intuitive Intellect and Cognition" on my website (see Notes on the Text).

through *scientia simplicis intelligentiae* are the properties he has the power to cause to be instantiated. We represent those properties modally as *possibly* instantiated properties.

My aim all along has been to reconcile the contingency of the actions of noumenal wills with (P2) and (P3), but the model sketched so far only applies to the existence of objects created by God and the properties he has the power to cause to be instantiated in them. I will now attempt to extend this model to the noumenal correlates of modality for the choices of noumenal wills, which I will treat as properties noumenal wills cause to be instantiated in themselves (e.g., I cause myself to have the property of being radically evil). The noumenal will of a finite rational agent makes a single timeless choice *either* to subordinate morality to self-interest (radical evil) *or* to subordinate self-interest to morality (moral goodness).[39] The noumenal correlate of the possibility that I choose otherwise than I do cannot be God's power to *cause* me to choose otherwise (or to create me with a will that chooses otherwise)[40] for that would destroy my freedom. Likewise, if the noumenal correlate of the *actuality* of my choice (radical evil) were God's causing me to choose, or creating me as so choosing, then freedom would be an illusion.

We have two options. The first would be to claim that the noumenal correlate of the possibility of my choosing otherwise is also God's power; likewise, the noumenal correlate of the actuality of my choice is at least *partial* dependence on God's will (God is either the partial or complete cause of my choice). This could take the form of Malebranchean occasionalism, Spinozistic monism, or divine concurrence (I do not have the power to do anything, including to determine my own will, without God's participation or 'concurrence'). Since Kant rejects the first two and expresses an at-best ambivalent attitude towards divine concurrence,[41] I think we should look elsewhere. The second option is that the noumenal correlates of the actuality and possibility of determination in my will are, respectively, what my will chooses and my power to choose otherwise. I choose to subordinate morality to self-interest—my will is the cause of my radical evil[42]—but I retain the power to choose otherwise. Restricting ourselves to properties that are within the power of my noumenal will,[43] the

[39] *Rel.*, Ak. 6: 25, 36, and the rest of Part 1 of *Rel.* Obviously, there are serious problems with interpreting Kant's 'rigorism' (e.g. how is moral conversion possible? Cf. Ak. 6:48), but I cannot address them here.

[40] In the former case, I do not essentially choose radical evil, but God creates me and causes me to so choose; in the latter case, I essentially choose radical evil, so in creating me God is creating me choosing evil.

[41] In *Pöl.RT.* he describes divine concurrence in free acts as an incomprehensible miracle (Ak. 28: 1105–7), but "quite possible" (Ak. 28: 1110). He is more skeptical of divine concurrence in *Danz.RT* (Ak. 28: 1039) and *Volck.RT* (Ak. 28: 1209). Cf. Hogan (2014).

[42] *Rel.*, Ak. 6: 31, 36.

[43] Some of my properties (e.g., my transcendental freedom, my finiteness) are direct consequences of God's creative act; since I *actually* possess these properties, we need some non-modal noumenal correlate of their actuality. See section 7.

noumenal correlate of actuality is the set of properties my will causes to be instantiated (in the cases of all human beings, radical evil),[44] and the noumenal correlate of possibility is the set of properties my will has the power to cause to be instantiated (on Kant's view, this set contains only two properties: radical evil and moral goodness).

My solution involves replacing the original (P1) with (P1*), or, more specifically, replacing the claim that noumenal wills are only contingently evil with the more complicated claim that noumenal wills have the non-modal noumenal correlate of *actually* willing evil (they cause themselves to be evil, i.e. they choose to subordinate morality to self-interest) and the non-modal noumenal correlate of *possibly* not being evil (they have the power to choose otherwise, i.e. to subordinate self-interest to morality). This requires developing a non-modal characterization of freedom, a 'noumenal correlate' of freedom that we then represent modally as *possibly* willing otherwise than it *actually* wills.

In fact, Kant has the materials to do precisely that. At least since the mid-1760s Kant has held the Crusian view that free actions of finite rational agents (noumenal wills) lack determining grounds.[45] In doing so, he is drawing on Crusius's distinction between a determining ground and a merely sufficient ground.[46] A merely sufficient ground brings about its effect but retains the power to omit that effect in the same circumstances (*Umstände*); a determining ground always brings about the very same effect in the same circumstances. While Crusius and Kant often characterize this distinction in *modal* terms (a merely sufficient ground possibly has a different consequence in the same circumstances, a determining ground necessarily has the same consequence in the same circumstances), it is possible to define them non-modally and use that distinction (between merely sufficient and determining grounds) to understand modality, as follows.[47] Crusius refers to a power that is indifferent in its nature between two outcomes as a "power of freedom": when a substance with a power of freedom acts in a given circumstance and posits a consequence it retains the power to posit the opposite consequence in the same

[44] *Rel.*, Ak. 6: 32–3.

[45] This shift from the compatibilist theory of freedom in *ND* (Ak. 1: 396–410) to a libertarian theory first appears in a series of *Reflexionen* from the 1760s: *Refl.* 3717 (Ak. 17: 260), 4034 (Ak. 17: 392), 4338 (Ak. 17: 510), 4544 (Ak. 17: 588), 4693 (Ak. 17: 677), and 5251 (Ak. 18: 132). See Hogan (2005) and (2009*a*) for critical discussion.

[46] *Ent.* §§82, 84 (148); *De Usu* §§II, V, XXII.

[47] Crusius: "to determine means nothing more than to leave a single possibility for how a thing in these circumstances is constituted or could be constituted" (*De Usu* §II; cf. §XVI). Kant: "[that] everything which happens has a determining ground, that is, something else which necessitates it, is the principle of the changes of all passive substances" (*Refl.* 4338, Ak. 17: 510). Kant cites Crusius's distinction in *ND*, but his explication of it, while non-modal, is not entirely perspicuous: "since, however, to determine is to posit in such a way that every opposite is excluded, the term 'determine' designates that which is certainly sufficient to conceive the thing in such and such a way, and in no other" (Ak. 1: 393). Nonetheless, in that early work he maintains that even the free actions of rational agents have determining grounds.

circumstances.[48] A merely sufficient ground of a consequence is the activation of a power of freedom. A determining ground of a consequence is the activation of a power that is not a power of freedom. Consequently, a consequence that has a determining ground has a completely intelligible explanation: the ground had no power to posit the opposite. By contrast, a consequence with only a sufficient ground cannot be made fully intelligible: the ground had the power to posit its opposite in the very same circumstances, and did not.

Starting in the mid-1760s Kant converts to the Crusian view that the free acts of finite rational agents (noumenal wills) lack determining grounds. In other words, our acts are the products of Crusian 'powers of freedom'; we retain the power to do the opposite under the same circumstances. Since the notion of a power of freedom can be defined non-modally (as I did above) and the notions of determining and sufficient grounds can be defined in terms of a power of freedom, they can also be defined non-modally. This means that freedom (or at least the 'leeway' requirement on freedom)[49] can be characterized non-modally: I am free because my acts are the consequence of a power of freedom, meaning, I retain the power to have acted differently under the very same circumstances.[50]

This will strike some readers as arguing in a very tight circle. Freedom is characterized in terms of the power to have done otherwise under the very same circumstances. But isn't the power to act otherwise to be understood modally as *possibly* acting otherwise? It is important to realize that this is not Kant's view. In *CJ* §76 and the Pölitz lectures he claims that an intuitive intellect would not cognize modally, but he does not claim it would not cognize powers, grounds, etc. In fact, he *explicitly* claims that God is aware of himself as the *cause* of the world.[51] Since God is omniscient and noumenal wills possess transcendental freedom, then either transcendental freedom does not involve power, or noumenal wills do have power and in cognizing this power God is not cognizing anything intrinsically modal. The representation of powers in modal terms (e.g., the possibility of acting otherwise than they do) is due to us, as I will explain in detail in the next section.

[48] *Ent.* §82.

[49] I am not claiming that the lack of a determining ground is sufficient for freedom. Kant still maintains that spontaneity (being the ground of one's action) and intelligence (being the ground of one's action through representing it) are necessary, but not sufficient, conditions on freedom (*CPrR*, Ak. 5: 96–7, 101). There is no reason to think these need to be understood modally. I am arguing that a non-modal characterization can be given of the 'leeway' requirement, which Kant sometimes characterizes in terms of absolute contingency: the free acts of a rational agent must lack determining grounds.

[50] In the Kantian context, we are talking about the free choices of *noumenal* wills, so circumstances will be non-spatial and non-temporal; they may include, for instance, facts about which noumenal wills God has created, which properties he has created in them (e.g., finiteness), and whether they causally interact. See A541/B569, A551f/B579f, A553/B581, and *CPrR* (Ak. 5: 100–3). Thanks to Hogan (2014) for drawing my attention to this point.

[51] *Pöl.RT*, Ak. 28: 1053–5.

Several difficulties still remain. First of all, there is the difficulty of reconciling God's creation of my noumenal will with my free choice.[52] Secondly, there is the difficulty of how God can be said to have *a priori*$_\text{J}$ cognition[53] of my choice if he is not himself the ground of my choice and there is no completely determining ground of that choice. Since God is not passive he cannot 'wait and see' how I will choose to act. However, my action has no determining ground, so there is nothing 'prior' (in the order of explanation, not temporally, since my noumenal will is atemporal)[54] to my choice from which God could 'predict' it.[55] I do not want to minimize these difficulties, but merely to point out that they are difficulties with Kant's theory of freedom, not with reconciling that theory with the amodalist consequences of (P2) and (P3). Regarding the first problem, Kant claims that it is a mystery how freedom is compatible with divine creation, but this should not weaken our credence in either (practical cognition and moral faith, respectively).[56] The second problem is really the problem of how to reconcile Kant's libertarian theory of freedom (free acts lack determining grounds) with the doctrine of divine non-passivity; this is a problem that Kant faces as a libertarian theist, independently of his doctrine of noumenal amodalism.[57]

10.6. Representing Noumenal Modality

In this section I am going to systematize the remarks I have made over the course of sections 4 and 5 about noumenal correlates and how we represent them using unschematized modal concepts.

In the previous chapter I gave an account of the content of the unschematized concept of real possibility *überhaupt*, of which various different kinds of real possibility are specifications. I have also discussed a specific such kind of real possibility,

[52] Kant admits as much in *Rel.*: "it is however, totally incomprehensible to our reason how beings can be *created* to use their powers freely, for according to the principle of causality we cannot attribute any other inner ground of action to a being, which we assume to have been produced, except that which the producing cause has placed in it. And, since through this ground (hence through an external cause) the being's every action is determined as well, the being itself cannot be free. So through our rational insight we cannot reconcile the divine and holy legislation, which only applies to free beings, with the concept of the creation of these beings, but must simply presuppose the latter as already existing free beings" (*Rel.*, Ak. 6: 142; cf. 44, 191). Cf. *CPrR*, Ak. 5: 100–3; and *Pöl.RT*, Ak. 28: 1105–7, 1110). See Brewer and Watkins (2012) for further discussion.

[53] For reasons I noted in §2 it is not strictly correct to talk of God's 'knowledge' [*Wissen*] because knowledge is a mode of holding a judgment to be true [*Fürwahrhalten*] and judgment is a discursive form of representation.

[54] *CPrR* (Ak. 5: 101–2), *Pöl.RT* (Ak. 28: 1095).

[55] Kant is aware of this problem, but it is unclear he has a consistent solution; cf. *MK*$_2$ (Ak. 28: 803) and *Pöl.RT* (Ak. 28: 1055). See Hogan (2014) for discussion.

[56] *Rel.*, Ak. 6: 142. See previous discussion.

[57] Cf. Leibniz's critique of libertarianism in *Theodicy* (*Th.* §§41, 42). This was a point Kant was well aware of, for he himself wielded it against libertarians in *ND*: "divine foreknowledge is only possible in respect of free actions if it is conceded that their futurition is determined by their own grounds" (Ak. 1: 405).

noumenal real possibility. Because noumenal real possibility is a kind of real possibility it instantiates this scheme:

(*Noumenal Possibility*$_N$) Noumenal possibility $\Diamond_N p$ is a kind of *real possibility* only if

(iii) *Non-logicality*: it is not a conceptual truth that $\Diamond_L p \supset \Diamond_N p$ (equivalently, it is not a conceptual truth that $\Box_N p \supset \Box_L p$), and

(iv) *Groundedness*: if $\Diamond_N p$ then the fact that $\Diamond_N p$ has a real ground in some actual object or principle.

(v) *Absolute necessity:* there is an associated notion of absolute necessity (absolute-$\Box_N p$) defined as follows: absolute-$\Box_N p =_{def} \neg p \Rightarrow$ (for all q, $\neg \Diamond_N p$). There is some actual p about the grounds of all noumenal possibility such that absolute-$\Box_N p$.

This is the concept whose non-modal noumenal correlates I will attempt to identify.

Up to this point, I have not discussed the concept <*actuality*> in much detail. I will offer a more complete interpretation of the unschematized concept of <*actuality*>[58] in the next section, but for now I just want to point out that we need some principles to guide our use of this concept in thinking about the difference between what is actually the case with noumenal wills (e.g., that I am radically evil) and what is merely possibly the case (e.g., that I am morally good). Just as with possibility, we need to distinguish different *kinds* of actuality: the different concepts of actuality will differ according to the different objects to which they are applied (e.g., actuality for phenomena will be different than actuality for noumena). We can think of the principles of actuality *überhaupt* as principles any lower-level concept must satisfy to be a concept of *actuality*. The principles of the unschematized concept of actuality (the minimal conditions any concept must meet to be a concept of actuality) are as follows:

(*Actuality*) A sentence operator actually(p) and associated object-level predicate actual(x) are concepts of actuality only if:

(i) *Is-Act*. What is, is actual: (a) for all propositions p if p then actually(p), and (b) for all objects x, actual(x).

(ii) *Alethic*. Actuality entails possibility: for all propositions p if actually(p) then possibly(p).

Strictly speaking, we should subscript the relevant notions of actuality and possibility and restrict the scope of objects and propositions accordingly: empirical-causal actuality (being an actual alteration) entails empirical-causal possibility, etc. but for simplicity I will suppress that complication. Each principle has two parts because we have two concepts of actuality: as an operator on propositions (what is actually the case) and a predicate of objects (which objects are actual). Note that clause (ii) holds

[58] As well as some reasons to think that <*actuality*> is indeed a category distinct from <*existence*>; cf. *MvS*, Ak. 28: 493.

only for the relation between actuality and *alethic* kinds of possibility; in particular, it does not apply to the deontic notions of the *permissible* and the *forbidden*, the "categories of freedom" discussed in section 3. Consequently, it is not part of the nominal definitions of actuality and possibility *überhaupt*, but it is part of the nominal definition of all of the kinds of real possibility I have been discussing in this book so far (with the exception of permissibility in section 3).[59] In order to deploy the unschematized concept of actuality appropriately we also need to respect this principle.

There is a very close structural isomorphism between, on the one hand, the concepts of noumenal real possibility and actuality (what is actual for noumena) and (what I have claimed are) their noumenal correlates. Let p range over logically contingent monadic propositions about noumena that are not themselves modal and existential generalizations thereof (i.e. of the form Fa or $\exists x$Fx); let x range over noumena.[60] Consider these principles about the noumenal correlates and the corresponding principle about possibility and actuality (listed in parentheses):

(i) It is not a conceptual truth that if $\Diamond_{L}p$ then some noumenon has the power to make it the case that p. (*Non-logicality.*)

(ii) If some noumenon has the power to make it the case that p this is grounded in a fact about God or some object God creates (the noumenon with that power). (*Groundedness.*)

(iii) For all p that some noumenon has the power to make the case, if God did not exist, nothing would have the power to make it the case that p. (*Absolute necessity.*)

(iv) If p is true, then either p is about God himself, or p is made true by the will of God or some other noumenon. For any x, x is either God, or x exists because God creates it. (*Is-Act.*)

(v) If p is true, then either p is about God or God or some other noumenon has the power to make it the case that p. (*Alethic*).[61]

Clearly, the non-modal principles (i)–(iv) correspond to the indicated modal principles. The reason for this is that when we map the noumenal correlates onto our modal concepts in the right way, these non-modal principles are transformed into

[59] I do not think that *Alethic* is part of the most general concept of real possibility. In the *CPrR* (Ak.5: 65–6, see section 3) Kant says that the modal category <*possibility*> can be used to think about permissibility. Permissibility is not a kind of alethic possibility. I argued in Chapter 9 that the unschematized modal category <*possibility*> is the *most general* concept of real possibility. It follows that real possibility *überhaupt* is not specifically a concept of alethic possibility: it is used to think about alethic possibilities (e.g., formal possibility) and non-alethic ones (e.g., permissibility).

[60] Intuitively, the idea behind this restriction is that we want to consider the grounds of real non-logical possibilities for noumena (hence the restriction to logically contingent propositions) and we want to ignore iterated modalities and logically complex properties.

[61] It is important to note that God himself is a noumenon (he is not a possible object of sensible intuition).

Table 10.1. The non-modal noumenal correlates of possibility and actuality

Non-modal noumenal correlate	*Represented modally as*
(a) God's powers and the powers of other noumenal wills	(a*) God's essence and the essences of other noumenal wills
(b) Properties instantiated by God and properties he has the power to cause other noumena to instantiate	(b*) A subset of noumenally really possible properties
(c) Properties finite noumenal wills have the power to cause themselves to instantiate	(c*) A subset of noumenally really possible properties (for finite noumenal wills)
(d) the relation between (b) and God's power and between (c) and the powers of finite noumenal wills	(d*) The grounding relation *making really noumenally possible*
(e) God himself and all the other noumena he has created	(e*) Actuality as a predicate of (noumenal) objects
(f) True propositions about God and true propositions about other noumena he has caused to be true	(f*) Actuality as an operator for propositions about God and his creative acts
(g) Propositions that attribute to finite noumenal wills properties they have caused themselves to instantiate	(g*) Actuality as an operator for propositions about the choices of noumenal wills
(h) The relation between (f) and God's will and (g) and finite noumenal wills	(h*) The relation of *making actual*

the corresponding modal principles. Table 10.1 gives the non-modal noumenal correlates in the left-hand column, and how we represent them modally in the right-hand column.

It can easily be checked that if we map non-modal noumenal correlates onto modal concepts according to Table 10.1 then the non-modal principles (i)-(iv) are transformed into the modal principles that correspond to them (in parentheses). For instance, if we represent properties God has the power to cause to be instantiated (b) and properties finite noumenal wills have the power to cause themselves to instantaite (c) as *possible* ((b*) and (c*)), the fact that (i) it is not a conceptual truth that if *p* is logically possible then some noumenon has the power to cause it to be the case that *p* is represented modally as *Non-logicality*: it is not a conceptual truth that if *p* is logically possible then it is noumenally really possible. Likewise, if we represent true propositions about God and propositions he or some other noumenon causes to be true ((f) and (g)) as actual ((f*) and (g*)), the fact that everything God or other noumena cause to be the case is within their power is represented modally by *Alethic*. What this means is that we use our unschematized concepts of possibility and actuality to represent the structurally isomorphic relations among the powers of God and finite noumenal wills and their actions. At the noumenal level considered 'in itself' there is no modality; these objects and properties (and propositions about them) have a modal status (possible, actual, contingent) only relative to our modalized thought about them.

Before continuing, I would like to respond to one lingering issue. From our modal perspective we can say that God is the ground of the possibility of the existence of noumenal objects in virtue of having the power to create them, but noumenal wills are the ground of the possibility of their free choices (in virtue of having the power to so choose). Some readers will object that this inappropriately limits the dependence of all noumenal possibility on God, but in the Pölitz lectures Kant himself reiterates that God is (represented by us as) the ground of all possibility *whatsoever*.[62] But, within my interpretation, we can nonetheless recover this absolute dependence of all of (what we represent as) possibility upon God. Take my noumenal will and some property (moral goodness) it has the power to produce in itself. My noumenal will would not have this power unless God created it with that power, and this would not be the case without God having the power to create it with that power. So the moral goodness that is within my power is mediately related to God's (actual) will that creates this power, and yet further mediately related to the divine power of which that creative act is an activation. Representing these noumenal correlates in modal terms, the possibility of my moral goodness is grounded in my power to so choose. I actually possess this power (the immediate ground of possibility) because of God's creation of me as a transcendentally free finite rational will. But this creative act was possible in virtue of God's power. This power has no further ground of possibility. So the mediate but *final* ground of the possibility of my moral goodness is God's power. By parity of reasoning, this will apply to any property possibly possessed by any finite being. Consequently, we can retain the dependence of all really possible properties of noumena upon God (specifically, on his powers).

10.7. Absolute Positing, Existence, and Actuality

Up to this point I have mainly been concerned with possibility and whether it figures in the intuitive intellect's cognitions. But Kant does not say merely that possibility does not figure in its cognitions; he says that every object of an intuitive intellect would exist and would be actual:

> That is, if our understanding were intuitive, it would have no objects except what is actual. (*CJ*, Ak. 5: 402)
>
> For an understanding to which this distinction [between thinking and intuiting] did not apply, all objects that I cognize would be (exist), and the possibility of some that did not exist, i.e. their contingency if they did exist, as well as the necessity that is to be distinguished from that, would not enter into the representation of such a being at all. (*CJ*, Ak. 5: 403)[63]

[62] Ak. 28: 1034, 1054.

[63] In *Refl.* 6020 Kant claims that for an intuitive intellect "only what is actual is possible" (Ak. 18: 426). But this means either that the intuitive intellect cognizes no difference between the actual and the possible,

In this section I want to clarify how the intuitive intellect represents what exists and what is actual, and whether and how these are different. As I have noted before, Kant sometimes identifies the second category of modality as 'existence-nonexistence' (*Dasein-nichtsein*), sometimes simply as 'existence' (*Dasein*), and sometimes as 'actuality' (*Wirklichkeit*).[64] It has been a running theme of this book that Kant uses 'existence' and 'actuality' in a number of different ways, and with the idea of an intuitive intellect in place we finally have the resources to systematize and explain those different uses.

1. *Existence as absolute positing (what there is).* The first notion we need to distinguish is existence as 'absolute positing': there being an object. This is an unschematized use of the concept because we can use it to think about noumena; the question of whether God exists is the question of whether there is an object that instantiates the concept <*God*>. This notion of existence was the theme of Chapters 1 and 2. This is the notion of existence that is not a 'real predicate' or a determination. Intuitions give us objects that exist in this sense: objects that are absolutely posited (the objects *there are*, q-objects). Since the intuitive intellect only intuits, and what is more, must intuit itself, it does represent objects as existing in this sense: it absolutely posits its objects. This is what Kant means when he writes: "for an intellect to which this distinction [between thinking and intuiting] did not apply, all objects that I cognize would **be** (exist)" (*CJ*, Ak. 5: 403).
2. *Existence as causal efficacy.* This is the concept I have previously dubbed existence*: the concept of an object possessing causal powers. This concept is a determination, for some objects do not exist*, e.g., mathematical objects.[65] Admitting that existence* is a real predicate does not, however, open any room for ontological arguments, for, from the analytic judgment *God exists** (<exists*> is one of the marks of the concept <God>) we cannot prove that there is a God (that God exists in the 'absolute positing' sense above). This is not properly even a modal category; the concept of existence* is really understood through the categories of relation (esp. <*cause–effect*>). Since we think about noumena as having causal powers (e.g., transcendental freedom), there is

in which case it is exactly what he says in §76, or it means that the intuitive intellect cognizes only actual things and cognizes them *as actual*, in which case it is inconsistent with his considered view in §76. Thanks to Ralf Bader for calling my attention to this *Reflexion*.

[64] In the initial table of the categories in the first *Critique* he gives '*Dasein–Nichtsein*' as the second category of modality (A80/B106), but in the Postulates of Empirical Thought, the later section devoted to the modal categories, the second modal category is now the category of '*Wirklichkeit*' (A218/B266). See also *MvS*, where Kant writes: "real actuality is here the category of existence [*Existenz*], in contrast to the possibility of a thing" (Ak. 28: 493). Cf. *ND* (Ak. 1: 396), *OPG* (2: 72, 75), *Volck.RT* (28: 1151), and *Danz.RT* (Ak. 28: 1256, 1291). See, however, MK_3 (Ak. 29: 986) as well as *Refl.* 6324 (Ak. 18: 644), where Kant explicitly distinguishes actuality from existence.

[65] See *MFNS*, Ak. 4: 468.

an unschematized concept of existence* (thought using the unschematized categories of relation). Since (if there is an intuitive intellect) the intuitive intellect exists* and it cognizes itself, it must cognize at least one existing object.* Presumably it must also cognize itself *as* existing*, but I do not want to get involved here in the puzzle of how the intuitive intellect represents properties of its objects (see section 2), including its own properties. It is clear, though, that God cognizes himself as existing* since he cognizes himself as the author of the world.[66]

3. *Actuality as a modal concept*. Finally, when we use the unschematized modal category of *<actuality>* we represent all the objects *there are* (existence as absolute positing) and everything that is the case about them as actual to distinguish it from what is *merely* possible (e.g., I actually subordinate the moral law to my self-interest, but possibly do otherwise). Consequently, the unschematized modal categories *<possibility>* and *<actuality>* are interdependent; we cannot use one without using the other. Their interconnection is given by two principles about actuality (see section 6):

    (i) *Is-Act.* If p then actually p. For all objects x, x is actual.
    (ii) *Alethic.* If actually p then possibly p.[67]

The first principle means that we represent everything that is the case as being actually the case and we represent all objects as actual. This may seem trivial, but it is a crucial constituent of applying our concept of actuality and thus modalizing what were otherwise non-modal concepts: what is the case, and what objects there are. The second principle means that (working within an alethic modality) we represent all actual truths as possibly true, and all actual objects as possible.

The possibility–actuality connection is also given by two principles about possibility:

(iii) *Groundedness*: if $\Diamond p$ then the fact that $\Diamond p$ has a real ground in some actual object object or principle.
(iv) *Absolute necessity:* there is an associated notion of absolute necessity (absolute-$\Box p$) defined as follows: absolute-$\Box p =_{def} \neg p \Rightarrow$ (for all q, $\neg\Diamond p$). There is some actual p such that absolute-$\Box p$.

The idea is this. If we are thinking about some objects, some concepts of them, and some propositions about them, then in order to think them under the (unschematized) modal categories we have to represent all of the objects there are as *actual* and all of the propositions that are true of them as *actually true*. By using the concepts *<possible>* and *<actual>* according to these principles, and the other principles of real possibility *überhaupt* (see section 6), we come to represent those objects and propositions using

[66] *Pöl.RT*, Ak. 28: 1054.

[67] The second principle only applies where the relevant notion of possibility is alethic (see sections 3 and 5).

the unschematized modal categories *<possibility>* and *<actuality>*. To think about them using various more specific modal notions (e.g., formal possibility) we need to employ the more specific principles described in Chapters 7 and 8. This notion of actuality—defined in terms of its relation to the unschematized category of possibility—is the most general *modal* notion of actuality/existence. It represents, so to speak, the 'modalization' of existence in the 'absolute positing' sense, for it requires us to represent all the objects *there are* as actual and not merely possible. For the sake of clarity I will refer to it as *actuality* to distinguish it from *existence* (absolute positing) and *existence**.

Kant claims in *CJ* §76 that "if our understanding were intuitive, it would have no objects except what is actual" (*CJ*, Ak. 5:402). Some readers might conclude from this that the intuitive intellect does represent *actuality* (as I am using this term). However, I do not think this can be correct. One cannot have the concept of actuality (in this sense) unless one represents it in contrast to mere possibility; this notion of actuality is an inherently modal concept. Kant denies that the intuitive intellect represents anything *as* possible, so I do not think that he is claiming that the intuitive intellect cognizes its objects as actual. Instead, I think Kant is speaking from our modal perspective here. All of the objects of the intuitive intellect's cognition fall under our concept of actuality because they include: the intuitive intellect itself (God), its various powers, and the noumenal objects it creates. The intuitive intellect represents itself as having the power to instantiate various properties but does not, in doing so, intuit anything that is non-actual (from our perspective); the object of its intuition is its power to cause that property to be instantiated. The more precise Kantian claim is expressed later in the same paragraph:

> For an understanding to which this distinction [between thinking and intuiting] did not apply, all objects that I cognize would **be** (exist), and the possibility of some that did not exist, i.e. their contingency if they did exist, as well as the necessity that is to be distinguished from that, would not enter into the representation of such a being at all. (*CJ*, Ak. 5: 403)

I take this to mean that (prescinding from our modalized concepts) all of the objects of an intuitive intellect exist in the 'absolute positing' sense. This is not the case for discursive intellects like ours because we form thoughts involving concepts that correspond to no existing objects (no objects given in intuition).

In the previous section I identified the noumenal correlate of *actuality* (as a predicate of objects) as follows: God and any noumenon that depends upon God's will. This may appear ad hoc to some readers; I have picked the noumenal correlate of the unschematized category *<actuality>*, they might object, precisely so that the structural isomorphism between the categories *<possibility>* and *<actuality>* and their noumenal correlates would obtain. The discussion of this section, though, alleviates this appearance of being ad hoc. Any noumenon is either God himself or exists because God has willed it to exist (according to our practical and theoretical belief in the existence of God). God's will is the ground of the existence of noumena in the 'absolute positing' sense of existence; for instance, *there is* my noumenal will because God wills it to be. We then represent *what there is* (according to our practically and theoretically grounded representation of God) modally by subsuming it under the principle *Is-Act* and thereby representing those objects as *actual*. Since there are no noumenal objects other than God and what God has willed to exist, nothing else falls under the noumenal correlate of actuality.

10.8. Back to *Beweisgrund*

Some readers will wonder how my interpretation in this chapter, on which noumena possess only non-modal correlates of our unschematized modal concepts, is compatible with my argument in Chapter 9 that we can possess theoretical rational belief [*Glaube*] in an absolutely necessary ground of all noumenal real possibility. If noumenal modality is merely a 'projection' of our discursive intellect and its modal concepts then why is theoretical reason engaged in a regressive search for noumenal grounds of noumenal real possibility?

It is clear, though, that Kant takes his noumenal amodalism ((P2) and (P3)) to be compatible with the subjective rational necessity of positing an absolutely necessary ground of noumenal real possibility, for he reiterates the latter claim in *CJ* §76, the *locus classicus* of the former doctrine:

> For that [actuality] cannot be inferred from [possibility], hence that those propositions [e.g., that not all possibilities are actual] are certainly valid of objects insofar as our cognitive faculty, as sensibly conditioned, is concerned with objects of these senses, but are not valid of objects in general, is evident from the unremitting demand of reason to assume some sort of thing (the original ground) as existing absolutely necessarily, in which possibility and actuality can no longer be distinguished at all, and for which idea our understanding has absolutely no concept, i.e. can find no way in which to represent such a thing and its way of existing [...] (*CJ*, Ak. 5: 402)

It is precisely because we have a discursive intellect, which gives rise to our representation of the distinction between between possibility and actuality, that we are rationally required to postulate an absolutely necessary ground of possibility:

> Hence the concept of an absolutely necessary being is an indispensable idea of reason but an unattainable problematic concept for the human understanding. It is still valid, however, for the use of our cognitive faculties in accordance with their special constitution, thus not for objects and thereby for every cognitive being [...] Now here this maxim is always valid, that even where the cognition of them outstrips the understanding, we should conceive all objects in accordance with the subjective conditions for the exercise of our faculties necessarily pertaining to our (i.e. human) nature [...] Just as in the theoretical consideration of nature reason must assume the idea of an unconditioned necessity of its primordial ground [...] (*CJ*, Ak. 5: 402–3)

Here Kant reiterates the very claim he makes in the Pölitz theology lectures. In the very same section of the third *Critique* where Kant denies that an intuitive intellect cognizes its objects modally he reiterates his claim that we are rationally required to postulate that there is a being that grounds all real possibility. What is more, we know from the Pölitz lectures that this being (God) has an omniscient intuitive intellect. We represent it as grounding all real possibility, and it cognizes itself as grounding all objects but not *modally*.

In Chapter 9 I reconstructed the rational necessity of this postulate as a consequence of reason's regressive search for the conditions (grounds) of noumenal real possibility. The idea of God as the ground of all real possibility is the idea of an unconditioned condition of all real possibility. However, the very same line of reasoning applies if we abstract from our modal concepts and think about non-modal noumenal correlates. The non-modal noumenal correlates we represent as grounding the possibility of our noumenal choices are our powers. Through practical reason we represent ourselves as having the power to make either of two noumenal choices (radical evil or moral goodness) but reason requires us to think about a condition of these powers, since these powers are not themselves thought as unconditioned. This in turns requires us to find a condition of that condition, and so on, until we form the idea of an unconditioned condition of all powers of noumena. This unified ground has powers that are activated when it creates us with a power of choice (represented as a ground of possibility of choice). This unconditioned condition of our powers is the non-modal noumenal correlate we represent as the unified ground of all noumenal real possibility.

In Chapter 4 we examined in detail Kant's theory of God as the ground of all real possibility in *Beweisgrund.* Various models have been proposed in the secondary literature for how God grounds real possibilities: through his intellect (Yong 2014), through his power (Stang 2010), or by instantiating them (Chignell 2009*a*). However, as I pointed out in Chapter 4, Kant not only explicitly states that *none* of these views is correct, but notably fails to identify what it is in God by which he grounds all real possibilities.[68] God's essence grounds all real possibilities, including his own power, wisdom (intellect), and goodness; consequently, neither his power, nor his intellect, nor his goodness can itself be the ground of real possibilities. Not only do we not understand what it is in God through which he grounds all real possibilities, but Kant denies that we could ever understand or even identify it:

> Though one can form, through an analogy with human action, some concept of how a being can be the cause of everything actual, one can form no concept of how such a being could contain the ground of the inner possibility of other things. It appears that this thought rises far higher than a created being can reach. (*OPG*, Ak. 2: 152–3)

In *Beweisgrund* Kant lacks not only an adequate epistemology of how we can know this metaphysical theory of real possibility to be true (something he will come to realize in the Critical period); as he himself admits, he lacks an adequate explanation

[68] *OPG*, Ak. 2: 125.

of how its content could even be among the contents represented by our discursive intellects.

In a reversal of the interpretive strategy that has been popular in recent Kant scholarship—in which the pre-Critical works are used to shed light on the Critical philosophy—Kant actually develops an account of the modal content of our representation of God in the Critical philosophy that fills the gap he himself noticed in *Beweisgrund*, thus retrospectively clarifying that philosophical work, but from a specifically Critical point of view. We have already seen, in previous chapters, Kant's Critical reconstruction of the 'only possible ground of proof' itself as an expression of the 'only possible way' to satisfy reason's regressive demand for an explanation of why anything is really (noumenally) possible at all. But now we can see that Kant's Critical theory also retrospectively answers the question Kant himself raises, but does not answer, in *Beweisgrund*: how can we represent whatever it is in God by which he grounds all real possibilities?

We represent God as the ground of all real noumenal possibilities by the means sketched in the previous three sections: we use the unschematized category of actuality to think about God and everything posited by God's will and we use the unschematized category of possibility to think about everything God has the power to posit. In doing so we come to represent intrinsically non-modal noumenal correlates modally and we represent God as grounding all real noumenal possibilities (everything we represent as a noumenal real possibility) in virtue of his powers; if God were canceled, his powers would be canceled, so nothing would be really possible.

In the second passage quoted above Kant denies that we can use an "analogy" with human power and will to represent God as the ground of all real possibility (Ak. 2: 152–3). Kant does not explicitly say *why* we cannot use this analogy in this fashion, but a natural answer suggests itself: God's essence is the ground of all real possibility, including the possibility of his own powers (cf. Ak. 2: 125), so those powers themselves cannot be the grounds of all real possibility.

But this worry about representing God as grounding all real possibility through his powers disappears when we no longer take God to have modal properties 'in himself.' Once Kant has in place the doctrine that the divine intuitive intellect does not cognize modal properties, it follows (by (P2) and (P3) from section 1) that neither God nor any other noumena have modal properties 'in themselves,' so restricting attention to God and noumena 'in themselves,' no question arises about the grounds of the *possibility* of God's powers. The source of noumenal modality is the content of our thoughts about noumena, using the unschematized categories of possibility and actuality. Now that God and noumena have no modal properties 'in themselves' (but only relative to our modalized thought about them) there is no barrier to thinking that God grounds various noumena and their properties through his will, and has the power to directly produce certain properties, and indirectly to produce powers to produce other properties (e.g., the choices of noumenal wills), and this comes to be represented *by us* as the relation of real grounds to real possibilities "through an analogy with action." In Chapters 4 and 5 I argued that, although he claims we cannot represent God's grounding of all real possibility through his powers, Kant implicitly relies on such a model in his argument for a unique *GARP*. I referred to this as the 'tension' within Kant's pre-Critical modal metaphysics. My solution to the 'antinomy' of Kant's Critical modal metaphysics also shows us, retrospectively, how to resolve the tension: God's powers are what we represent modally as the grounds of all real possibility. The problem that arose from thinking that God grounds all real possibilities through his powers—powers presuppose possibility—is dissolved, since possibility (for noumena) is no longer something 'worldly' or 'noumenal' but the content of our modalized thoughts about noumena. God's powers do not ground real possibilities for noumena 'in themselves' so to speak; we think of those powers modally and thereby represent them as grounds of possibility of the very consequences they are the powers to bring about.

We have already seen that the modal theory of *Beweisgrund* raises (implicitly and explicitly) various questions in the metaphysics of modality, none of which receive satisfactory answers in that work:

(i) How are really possible predicates given to a discursive intellect?
(ii) How can a discursive intellect know what is really possible?
(iii) What is the content of modal thought? I.e. what, in general, are we as discursive intellects doing when we represent some predicate as really possible?
(iv) What are the real grounds of real possibility?
(v) Can we prove that there is a unique absolutely necessary real ground of real possibility?
(vi) What is the real grounding relation between God and really possible predicates? In the terms of Chapters 4 and 5, what are God's fundamental predicates and what is the derivation relation between them and derivative really possible predicates?

We can now see that Kant's Critical modal metaphysics constitutes a systematic set of answers to all of these questions. His answers to (i) and (ii) constitute his positive theory of *a priori* cognition (Chapter 6). His answer to (iii) constitutes his highly general theory of (the unschematized category of) real possibility *überhaupt* (Chapters 7 and 9). His answer to (iv) constitutes his theory of various notions of real possibility and their grounds, systematically unified by his answer to (iii) (Chapters 7, 8, and 9). His answer to (v) constitutes his Critical reinterpretation of the epistemic status of the argument in *Beweisgrund*: it is a subjectively necessary postulate of theoretical reason (Chapter 9). And, finally, his answer to (vi) is that the divine intuitive intellect cognizes no modal properties, so the relation between God and real possibility is a consequence of our modalized representation of the relation between God and what he has the power to do.

Note on Sources and Key to Abbreviations and Translations

Citations to the *Critique of Pure Reason* use the customary format of giving the page in the first edition of 1781 (A), followed by the page in the second edition of 1787 (B) (e.g., A327/B384). Citations to works of Kant other than the *Critique of Pure Reason* give the volume and page number in the Academy edition, *Kants gesammelte Schriften* [Ak.], edited by the Berlin-Brandenburg Academy of Sciences (Berlin: Walter de Gruyter, 1902–). Unless context makes it clear, I also refer to individual works by abbreviations of their English titles. Below I list those abbreviations, their complete (German) title, which Ak. volume they are found in, as well as dates of publication (for published works), of composition (for unpublished works), and, in the case of Kant's lectures, of the lectures themselves; where dates are uncertain I indicate this with a question mark (and, in some cases, a range of possible dates). I also list the translations from which I quote, often with slight modifications of my own (see Preface). Where no published translation is listed, either translations are my own or the work is cited but not quoted.

A	Leibniz, G. W. (1923–). *Sämtliche Schriften und Briefe*. Darmstadt: Berlin Academy of the Sciences. Cited by series, volume, and page.
AG	Leibniz, G. W. (1989). *Philosophical Essays*. R. Ariew and D. Garber (trans.). Indianapolis: Hackett.
Ak.	(1902–). *Kants gesammelte Schriften* (vols. 1–29). Berlin-Brandenburg (formerly: Royal Prussian) Academy of Sciences (eds.). Berlin: Walter de Gruyter. Cited by volume and page number.
Anthr.	*Anthropologie in pragmatischer Hinsicht* (1798). Ak. 7: 117–334.
AT	Descartes, R. (1964–76). *Oeuvres* (vols. 1–12). C. Adam and P. Tannery (eds.). Paris: Vrin/CNRS. Cited by volume and page number.
AzDM	*Anmerkungen zur Deutschen Metaphysik*. In *Werke* (Wolf (1965–)), Abt. I, Bd. 3.
CJ	*Kritik der Urtheilskraft* (1790). Ak. 5: 165–485. *Critique of the Power of Judgment*, P. Guyer and E. Matthews (trans.). Kant (2000).
Corr.	*Kants Briefwechsel*. Ak. 10–13. Selected translations in Kant (1999).
CPR	*Kritik der reinen Vernunft* (A:1781, B:1787). Ak. 3 (B) and 4: 1–252 (A). I have also consulted Kant (1998*b*). *Critique of Pure Reason*, P. Guyer and A. Wood (trans.). Kant (1998*a*). I have also occasionally consulted the Kemp Smith translation, Kant (1929).
CPrR	*Kritik der praktischen Vernunft* (1788). Ak. 5: 1–164. *Critique of Practical Reason*, M. J. Gregor (trans.). In Kant (1996*a*).

CSM — Descartes, R. (1984–1991). *The Philosophical Writings of Descartes* (vols. 1–3), J. Cottingham, R. Stoothoff, and D. Murdoch (trans. and ed.). Cambridge: Cambridge University Press.

Danz.RT — *Danziger Rationaltheologie* (1783/4), Ak. 28: 1226–1319.

De Usu — Crusius, C. A. (1743). *Dissertatio de usu et limitibus principii rationis determinantis, vulgo sufficientis.* In (1964). *Die Philosophischen Hauptwerke*, G. Tonelli (ed.). Hildesheim: Georg Olms, Vol. 4.
Ausfürhliche Abhandlung von dem rechten Gebrauche und der Einschränkung des sogenannten Satzes vom zureichenden oder besser determinierenden Gründe (1744), Christian Friedrich Krausen (trans.). Leipzig. Cited by section number.

Disc. — *Über eine Entdeckung, nach der alle neue Kritik der reinen Vernunft durch eine ältere entebehrlich gemacht werden soll* (1790). Ak. 8: 185–252.
On a Discovery Whereby any New Critique of Pure Reason is to be Made Superfluous by an Older One, H. Allison (trans.). In Kant (2002).

DM — Leibniz, G.W. *Discourse on Metaphysics.* In AG 35–68.

DSR — Leibniz, G. W. (1992). *De Summa Rerum*, G. H. R. Parkinson (trans. and ed.). New Haven: Yale University Press.

Dt.Log. — Wolff, C., *Vernünftige Gedanken von den Kräften des menschlichen Verstandes und ihrem richtigen Gebrauche im Erkenntnis der Warhheit.* In *Werke* (Wolff (1965–)), Abt. I, Bd. 1. Cited by chapter and paragraph number (e.g., §3.3 refers to §3 of Chapter 3; Chapter 0 corresponds to '*Vorbericht von der Welt-Weisheit*'.)

Dt.Met. — Wolff, C., *Verunünftige Gedancken von Gott, der Welt und der Seele des Menschen, auch allen Dingen überhaupt.* In *Werke* (Wolff (1965–)), Abt. I, Bd. 2.

E — Erdmann, B. (1881). *Nachträge zu Kants Kritik der reinen Vernunft.* Kiel: Lipsius & Tischer. Reproduced in Ak. 14–23. Cited by Roman numeral in Erdmann's edition and volume and page number in Ak.

Ent. — Crusius, C. A. (1745). *Entwurf der nothwendigen Verunft-Wahrheiten* (Leipzig). In (1964). *Die Philosophischen Hauptwerke*, G. Tonelli (ed.). Hildesheim: Georg Olms, Vol. 2. Cited by section number.

GMM — *Grundlegung zur Metaphysik der Sitten* (1785). Ak. 5: 385–464.
Groundwork of the Metaphysics of Morals, M. J. Gregor (trans.). In Kant (1996*a*).

H.*Werke* — Hegel, G.W.F. (1969–1971). *Werke*, E. Moldenhauer and K.M. Michel (eds.). Frankfurt a.M.: Suhrkamp. 20 volumes. Cited by volume and page number.

ID — *De mundi sensibilis atque intelligibilis forma et principiis* (1770). Ak. 2: 385–420. [Inaugural Dissertation.]
On the Form and Principles of the Sensible and Intelligible World, D. Walford and R. Meerbote (trans.). In Kant (1992*a*).

JL — *Logik*, herausgegeben von Gottlob Benjamin Jäsche (1800). Ak. 9: 1–150.
J. M. Young (trans.). In Kant (1992*b*).

L	Leibniz, G. W. (1969). *Philosophical Papers and Letters*, L. Loemker (trans. and ed.). Dordrecht: D. Reidel.
LB	*Logik Blomberg* (1771?). Ak. 24: 7–301. J. M. Young (trans.). In Kant (1992*b*).
LBu	*Logik Busolt* (1789/90?). Ak. 24: 604–86.
LDW	*Logik Dohna-Wundlacken* (1792), Ak. 24: 697–784. J. M. Young (trans.). In Kant (1992*b*).
LF	*Gedanken von der wahren Schätzung der lebendigen Kräfte* (1747). Ak. 1: 1–182.
Log.	Wolff, C., *Philosophia rationalis sive Logica*. In *Werke* (Wolf (1965–)), Abt. II, Bd. 1.2–3.
LP	*Logik Philippi* (1772?). Ak. 24: 303–496.
LPö	*Logik Pölitz* (1780?). Ak. 24: 499–602.
MD	*Metaphysik Dohna* (1792–3). Ak. 28: 610–704.
Meta.	Baumgarten, A. G. (1757). *Metaphysica* (4th edn.). Halle. Baumgarten, A. G. (2013). *Metaphysics: A Critical Translation with Kant's Elucidations, Selected Notes and Related Materials*, C. Fugate and J. Hymers (trans. and ed.). London: Bloomsbury.
MFNS	*Metaphysische Anfangsgründe der Naturwissenschaft* (1786). Ak. 4: 465–566. *Metaphysical Foundations of Natural Science*, M. Friedman (trans.). In Kant (2002).
MH	*Metaphysik Herder* (1762–4?). Ak. 28: 1–181. Selections trans. K. Ameriks and S. Naragon in Kant (1997).
MK_2	*Metaphysik K_2* (early 1790s). Ak. 28: 705–816. Selections trans. K. Ameriks and S. Naragon in Kant (1997).
MK_3	*Metaphysik Vigilantius* (1794–5). Ak. 28: 817–838, 941–1047. K. Ameriks and S. Naragon (trans.). In Kant (1997).
ML_1	*Metaphysik L_1* (1770–5?). Ak. 28: 167–250. Selections trans. K. Ameriks and S. Naragon in Kant (1997).
ML_2	*Metaphysik L_2* (1780?), Ak. 28: 531–604. K. Ameriks and S. Naragon (trans.). In Kant (1997).
MM	*Metaphysik Mrongovius* (1782/3), Ak. 29: 747–940. K. Ameriks and S. Naragon (trans.). In Kant (1997).
Mon.	Leibniz, G. W. (1714). *Principes de philosophie ou Monadologie.* *The Principles of Philosophy, or Monadology*. AG 213–24.
MV	*Metaphysik Volckmann* (1784/5). Ak. 28: 351–459.
MvS	*Metaphysik von Schön* (1785–90?). Ak. 28: 460–524.
ND	*Principiorum primorum cognitionis metaphysicae nova delucidatio* (1755). Ak. 1: 385–416. *New Elucidation of the First Principles of Metaphysical Cognition*, D. Walford and R. Meerbote (trans.). In Kant (1992*a*).

NE	Leibniz, G. W. (1704). *Nouveaux Essais sur l'entendement humain.* Leibniz, G.W. (1966). *New Essays on Human Understanding*, P. Remnant and J. Bennett (trans. and ed.). Cambridge: Cambridge University Press.
NG	Wolff, C. *Natürliche Gottesgelahrheit nach beweisender Lehrart abgefasst.* In *Werke* (Wolf (1965–)), Abt. 1, Bd. 23.1–4.
NM	*Versuch den Begriff von Negativen Grössen in der Philosophie einzuführen* (1763). Ak. 2: 165–204. *Attempt to Introduce the Concept of Negative Magnitudes into Philosophy*, D. Walford and R. Meerbote (trans.). In Kant (1992).
Ont.	Wolff, C., *Philosophia prima sive Ontologia* . In *Werke* (Wolff (1965–)), Abt. 2, Bd. 3.
OPG	*Der einzig Mögliche Beweisgrund zu einer Demonstration des Dasein Gottes* (1763). Ak. 2: 63–164. I have also consulted Kant (2011). [*Beweisgrund*] *The Only Possible Argument in Support of a Demonstration of the Existence of God*, D. Walford and R. Meerbote (trans. and ed.). In Kant (1992*a*).
P	Leibniz, G. W. (1966). *Logical Papers*, G. H. R. Parkinson (trans. and ed.). Oxford: Clarendon Press.
PE	*Vorlesungen über Philosophische Enzyklopädie.* Ak. 29: 3–45.
Pöl.RT	*Philosophische Religionslehre nach Pölitz* (1783–6?). Ak. 28: 989–1126. *Lectures on the Philosophical Doctrine of Religion*, A. Wood (trans.). In Kant (1996*b*).
Prize	*Untersuchung über die Deutlichkeit der Grundsätze der natürlichen Theologie und Moral* (1764). Ak. 2: 273–302. *Inquiry Concerning the Distinctness of the Principles of Natural Theology and Morality*, D. Walford and R. Meerbote (trans.). In Kant (1992*a*). [Prize Essay]
Prog.	*Welches sind die wirklichen Fortschritte, die die Metaphysik seit Leibnitzens und Wolffs Zeiten in Deutschland gemacht hat?* (1791/1804), Ak. 20: 253–332. *What Real Progress has Metaphysics Made in Germany since the Time of Leibniz and Wolff?*, P. Heath (trans.). In Kant (2002).
Prol.	*Prolegomena zu einer jeden künftigen Metaphysik die als Wissenschaft wird auftreten können* (1783). Ak. 4: 253–384. *Prolegomena to any Future Metaphysics that Will be Able to Present itself as a Science*, G. Hatfield (trans.). In Kant (2002).
Refl.	*Kants handschriftlicher Nachlaß.* Ak. 14–23. Cited by four-digit number and volume and page number. Also referred to as *Reflexionen*.
Rel.	*Die Religion innerhalb der Grenzen der bloßen Vernunft* (1793). Ak. 6: 1–202. *Religion within the Boundaries of Mere Reason*, G. di Giovanni (trans.). In Kant (1996*b*).
Search	Malebranche, N. (1980). *The Search after Truth*, M. Lennon and P. J. Olscamp (trans. and ed.). Columbus, Ohio: Ohio University Press. Repr. (1997), Cambridge: Cambridge University Press. Cited by Book, Part, and Chapter.

Th.	Leibniz, G. W. (1710). *Théodicée.* Leibniz, G. W. (1985). *Theodicy*, E. M. Huggard and A. Ferrer (trans. and ed.). Chicago and La Salle, Ill.: Open Court.
TN	Wolff, C., *Theologia naturalis.* In *Werke* (Wolff (1965–)). Abt. II, Bd. 7–8.
Tone	*Von einem neuerdings erhobenen vornehmen Ton in der Philosophie* (1796). Ak. 8: 387–406. *On a Recently Prominent Tone of Superiority in Philosophy*, P. Heath (trans.). In Kant (2002).
Volck.RT	*Natürliche Theologie Volckmann* (1783/4). Ak. 28: 1127–1225.
Weg	Crusius, C. (1747). *Weg zur Gewißheit und Zuverläßigkeit der menschlichen Erkenntniß.* In (1964). *Die Philosophischen Hauptwerke*, G. Tonelli (ed.). Hildesheim: Georg Olms, Vol. 3. Cited by section number.
Werke	Wolff, C. (1965–). *Gesammelte Werke*, J. École, H. W. Arndt, Ch. A. Corr, J. E. Hoffmann, and M. Thomann (eds.). Hildesheim, Zürich, and New York: Georg Olms. Cited by division (Abt.) and volume (Bd.).
WL	*Wiener Logik* (1780). Ak. 24: 785–940. J. M. Young (trans.). In Kant (1992*b*).

Bibliography

Works of Kant

(1902–). *Kants gesammelte Schriften* (vols. 1–29). Berlin-Brandenburg (formerly: Royal Prussian) Academy of Sciences (eds.). Berlin: Walter de Gruyter.

(1929). *Immanuel Kant's Critique of Pure Reason*, N. Kemp Smith (trans.). London: Macmillan.

(1976). *Prolegomena; zu einer jeden künftigen Metaphysik, die als Wissenschaft wird auftreten können*, K. Vorländer (ed.). Hamburg: Felix Meiner.

(1977). *Prolegomena to any Future Metaphysics*, P. Carus and J. Ellington (trans.). New York: Hackett.

(1992*a*). *Theoretical Philosophy, 1775–1770*, D. Walford and R. Meerbote (trans. and ed.). Cambridge: Cambridge University Press.

(1992*b*). *Lectures on Logic*, J. M. Young (trans. and ed.). Cambridge: Cambridge University Press.

(1996*a*). *Practical Philosophy*, M. J. Gregor (trans. and ed.). Cambridge: Cambridge University Press.

(1996*b*). *Religion and Rational Theology*, A. Wood and G. di Giovanni (trans. and ed.). Cambridge: Cambridge University Press.

(1997). *Lectures on Metaphysics*, K. Ameriks and S. Naragon (trans. and ed.). Cambridge: Cambridge University Press.

(1998*a*). *Critique of Pure Reason*, P. Guyer and A. Wood (trans. and ed.). Cambridge: Cambridge University Press.

(1998*b*). *Kritik der reinen Vernunft. Nach der 1. And 2. Original Ausgabe*, J. Timmerman (ed.). Hamburg: Felix Meiner.

(1999). *Correspondence*, A. Zweig (trans. and ed.). Cambridge: Cambridge University Press.

(2000). *Critique of the Power of Judgment*, P. Guyer and E. Matthews (trans. and ed.). Cambridge: Cambridge University Press.

(2002). *Theoretical Philosophy after 1781*, H. Allison and P. Heath (trans. and ed.). Cambridge: Cambridge University Press.

(2011). *Der einzig mögliche Beweisgrund zu einer Demonstration des Dasein Gottes. Historisch-kritische Edition*, L. Kreimendahl and M. Oberhausen (eds.). Hamburg: Felix Meiner.

Other Primary Sources

Baumgarten, A. G. (1757). *Metaphysica* (4th edn.). Halle. Repr. in Ak. 17.

Baumgarten, A. G. (1783). *Metaphysik* (2nd German edn.), G. F. Meier and J. E. Eberhard (trans. and ed.). Halle. Repr. (2004). *Metaphysik*, D. Mirbach (ed.). Jena: Dietrich Schlegelmann.

Baumgarten, A. G. (2011). *Metaphysica = Metaphysik: historisch-kritische Ausgabe*, G. Gawlick and L. Kreimendahl (trans.). Stuttgart: Fromann-Holzboog.

Baumgarten, A. G. (2013). *Metaphysics: A Critical Translation with Kant's Elucidations, Selected Notes and Related Materials*, C. Fugate and J. Hymers (trans. and ed.). London: Bloomsbury.

Crusius, C. A. (1964). *Die philosophischen Hauptwerke* (vols. 1–4), G. Tonelli (ed.). Hildesheim: Georg Olms.

Descartes, R. (1964–76). *Oeuvres* (vols. 1–12), C. Adam and P. Tannery (eds.). Paris: Vrin/CNRS.

Descartes, R. (1984–1991). *The Philosophical Writings of Descartes* (vols. 1–3), J. Cottingham, R. Stoothoff, and D. Murdoch (trans. and ed.). Cambridge: Cambridge University Press.

Eberhard, J. E. (1789). Über die Unterscheidung der Urteile in analytische und synthetische. *Philosophisches Magazin*, 1: 307–32.

Hegel, G. W. F. (1969–71). *Werke* (vol. 1–20), E. Moldenhauer and K.M. Michel (eds.). Frankfurt a.M.: Suhrkamp.

Leibniz, G. W. (1969). *Philosophical Papers and Letters*, L. Loemker (trans. and ed.). 2nd edition. Dordrecht: D. Reidel.

Leibniz, G. W. (1985). *Theodicy*, E. M. Huggard and A. Farrer (trans. and ed.). Chicago and La Salle, Ill.: Open Court.

Leibniz, G. W. (1989). *Philosophical Essays*, R. Ariew and D. Garber (trans. and ed.). Indianapolis: Hackett.

Leibniz, G. W. (1996). *New Essays on Human Understanding*, P. Remnant and J. Bennett (trans. and ed.). Cambridge: Cambridge University Press.

Leibniz, G. W. (2008). *Protogaea*, C. Cohen and A. Wakefield (trans. and ed.). Chicago: University of Chicago Press.

Maaß, J. G. (1789). Über den höchsten Grundsatz der synthetischen Urtheile, in Beziehung auf die Theorie von der mathematischen Gewißheit. *Philosophisches Magazin*, 2(2): 186–231.

Malebranche, N. (1980). *The Search after Truth*, M. Lennon and P. J. Olscamp (trans. and ed.). Columbus, Ohio: Ohio University Press. Repr. (1997). Cambridge: Cambridge University Press.

Malebranche, N. (1997). *Dialogues on Metaphysics and on Religion*, N. Jolley and D. Scott (trans. and ed.). Cambridge: Cambridge University Press.

Mendelssohn, M. (1997). *Philosophical Writings*, D. Dahlstrom (trans. and ed.). Cambridge: Cambridge University Press.

Suárez, F. (1983). *On the Essence of Finite Being as Such, on the Existence of that Essence and their Distinction*, N. Wells (trans. and ed.). Milwaukee, Wisc.: Marquette University Press.

Valentini, M. B. (1709). *Armamentarium artis & naturae*. Gissae-Hassorum.

Warda, A. (1922). *Immanuel Kants Bücher*. Berlin: Martin von Breslauer.

Watkins, E. (2009). *Kant's Critique of Pure Reason: Background Source Materials*. Cambridge: Cambridge University Press.

Wolff, C. (1965–). *Gesammelte Werke*, J. École, H.W. Arndt, Ch. A. Corr, J. E. Hoffmann, and M. Thomann (eds.). Hildesheim, Zürich, and New York: Georg Olms.

Wolff, C. (2005). *Erste Philosophie oder Ontologie* §§1–78, D. Effertz (trans. and ed.). Hamburg: Felix Meiner.

Secondary Literature

Abaci, U. (2008). Kant's Theses on Existence. *British Journal for the History of Philosophy*, 16 (3): 559–93.

Abaci, U. (2013). The Coextensiveness Thesis and Kant's Modal Agnosticism in the 'Postulates.' *European Journal of Philosophy* 23(2).

Abaci, U. (2014). Kant's *Only Possible Argument* and Chignell's Real Harmony. *Kantian Review*, 19(1): 1–25.

Adams, R. M. (1974). Theories of Actuality. *Noûs*, 8(3): 211–31.

Adams, R. M. (1994). *Leibniz: Determinist, Idealist, Theist.* Oxford: Oxford University Press.

Adams, R. M. (1997). Things in Themselves. *Philosophy and Phenomenological Research*, 57 (4): 801–25.

Adams, R. M. (2000). God, Possibility and Kant. *Faith and Philosophy*, 17(4): 425–40.

Adickes, E. (1924). *Kant und das Ding an sich.* Berlin: Pan Verlag Rolf Heisse. Repr. (1977). Hildesheim: Georg Olms.

Allais, L. (2004). Kant's One World: Interpreting 'Transcendental Idealism.' *The British Journal for the History of Philosophy*, 12(4): 655–84.

Allais, L. (2006). Intrinsic Natures: A Critique of Langton on Kant. *Philosophy and Phenomenological Research*, 73(1): 144–69.

Allais, L. (2007). Kant's Idealism and the Secondary-Quality Analogy. *Journal of the History of Philosophy*, 45(3): 459–84.

Allais, L. (2009). Kant, Non-conceptual Content, and the Representation of Space. *Journal of the History of Philosophy*, 47(3): 383–413.

Allais, L. (2015). *Manifest Reality: Kant's Idealism and his Realism.* Oxford: Oxford University Press.

Allison, H. (1973). *The Kant–Eberhard Controversy.* Baltimore: Johns Hopkins University Press.

Allison, H. (1983). *Kant's Transcendental Idealism: An Interpretation and Defense.* New Haven: Yale University Press.

Allison, H. (1994). Causality and Causal Laws in Kant: A Critique of Michael Friedman. In P. Parrini (ed). *Kant and Contemporary Epistemology.* Dordrecht: Kluwer, 291–307.

Allison, H. (2000). Is the Critique of Judgment 'Post-Critical?'. In S. Sedgwick (ed.). *The Reception of Kant's Critical Philosophy: Fichte, Schelling, and Hegel.* Cambridge: Cambridge University Press, 78–92.

Allison, H. (2004). *Kant's Transcendental Idealism: An Interpretation and Defense.* 2nd edn. New Haven: Yale University Press.

Allison, H. (2012). *Essays on Kant.* Oxford: Oxford University Press.

Ameriks, K. (1981). Kant's Deduction of Freedom and Morality. *Journal of the History of Philosophy*, 19(1): 53–79.

Ameriks, K. (1982*a*). Recent Work on Kant's Theoretical Philosophy. *American Philosophical Quarterly*, 19(1): 1–24.

Ameriks, K. (1982*b*). *Kant's Theory of Mind: An Analysis of the Paralogisms of Pure Reason.* Oxford: Clarendon Press.

Ameriks, K. (1992). The Critique of Metaphysics: Kant and Traditional Ontology. In P. Guyer (ed.). *The Cambridge Companion to Kant.* Cambridge: Cambridge University Press, 249–79.

Ameriks, K. (2003). Kant and Short Arguments to Humility. In *Interpreting Kant's Critiques.* Oxford: Oxford University Press, 135–57.

Anderson, E. (1994). Kant, Natural Kind Terms, and Scientific Essentialism. *History of Philosophy Quarterly*, 11(4): 355–73.

Anderson, R. L. (2004). It Adds Up After All: Kant's Philosophy of Arithmetic in Light of the Traditional Logic. *Philosophy and Phenomenological Research*, 69(3): 501–40.

Aquila, R. (1979). Things in Themselves and Appearances: Intentionality and Reality in Kant. *Archiv für Geschichte der Philosophie*, 61(3): 293–308.

Aquila, R. (1983). *Representational Mind: A Study of Kant's Theory of Knowledge*. Bloomington, Ind.: Indiana University Press.

Armstrong, D. (1989). *Universals: An Opinionated Introduction*. Boulder, Colo.: Westview Press.

Ariew, R. (1998). Leibniz on the Unicorn and Various Other Curiosities. *Early Modern Science and Medicine*, 3(4): 267–88.

Audi, P. (2012*a*). Grounding: Toward a Theory of the In-Virtue-Of Relation. *Journal of Philosophy*, 109(12): 685–711.

Audi, P. (2012*b*). A Clarification and Defense of the Notion of Grounding. In F. Correia and B. Schnieder (eds.). *Metaphysical Grounding: Understanding the Structure of Reality*. Cambridge: Cambridge University Press, 101–21.

Ayer, A. J. (1936). *Language, Truth, and Logic*. London: Gollancz.

Bader, R. M. (2009). Kant and the Categories of Freedom. *British Journal for the History of Philosophy*, 17(4): 799–820.

Baumgardt, D. (1915). *Das Möglichkeitsproblem der Kritik der reinen Vernunft, der modernen Phänomenologie und der Gegenstandstheorie*. Kant-Studien Ergänzungsheft 51. Berlin: Walter de Gruyter.

Barnes, J. (1972). *The Ontological Argument*. New York: St Martin's Press.

Beck, L. W. (1960). *A Commentary on Kant's Critique of Practical Reason*. Chicago: University of Chicago Press.

Beck, L. W. (1967). Can Kant's Synthetic Judgments Be Made Analytic? In M. Gram (ed.). *Kant: Disputed Questions*. Chicago: Quadrangle Books, 228–46.

Beck, L. W. (1969). *German Philosophy before Kant*. Cambridge, Mass.: Harvard University Press.

Beck, L.W. (1978). A Prussian Hume and a Scottish Kant. In *Essays on Kant and Hume*. New Haven: Yale University Press, 111–29.

Beck, L. W. (1989). Two Ways of Reading Kant's Letter to Herz: Comments on Carl. In E. Förster (ed.). *Kant's Transcendental Deductions*. Stanford: Stanford University Press, 21–6.

Beiser, F. (2002). *German Idealism: The Struggle against Subjectivism, 1781–1801*. Cambridge, Mass.: Harvard University Press.

Bird, A. (2005). Laws and Essences. *Ratio*, 18(4): 437–61.

Bird, G. (1962). *Kant's Theory of Knowledge*. London: Routledge & Kegan Paul.

Bissinger, A. (1970). *Die Struktur der Gotteserkenntnis: Studien zur Philosophie Christian Wolffs*. Bonn: H. Bouvier.

Bliss, R. and Trogdon, K. (2014). Metaphysical Grounding. *The Stanford Encyclopedia of Philosophy*. E.N. Zalta (ed.). <http://plato.stanford.edu/archives/win2014/entries/grounding/>.

Blumenfeld, D. (1973). Leibniz's Theory of the Striving Possibles. *Studia Leibnitiana*, 5(2): 163–77.

Boghossian, P. (1996). Analyticity Reconsidered. *Noûs*, 30(3): 360–91.

Brandt, R. (1991). *Die Urteilstafel: Kritik der reinen Vernunft A67–76/B92–201*. Hamburg: Felix Meiner.

Breazeale, D. (1998). Fichte's *nova methodo phenomenologica*: On the Methodological Role of 'Intellectual Intuition' in the Later Jena *Wissenschaftslehre*. *Revue internationale de philosophie*, 52(206): 587–616.

Brewer, K. and Watkins, E. (2012). A Difficulty Still Awaits: Kant, Spinoza, and the Threat of Theological Determinism. *Kant-Studien*, 103(2): 163–87.

Brittan, G. (1978). *Kant's Philosophy of Science*. Princeton: Princeton University Press.

Brogaard, B. and Salerno, J. (2013). Remarks on Counterpossibles. *Synthèse*, 19(4): 639–60.

Buchdahl, G. (1965). Causality, Causal Laws and Scientific Theory in the Philosophy of Kant. *British Journal for the Philosophy of Science*, 16(63): 187–208.

Buchdahl, G. (1969). The Kantian 'Dynamic of Reason', with Special Reference to the Place of Causality in Kant's System. In L. W. Beck (ed.). *Kant Studies Today*. La Salle, Ill.: Open Court, 341–74.

Buchdahl, G. (1986). Kant's 'Special Metaphysics' and *The Metaphysical Foundations of Natural Science*. In R. Butts (ed.). *Kant's Philosophy of Physical Science: Metaphysische Anfangsgründe der Naturwissenschaft 1786–1986*. Dordrecht: D. Reidel, 127–62.

Buchdahl, G. (1992). *Kant and the Dynamics of Reason*. Cambridge, Mass.: Blackwell.

Burge, T. (1988). Individualism and Self-Knowledge. *Journal of Philosophy*, 85(11): 649–63.

Burgess, J. (2013). *Saul Kripke*. Cambridge: Polity.

Butts, R. (1986). The Methodological Structure of Kant's Metaphysics of Science. In R. Butts (ed.). *Kant's Philosophy of Physical Science: Metaphysische Anfangsgründe der Naturwissenschaft 1786–1986*. Dordrecht: D. Reidel, 163–200.

Byrne, A. and Hilbert, D. R. (1997). *Readings on Color, Vol. 1: The Philosophy of Color*. Cambridge, Mass.: MIT Press.

Carl, W. (1989*a*). *Der schweigende Kant: die Entwürfe zu einer Deduktion der Kategorien vor 1781*. Göttingen: Vandenhoek & Ruprecht.

Carl, W. (1989*b*). Kant's First Draft of the Deduction of the Categories. In E. Förster (ed.). *Kant's Transcendental Deductions*. Stanford: Stanford University Press, 3–20.

Cassirer, E. (1981). *Kant's Life and Thought*. (James Haden, trans.). New Haven: Yale University Press.

Casullo, A. (2002). A Priori Knowledge. In P. Moser (ed.). *The Oxford Handbook of Epistemology*. Oxford: Oxford University Press, 95–143.

Chalmers, D., Manley, R., and Wasserman, R. (eds.). (2009). *Metametaphysics*. Oxford: Oxford University Press.

Chignell, A. (2003). Kant's Ethics of Assent: Knowledge and Belief in the Critical Philosophy. Unpublished Ph.D dissertation. Cornell University.

Chignell, A. (2007). Belief in Kant. *Philosophical Review*, 116(3): 323–60.

Chignell, A. (2009*a*). Kant, Modality, and the Most Real Being. *Archiv für Geschichte der Philosophie*, 91(2): 157–92.

Chignell, A. (2009*b*). Are Supersensibles Really Possible? The Evidential Role of Symbols. In V. Rhoden, T. Terra, and G. Almeida (eds.). *Recht und Frieden in der Philosophie Kants*. Berlin: Walter de Gruyter, vol. 4, 99–110.

Chignell, A. (2010). Real Repugnance and Belief about Things-in-Themselves: A Problem and Kant's Three Solutions. In J. Krueger and B. B. Lipscomb (eds.). *Kant's Moral Metaphysics*. Berlin: Walter de Gruyter, 177–210.

Chignell, A. (2011). Real Repugnance and Our Ignorance of Things-in-Themselves: A Lockean Problem in Kant and Hegel. *Internationales Jahrbuch des Deutschen Idealismus*, 7: 135–59.

Chignell, A. (2012). Kant, Possibility, and the Threat of Spinoza. *Mind*, 121(483): 635–75.

Chignell, A. (2014*a*). Kant and the 'Monstrous' Ground of Possibility: A Reply to Abaci and Yong. *Kantian Review*, 19(1): 53–69.

Chignell, A. (2014*b*). Modal Motivations for Noumenal Ignorance: Knowledge, Cognition, and Coherence. *Kant-Studien* 105(4): 573–597.

Cicovacki, P. (1991). An Aporia of *A priori* Knowledge. *Kant-Studien*, 82(3): 349–60.

Cohen, C. (2002). *The Fate of the Mammoth*. Chicago: University of Chicago Press.

Corr, E. H. (1973). The Existence of God, Natural Theology and Christian Wolff. *International Journal for Philosophy of Religion*, 4(2): 105–18.

Correia, F. and Schnieder, B. (eds.) (2012). *Metaphysical Grounding: Understanding the Structure of Reality*. Cambridge: Cambridge University Press.

Cover, J. and O'Leary-Hawthorne, J. (1999). *Substance and Individuation in Leibniz*. Cambridge: Cambridge University Press.

Cunning, David (2008). Descartes' Modal Metaphysics. *The Stanford Encyclopedia of Philosophy* (Winter 2008 edn.). E.N. Zalta (ed.). <http://plato.stanford.edu/archives/spr2014/entries/descartes-modal/>.

Curley, E. (1984). Descartes on the Creation of the Eternal Truths. *Philosophical Review*, 93(4): 569–97.

Dell'Oro, R. (1994). *From Existence to the Ideal: Continuity and Development in Kant's Theology*. New York: Peter Lang.

De Vleeschauwer, H. J. (1962). *The Development of Kantian Thought*, trans. A. R. C. Duncan. Edinburgh: Thos. Nelson.

Di Bella, S. (2005). *The Science of the Individual*. Dordrecht: Kluwer.

Di Sala, G. (1990). *Kant und die Frage nach Gott*. Kant-Studien Ergänzungsheft 122. Berlin: Walter de Gruyter.

Divers, J. (2002). *Possible Worlds*. New York: Routledge.

Dummett, M. (1983). Could There Be Unicorns? In *The Seas of Language*. Oxford: Oxford University Press, 328–48.

Düsing, K. (1986). *Die Teleologie in Kants Weltbegriff*. Kant-Studien Ergänzungsheft 80. Bonn: H. Bouvier.

Easton, P. (2009). What is at Stake in the Cartesian Debates on the Eternal Truths? *Philosophy Compass*, 4(2): 348–62.

École, J. (1990). *La Métaphysique de Christian Wolff*. In Wolff, C. *Werke*, Abt. III, Bd. 12.1–2.

Ellis, B. and Lierse, C. (1994). Dispositional Essentialism. *Australasian Journal of Philosophy*, 72(1): 27–45.

Falkenstein, L. (1995). *Kant's Intuitionism: A Commentary on the Transcendental Aesthetic*. Toronto: University of Toronto Press.

Ferrari, J. (1998). Das Ideal der reinen Vernunft. In G. Mohr and M. Willaschek (eds.). *Immanuel Kant, Kritik der reinen Vernunft*. Berlin: Academie Verlag, 491–523.

Fine, K. (1994*a*). Essence and Modality. *Philosophical Perspectives*, 8: 1–16.

Fine, K. (1994*b*). Compounds and Aggregates. *Noûs*, 28(2): 137–58.

Fine, K. (1995). Senses of Essence. In W. Sinnott-Armstrong, D. Raffman, and N. Asher (eds.). *Modality, Morality and Belief: Essays in Honor of Ruth Barcan Marcus*. Cambridge: Cambridge University Press, 53–73.

Fine, K. (1999). Things and Their Parts. *Midwest Studies in Philosophy*, 23(1): 61–74.

Forgie, J. W. (2007). Kant and Gassendi on Existence. *Journal of the History of Philosophy*, 45 (4): 511–23.

Förster, E. (2011). *Die 25 Jahre der Philosophie*. Frankfurt a.M.: Vittorio Klostermann.

Friedman, M. (1986). The Metaphysical Foundations of Newtonian Science. In R. Butts (ed.). *Kant's Philosophy of Physical Science: Metaphysische Anfangsgründe der Naturwissenschaft 1786–1986*. Dordrecht: D. Reidel, 25–60.

Friedman, M. (1992*a*). *Kant and the Exact Sciences*. Cambridge, Mass.: Harvard University Press.

Friedman, M. (1992*b*). Causal Laws and the Foundation of Natural Science. In P. Guyer (ed.). *The Cambridge Companion to Kant*. Cambridge: Cambridge University Press, 161–99.

Friedman, M. (2013). *Kant's Construction of Nature: A Reading of the Metaphysical Foundations of Natural Science*. Cambridge: Cambridge University Press.

Garber, D. (1985). Leibniz and the Foundations of Physics: The Middle Years. In K. Okruhlik and J. R. Brown (eds.). *The Natural Philosophy of Leibniz*. Dordrecht: D. Reidel, 27–130.

Garber, D. (2005). Leibniz: Physics and Philosophy. In Donald Rutherford (ed.). *The Cambridge Companion to Leibniz*. Cambridge: Cambridge University Press, 270–352.

Garber, D. (2009). *Leibniz: Body, Substance, Monad*. Oxford: Oxford University Press.

Gardner, S. (1999). *Kant and the Critique of Pure Reason*. New York: Routledge, 1999.

Gibson, M. (2011). A Revolution in Method, Kant's 'Copernican Hypothesis,' and the Necessity of Natural Laws. *Kant-Studien*, 102(1): 1–21.

Gloy, K. (1976). *Die Kantische Theorie der Naturwissenschaft: eine Strukturanalyse ihrer Möglichkeit, ihres Umfangs und ihrer Grenzen*. Berlin: Walter de Gruyter.

Goodman, N. (1954). *Fact, Fiction, and Forecast*. London: Athlone Press.

Gram, M. S. (1981). Intellectual Intuition: The Continuity Thesis. *Journal of the History of Ideas*, 42(2), 287–304.

Grier, M. (2001). *Kant's Doctrine of Transcendental Illusion*. Cambridge: Cambridge University Press.

Griffin, M. (2013). *Leibniz, God, and Necessity*. Cambridge: Cambridge University Press.

Grüne, S. (2009). *Blinde Anschauung: die Rolle von Begriffen in Kants Theorie sinnlicher Synthesis*. Frankfurt a.M.: Vittorio Klostermann.

Grüne, S. (2011). Is There a Gap in Kant's B Deduction? *International Journal of Philosophical Studies*, 19(3): 465–90.

Guyer, P. (1987). *Kant and the Claims of Knowledge*. Cambridge: Cambridge University Press.

Guyer, P. (1990*a*). Reason and Reflective Judgment: Kant on the Significance of Systematicity. *Noûs*, 24(1): 17–43.

Guyer, P. (1990*b*). Kant's Conception of Empirical Law. *Proceedings of the Aristotelian Society*, 64: 221–42.

Guyer, P. (ed.). (1992*a*). *The Cambridge Companion to Kant*. Cambridge: Cambridge University Press.

Guyer, P. (1992*b*). The Transcendental Deduction of the Categories. In P. Guyer (ed.). *The Cambridge Companion to Kant*. Cambridge: Cambridge University Press, 123–60.

Guyer, P. (1998). The Postulates of Empirical Thinking in General and the Refutation of Idealism. In G. Mohr and M. Willaschek (eds.). *Immanuel Kant: Kritik der reinen Vernunft*. Berlin: Akademie Verlag, 297–324.

Guyer, P. (2003). Kant on the Systematicity of Nature: Two Puzzles. *History of Philosophy Quarterly*, 20(3): 277–95.

Guyer, P. (2006). *Kant*. New York: Routledge.

Guyer, P. (ed.). (2010). *The Cambridge Companion to Kant's Critique of Pure Reason*. Cambridge: Cambridge University Press.

Hanna, R. (1998). A Kantian Critique of Scientific Essentialism. *Philosophy and Phenomenological Research*, 58(3): 497–528.

Hanna, R. (2001). *Kant and the Foundations of Analytic Philosophy*. Oxford: Oxford University Press.

Hanna, R. (2006). *Kant, Science, and Human Nature*. Oxford: Oxford University Press.

Hanna, R. (2008). Kantian Non-conceptualism. *Philosophical Studies*, 137(1): 41–64.

Hanna, R. (2011). Kant's Non-Conceptualism, Rogue Objects, and the Gap in the B Deduction. *International Journal of Philosophical Studies*, 19(3): 399–415.

Hardin, C. L. (1988). *Color for Philosophers*. Indianapolis: Hackett.

Hartmann, N. (1938). *Möglichkeit und Wirklichkeit*. Berlin: Walter de Gruyter.

Hawthorne, J. (2001). Causal Structuralism. *Philosophical Perspectives*, 15: 361–78.

Heimsoeth, H. (1956). *Metaphysik und Kritik bei. Chr. A. Crusius*. *Kant-Studien* Ergänzungsheft 71. Cologne: Kölner Üniversitats-Verlag.

Heidegger, M. (1975). *Die Grundprobleme der Phänomenologie*. Frankfurt a.M.: Vittorio Klostermann.

Henrich, D. (1967). Kants Denken 1762/3: Über den Ursprung der Unterscheidung analytischer und synthetischer Urteile. In H. Heimsoeth, D. Henrich, and G. Tonelli (eds.). *Studien zu Kants philosophischer Entwicklung*. Hildesheim: Olms, 9–38.

Henrich, D. (1968–9). The Proof-structure of Kant's Transcendental Deduction. *Review of Metaphysics*, 22(4): 640–59.

Henrich, D. (1976). *Identität und Objektivität: eine Untersuchung über Kants transzendentale Deduktion*. Heidelberg: C. Winter.

Hintikka, J. (1969). On Kant's Notion of Intuition. In T. Penelhum and J. J. MacIntosh (eds.). *The First Critique: Reflections on Kant's Critique of Pure Reason*. California: Wadsworth Publishing Company, 38–53.

Hoffer, Noam. (forthcoming). Was Kant a Spinozist? *Kantian Review*.

Hogan, D. (2005). Rationalism and Causal Realism in Kant's Metaphysics. Unpublished Ph.D Dissertation. Yale University.

Hogan, D. (2009*a*). Three Kinds of Rationalism and the Non-Spatiality of Things in Themselves. *Journal of the History of Philosophy*, 47(3): 355–82.

Hogan, D. (2009*b*). Noumenal Affection. *Philosophical Review*, 118(4): 501–32.

Hogan, D. (2009*c*). How to Know Unknowable Things in Themselves. *Noûs*, 43(1): 49–63.

Hogan, D. (2013). Metaphysical Motives of Kant's Analytic-Synthetic Distinction. *Journal of the History of Philosophy*, 51(2): 267–307.

Hogan, D. (2014). Kant on Foreknowledge of Contingent Truths. *Res Philosophica*, 19(1): 47–70.

Honnefelder, L. (1990). *Scientia transcendens: die formale Bestimmtheit der Seiendheit und Realität in der Metaphysik des Mittelalters und der Neuzeit*. Hamburg: Felix Meiner.

Howell, R. (1973). Intuition, Synthesis, and Individuation in the *Critique of Pure Reason*. *Noûs*, 7(3): 207–32.

Insole, C. (2011). Intellectualism, Relational Properties and the Divine Mind in Kant's Pre-critical Philosophy. *Kantian Review*, 16 (3): 399–427.

Jago, M. (2014). *The Impossible: An Essay on Hyperintensionality*. Oxford: Oxford University Press.

Jolley, N. (1990). *The Light of the Soul: Theories of Ideas in Leibniz, Malebranche, and Descartes*. Oxford: Clarendon Press.

Kain, P. (forthcoming). The Development of Kant's Conception of Divine Freedom. In B. Look (ed.). *Leibniz and Kant*. Oxford: Oxford University Press.

Kaufman, D. (2002). Descartes's Creation Doctrine and Modality. *Australasian Journal of Philosophy*, 80(1): 24–41.

Kauppi, R. (1960). Über die Leibnizsche Logik. *Acta Philosophica Fennica, XII*, 243–67.

Kemp Smith, N. (1918). *A Commentary to Kant's Critique of Pure Reason*. London: Macmillan. Repr. (1984). Atlantic Highlands, NJ: Humanities Press.

Kenny, A. (1968). *Descartes: A Study of His Philosophy*. New York: Random House.

Kitcher, P. (1975). Kant and the Foundations of Mathematics. *Philosophical Review*, 84: 23–50.

Kitcher, P. (1986). Projecting the Order of Nature. In R. Butts (ed.). *Kant's Philosophy of Physical Science: Metaphysische Anfangsgründe der Naturwissenschaft 1786–1986*. Dordrecht: D. Reidel, 201–37.

Kitcher, P. (1993). *Kant's Transcendental Psychology*. New York: Oxford University Press.

Kment, B. (2006). Counterfactuals and the Analysis of Necessity. *Philosophical Perspectives*, 20(1): 237–302.

Kohl, M. (2015). Kant on the Inapplicability of the Categories to Things in Themselves. *British Journal for the History of Philosophy*, 23(1): 1–25.

Korcz, K.A. (2010). The Epistemic Basing Relation. *The Stanford Encyclopedia of Philosophy* (Spring 2010 edn.). E. N. Zalta (ed.). <http://plato.stanford.edu/archives/spr2010/entries/basing-epistemic/>.

Koslicki, K. (forthcoming). The Coarse-Grainedness of Grounding. In K. Bennett and D. Zimmerman (eds.). *Oxford Studies in Metaphysics*. Oxford: Oxford University Press.

Kreimendahl, L. (1990). *Kant: der Durchbruch von 1769*. Cologne: J. Dinter.

Kreines, J. (2009). Kant on the Laws of Nature: Laws, Necessitation, and the Limits of our Knowledge. *European Journal of Philosophy*, 17(4): 527–58.

Kripke, S. (1980). *Naming and Necessity*. Cambridge, MA: Harvard University Press.

Kroon, F. and Nola, E. (1987). Kant, Kripke, and Gold. *Kant-Studien*, 78(4): 442–58.

Kuehn, M. (2001). *Kant: A Biography*. Cambridge: Cambridge University Press.

Laerke, M. (2008). *Leibniz lecteur de Spinoza: la genèse opposition complexe*. Paris: Champion.

Land, Thomas (2013). Intuition and Judgment: How Not to Think about the Singularity of Intuition. In S. Bacin, A. Ferrarin, C. LaRocca, M. Ruffing (eds.). *Akten des XI. Internationalen Kant-Kongresses*. Berlin: Walter de Gruyter, vol. 2, 221–32.

Lange, M. (2009). *Laws and Lawmakers: Science, Metaphysics, and the Laws of Nature*. Oxford: Oxford University Press.

Langton, R. (1998). *Kantian Humility*. Oxford: Oxford University Press.

Lapointe, S. (2012). Kant, Bolzano, and the Kantian Logicians. *Grazer Philosophische Studien*, 85(1): 11–32.

Laywine, A. (1993). *Kant's Early Metaphysics and the Origins of the Critical Philosophy*. North American Kant Society Studies in Philosophy, vol. 3. Atascadero, Calif.: Ridgeview.

Leech, J. (2012). Kant's Modalities of Judgment. *European Journal of Philosophy*, 20(2): 260–84.

Leech, J. (2014). Making Modal Distinctions: Kant on the Possible, the Actual, and the Intuitive Understanding. *Kantian Review*, 19(3): 339–65.

Lewis, D. (1973). *Counterfactuals*. Oxford: Basil Blackwell.

Lewis, D. (1983). New Work for a Theory of Universals. *Australian Journal of Philosophy* 61, 343–77. Repr. in D. Lewis (1999). *Papers in Metaphysics and Epistemology*. New York: Cambridge University Press, 8–55.

Lewis, D. (1986). *On the Plurality of Worlds*. Oxford: Blackwell Publishers. Cited by page numbers in 2nd edition (2001).

Lin, M. (2012). Rationalism and Necessitarianism. *Noûs*, 46(3): 418–48.

Logan, I. (2007). Whatever Happened to Kant's Ontological Argument? *Philosophy and Phenomenological Research*, 74(2): 346–63.

Longuenesse, B. (1998). *Kant and the Capacity to Judge*. Princeton: Princeton University Press.

Longuenesse, B. (2000). Point of View of Man or Knowledge of God: Kant and Hegel on Concept, Judgment, and Reason. In S. Sedgwick (ed.). *The Reception of Kant's Critical Philosophy: Fichte, Schelling, and Hegel*. Cambridge: Cambridge University Press, 253–82.

Longuenesse, B. (2005). The Transcendental Ideal and the Unity of the Critical System. In B. Longuenesse. *Kant and the Human Standpoint*. New York: Cambridge University Press, 210–35.

Look, B. (2005). Leibniz and the Shelf of Essences. *Leibniz Review*, 15: 27–47.

Malcolm, N. (1960). Anselm's Ontological Arguments. *Philosophical Review*, 69(1): 41–62.

Marshall, C. (2013). Kant's Appearances and Things in Themselves as Qua-Objects. *The Philosophical Quarterly*, 63(252): 520–45.

Mates, B. (1986). *The Philosophy of Leibniz: Metaphysics and Language*. New York: Oxford University Press.

McDaniel, K. (2013). A Philosophical Model of the Relation between Things in Themselves and Appearances. *Noûs*, 48(4): 1–22.

McFarlane, J. (2002). Kant, Frege, and the Logic in Logicism. *Philosophical Review*, 111(1): 25–65.

McLear, C. (2011). Kant on Animal Consciousness. *Philosophers' Imprint*, 11(15): 1–16.

McLear, C. (2014). The Kantian (Non)-conceptualism Debate. *Philosophy Compass*, 9(11): 769–90.

McLear, C. (2015). Two Kinds of Unity in the *Critique of Pure Reason*. *Journal of the History of Philosophy*, 53(1): 79–110.

McLear, C. (forthcoming). Kant on Perceptual Content. *Mind*.

McMichael, A. (1983). A Problem for Actualism about Possible Worlds. *The Philosophical Review*, 92(1983): 49–66.

Menne, A. (1982). Das unendliche Urteil Kants. *Philosophia Naturalis*, 19(1–2): 151–62.

Mondadori, F. (1973). Reference, Essentialism, and Modality in Leibniz's Metaphysics. *Studia Leibnitiana*, 5(1): 74–101.

Mondadori, F. (1975). Leibniz and the Doctrine of Inter-world Identity. *Studia Leibnitiana*, 7(1): 21–57.

Mondadori, F. (1985). Understanding Superessentialism. *Studia Leibnitiana*, 17(2): 162–90.

Morujão, C. (2008). Wolff, Kant und der Begriff der Existenz. In J. Stolzenburg, and O. P. Rudolph (eds.). *Christian Wolff und die europäischen Aufklärung*. Akten des 1. Internationalen Christian-Wolff Kongresses. Hildesheim: Georg Olms.

Nachtomy, O. (1997). *Possibility, Agency, and Individuality in Leibniz's Metaphysics*. Dordrecht: Springer.

Nachtomy, O. (2012). Leibniz and Kant on Possibility and Existence. *British Journal for the History of Philosophy*, 20(5): 953–72.

Nadler, S. (1992). *Malebranche and Ideas*. Oxford: Clarendon Press.

Newlands, S. (2010). The Harmony of Spinoza and Leibniz. *Philosophy and Phenomenological Research*, 81(1): 64–104.

Newlands, S. (2013). Leibniz on the Ground of Possibility. *Philosophical Review*, 122(2): 155–87.

Newlands, S. and Jorgensen, L. (eds.). (2009). *Metaphysics and the Good: Themes from the Philosophy of Robert Merrihew Adams*. Oxford: Oxford University Press.

Nolan, D. (1997). Impossible Worlds: A Modest Approach. *Notre Dame Journal of Formal Logic*, 38(4): 535–72.

Nolan, D. (2014). Hyperintensional Metaphysics. *Erkenntnis*, 171(1): 149–60.

Nolan, L. (1997). The Ontological Status of Cartesian Natures. *Pacific Philosophical Quarterly*, 78(2): 169–94.

Nolan, L. (2005). The Ontological Argument as an Exercise in Cartesian Therapy. *Canadian Journal of Philosophy*, 35(4): 521–62.

Nolt, J. (2014). Free Logic. *The Stanford Encyclopedia of Philosophy*. E.N. Zalta (ed.). <http://plato.stanford.edu/archives/win2014/entries/logic-free/>.

O'Neill, O. (1992). Vindicating Reason. In P. Guyer (ed.). *The Cambridge Companion to Kant*. Cambridge: Cambridge University Press, 280–308.

Oliver, A. (1996). The Metaphysics of Properties. *Mind*, 105(417): 1–80.

Oppy, G. (1995). *Ontological Arguments and Belief in God*. New York: Cambridge University Press.

Parrini, P. (1994). (ed). *Kant and Contemporary Epistemology*. Dordrecht: Kluwer.

Parsons, C. (1969). Kant's Philosophy of Arithmetic. In S. Morgenbesser, P. Suppes, and M. White (eds.). *Philosophy, Science, and Method*. New York: St. Martin's Press, 568–94. Repr. (with Postscript) in (1983). *Mathematics in Philosophy*. Ithaca, NY: Cornell University Press, 110–41.

Parsons, C. (1992). Transcendental Aesthetic. In P. Guyer (ed.). *The Cambridge Companion to Kant*. Cambridge: Cambridge University Press, 62–100.

Pasqualino, M. (1994). Möglichkeit, Grund und Prädikation: Zur Rezeption einer Leibnizischen Argumentation bei Wolff und Baumgarten. In *Leibniz und Europa*. Akten des VI. Internationaler Leibniz-Kongreß. Leibniz Gesellschaft: Hannover, 465–71.

Paton, H. J. (1936). *Kant's Metaphysic of Experience*. London: Allen & Unwin. Repr. (1951) New York: MacMillan.

Pereboom, D. (1991). Is Kant's Transcendental Philosophy Inconsistent? *History of Philosophy Quarterly*, 8(4): 357–72.

Pereboom, D. (2007). Kant on Transcendental Freedom. *Philosophy and Phenomenological Research*, 73(3): 537–67.

Pichler, H. (1910). *Über Christian Wolffs Ontologie*. Leipzig: Dürr'schen Buchhandlung.

Pichler, H. (1912). *Möglichkeit und Wiederspruchslosigkeit*. Leipzig: Johann Ambrosius Barth.

Plaass, P. (1965). *Kants Theorie der Naturwissenschaft. Eine Untersuchung zur Vorrede von Kants 'Metaphysischen Anfangsgründe der Naturwissenschaft.'* Göttingen: Vandenhoek & Ruprecht.

Plantinga, A. (ed.). (1965). *The Ontological Argument*. Garden City, NY: Doubleday.

Plantinga, A. (1967). *God and Other Minds*. Ithaca, NY: Cornell University Press.

Plantinga, A. (1966). Kant's Objection to the Ontological Argument. *Journal of Philosophy*, 63 (19): 537–46.

Plantinga, A. (1974). *God, Freedom, and Evil*. New York: Harper & Row.

Pollok, K. (2001). *Kant's 'Metaphysische Anfangsgründe der Naturwissenschaft': ein Kritischer Kommentar*. Hamburg: Felix Meiner.

Pollok, K. (2002). 'Fabricating a World in Accordance with Mere Fantasy'? The Origins of Kant's Critical Theory of Matter. *Review of Metaphysics*, 56(1): 61–97.

Poppe, B. (1907). *Alexander Gottlieb Baumgarten: seine Bedeutung und Stellung in der Leibniz-Wolffischen Philosophie und seine Bedeutung zu Kant*. Borna-Leipzig: R. Noske.

Prauss, G. (1974). *Kant und das Problem der Dinge an sich*. Bonn: Grundman.

Proops, I. (2015). Kant on the Ontological Argument. *Noûs* 49(1), 1–27.

Putnam, Hilary (1975). The Meaning of 'Meaning'. *Minnesota Studies in the Philosophy of Science*, 7: 131–93.

Quine, W.V. (1948). On What There Is. *Review of Metaphysics* 2(1), 21–38.

Raven, M. (forthcoming). Ground. In *Philosophy Compass*.

Reich, K. (1932). *Die Vollständigkeit des Kantischen Urteilstafel*. Berlin: Richard Schoetz.

Robinson, H. (1994). Two Perspectives on Kant's Appearances and Things-in-Themselves. *Journal of the History of Philosophy*, 32(3): 411–41.

Rohs, P. (1978). Kants Prinzip der durchgängigen Bestimmung alles Seienden. *Kant-Studien*, 69(1–4): 170–80.

Rosefeldt, T. (2007). Dinge an sich und sekundäre Qualitäten. In J. Stolzenberg (ed.). *Kant in der Gegenwart*. Berlin: Walter de Gruyter, 167–209.

Rosefeldt, T. (2008). Kants Begriff der Existenz. In V. Rohden, V. R. Terra, and G. Almeida (eds.). *Recht und Frieden in der Philosophie Kants* Berlin: Walter de Gruyter, vol. 2, 657–68.

Rosefeldt, T. (2011). Frege, Pünjer, and Kant on Existence. *Grazer Philosophische Studien*, 82: 329–51.

Rosefeldt, T. (2013). Subject-dependence and Trendelenburg's Gap. In S. Bacin, A. Ferrarin, C. LaRocca, M. Ruffing (eds.). *Akten des XI. Internationalen Kant-Kongresses*. Berlin: de Gruyter, vol. 2, 755–64.

Rosen, G. (2010). Metaphysical Dependence: Grounding and Reduction. In R. Hale and A. Hoffman (eds.). *Modality: Metaphysics, Logic, and Epistemology*. Oxford: Oxford University Press, 109–36.

Rosenkoetter, T. (2010). Absolute Positing, the Frege Anticipation Thesis, and Kant's Definitions of Judgment. *European Journal of Philosophy*, 18(4): 539–66.

Rosenkoetter, T. (2013). Towards a Non-Embarrassing Account of the Modal Functions of Judging. In S. Bacin, A. Ferrarin, C. LaRocca, M. Ruffing (eds.). *Akten des XI. Internationalen Kant-Kongresses*. Berlin: Walter de Gruyter, vol. 2, 383–94.

Russell, B. (1910–11). Knowledge by Acquaintance and Knowledge by Description. *Proceedings of the Aristotelian Society*, 11: 108–28.

Sassen, B. (trans. and. ed.). (2000). *Kant's Early Critics: The Empiricist Critique of the Theoretical Philosophy*. Cambridge: Cambridge University Press.

Schafer, K. (forthcoming). Kant's Conception of Cognition and Our Knowledge of Things in Themselves. In N. Stang and K. Schafer (eds.). *The Sensible and Intelligible Worlds: New Essays on Kant's Metaphysics and Epistemology*. Oxford: Oxford University Press.

Schafer, K. (unpublished). Practical Cognition and Knowledge of Things in Themselves.

Schaffer, J. (2009). On What Grounds What. In D. Chalmers, R. Manley, and R. Wasserman (eds.). *Metametaphysics*. Oxford: Oxford University Press, 247–83.

Schmaltz, T. (1996). *Malebranche's Theory of the Soul: A Cartesian Interpretation*. Oxford: Oxford University Press.

Schmucker, J. (1983). *Kants vorkritische Kritik der Gottesbeweise*. Mainz: Akademie der Wissenschaften und der Literatur.

Schneeberger, G. (1952). *Kants Konzeption der Modalbegriffe*. Basel: Verlag für Recht und Gesellschaft.

Schönfeld, M. (2000). *The Philosophy of the Young Kant*. Oxford: Oxford University Press.

Schwaiger, C. (2011). *Alexander Gottlieb Baumgarten*. Stuttgart-Bad Canstatt: Frommann Holzboog.

Sellars, W. (1968). *Science and Metaphysics: Variations on Kantian Themes*. New York: Humanities Press.

Sellars, W. (1976). Kant's Transcendental Idealism. In P. Laberge, F. Duchesneau, and B. E. Morrisey (eds.). *Proceedings of the Ottawa Congress on Kant*. Ottawa: University of Ottawa Press, 165–81.

Shabel, L. (2007). Kant's Philosophy of Mathematics. In P. Guyer (ed.). *The Cambridge Companion to Kant and Modern Philosophy*. Cambridge: Cambridge University Press, 94–128.

Shaffer, J. (1969). Existence, Predication and the Ontological Argument. In T. Penelhum (ed.). *The First Critique: Reflections on Kant's Critique of Pure Reason*. Belmont, Calif.: Wadsworth, 123–42.

Shoemaker, S. (1980). Causality and Properties. In P. van Inwagen (ed.). *Time and Cause*. Dordrecht: D. Reidel, 109–35.

Shoemaker, S. (1998). Causal and Metaphysical Necessity. *Pacific Philosophical Quarterly*, 79(1): 59–77.

Sider, T. (2012). *Writing the Book of the World*. Oxford: Oxford University Press.

Sleigh, R. (1990). *Leibniz and Arnauld: A Commentary on their Correspondence*. New Haven: Yale University Press.

Smit, H. (2000). Kant on Marks and the Immediacy of Intuition. *Philosophical Review*, 109(2): 235–66.

Smit, H. (2009). Kant on A Priority and the Spontaneity of Cognition. In S. Newlands and L. Jorgensen (eds.). *Metaphysics and the Good: Themes from the Philosophy of Robert Merrihew Adams*. Oxford: Oxford University Press, 188–251.

Smit, H. (2010). A Priority, Reason and Induction in Hume. *Journal of the History of Philosophy*, 48(3): 313–43.

Stadler, A. (1874). *Die Grundsätze der reinen Erkenntnistheorie in der Kantischen Philosophie*. Zürich: S. Hirzel.

Stalnaker, R. (1984). *Inquiry*. Cambridge, Mass.: MIT Press.

Stang, N. (2008). Kant's Modal Metaphysics. Unpublished Ph.D dissertation. Princeton University.

Stang, N. (2010). Kant's Possibility Proof. *History of Philosophy Quarterly*, 27(3): 275–99.

Stang, N. (2011). Did Kant Conflate the Necessary and the *A Priori*? *Noûs*, 45(3): 443–71.

Stang, N. (2012). Kant on Complete Determination and Infinite Judgment. *British Journal for the History of Philosophy*, 20(6): 1117–39.

Stang, N. (2014*a*). Kant, Bolzano, and the Formality of Logic. In C. Tolley and S. Lapointe (trans. and eds.). *New Anti-Kant*. New York: Palgrave-McMillan, 192–234.

Stang, N. (2014*b*). Review of Henry Allison, *Essays on Kant* (Oxford, 2012). *Notre Dame Philosophical Reviews*. https://ndpr.nd.edu/news/47450-essays-on-kant/.

Stang, N. (forthcoming). Kant's Transcendental Idealism. *The Stanford Encyclopedia of Philosophy*.

Stephenson, A. (forthcoming). Kant on the Object-Dependence of Intuition and Hallucination. *The Philosophical Quarterly*.

Stolzenburg, J. and Rudolph, O. P. (eds.). (2008). *Christian Wolff und die europäischen Aufklärung*. Akten des 1. Internationalen Christian-Wolff Kongresses. Hildesheim, Zürich, New York: Georg Olms.

Strawson, P. F. (1966). *The Bounds of Sense: An Essay on Kant's Critique of Pure Reason*. London: Methuen.

Theis, Robert (1994). *Gott: Untersuchung zur Entwicklung des theologischen Diskurses in Kants Schriften zur theoretischen Philosophie bis hin zum Erscheinen der Kritik der reinen Vernunft*. Stuttgart-Bad Canstatt: Frommann-Holzboog.

Thompson, M. (1973). Singular Terms and Intuition in Kant's Epistemology. *Review of Metaphysics*, 26(2): 314–43.

Tolley, C. (2011). Kant on the Content of Cognition. *European Journal of Philosophy*, 20(4): 200–28.

Tolley, C. (2013). The Non-Conceptuality of the Content of Intuitions: A New Approach. *Kantian Review*, 18(1): 107–36.

Tolley, C. (forthcoming). Kant on the Distinction between Perception and Experience.

Tüschling, B. (1990). Intuitiver Verstand, absolute Identität, Idee. Thesen zu Hegels frühe Rezeption der *Kritik der Urteilskraft*. In H. F. Fulda and R. P. Horstmann (eds.). *Hegel und die Kritik der Urtheilskraft*. Stuttgart: Klett-Cotta, 174–88.

Tüschling, B. (1992). The System of Transcendental Idealism: Questions Raised and Left Open in the *Kritik der Urtheilskraft*. In H. Robinson (ed.). *System and Teleology in Kant's Critique of Judgment* (Spindel Conference 1991). *Southern Journal of Philosophy*, 30, Supplement, 109–27.

Vaihinger, H. (1881). *Commentar zu Kants Kritik der reinen Vernunft*, 2 vols. Stuttgart: Spemann.

Van Cleve, J. (1979). Foundationalism, Epistemic Principles, and the Cartesian Circle. *Philosophical Review*, 88(1): 55–91.

Van Cleve, J. (1994). Descartes and the Destruction of the Eternal Truths. *Ratio*, 7(1): 58–62.

Van Cleve, J. (1999). *Problems from Kant*. New York: Oxford University Press.

Vander Laan, D. (2004). Counterpossibles and Similarities. In G. Priest and F. Jackson (eds.). *Lewisian Themes*. New York: Oxford University Press, 258–76.

Walker, R. (1990). Kant's Conception of Empirical Law. *Proceedings of the Aristotelian Society*, 64: 243–58.

Watkins, E. (1997). The Laws of Motion from Newton to Kant. *Perspectives on Science* 5(3): 311–48.

Watkins, E. (1998*a*). The Argumentative Structure of Kant's *Metaphysical Foundations*. *Journal of the History of Philosophy*, 36(4): 567–93.

Watkins, E. (2005). *Kant and the Metaphysics of Causality*. Cambridge: Cambridge University Press.

Watkins, E. (2006). On the Necessity and Nature of Simples: Leibniz, Wolff, Baumgarten and the Pre-Critical Kant. *Oxford Studies in Early Modern Philosophy*, 3: 261–314.

Watkins, E. and Fisher, M. (1998). Kant on the Material Grounds of Possibility. *Review of Metaphysics*, 52(2): 369–95.

Werkmeister, W. H. (1981). *Kant's Silent Decade*. Tallahassee, Fla.: University Press of Florida.

Westphal, K. (1995). Kant's Dynamic Constructions. *Journal of Philosophical Research* 20: 381–429.

Westphal, K. (2000). Kant, Hegel, and the Fate of 'the' Intuitive Intellect. In S. Sedgwick (ed.). *The Reception of Kant's Critical Philosophy: Fichte, Schelling, and Hegel*. Cambridge: Cambridge University Press, 283–305.

Westphal, K. (2004). *Kant's Transcendental Proof of Realism*. Cambridge: Cambridge University Press.

Willaschek, M. (1998). Phaenomena/noumena und die Amphibolie der Reflexionsbegriffe (A235/B294-A292/B349). In G. Mohr and M. Willaschek (eds.). *Immanuel Kant: Kritik der reinen Vernunft*. Berlin: Akademie Verlag, 324–52.

Williamson, T. (2013). *Modal Logic as Metaphysics*. Oxford: Oxford University Press.

Wilson, J. (2014). No Work for a Theory of Grounding. *Inquiry*, 57(5–6): 1–45.

Wilson, K. (1975). Kant on Intuition. *The Philosophical Quarterly*, 25(100): 247–65.

Wilson, M. (1978). Possible Gods. *Review of Metaphysics*, 32(4): 717–33.

Wilson, M. (1990). *Leibniz's Doctrine of Necessary Truth*. New York: Garland Publishing.

Wippel, J. (1982*a*). The Relationship between Essence and Existence in Late-Thirteenth-Century Thought: Giles of Rome, Henry of Ghent, Godfrey of Fontaines, and James of Viterbo. In P. Morewedge (ed.). *Philosophies of Existence: Ancient and Medieval*. New York: Fordham University Press, 131–64.

Wippel, J. (1982*b*). Essence and Existence. In N. Kretzmann, A. Kenny, and J. Pinborg (eds.). *The Cambridge History of Later Medieval Philosophy*. Cambridge: Cambridge University, 385–410.

Witt, J. (2010). Essence and Existence. In H. Lagerlund (ed.). *Encyclopedia of Medieval Philosophy*. New York: Springer, 304–9.

Wolff, M. (1995). *Die Vollständigkeit des Kantischen Urteilstafel*. Frankfurt a.M.: Vittorio Klostermann.

Wolfson, H. A. (1947). Infinite and Privative Judgments in Aristotle, Averroes and Kant. *Philosophy and Phenomenological Research*, 8(2): 173–87.

Wood, A. (1970). *Kant's Moral Religion*. Ithaca, NY: Cornell University Press.

Wood, A. (1978). *Kant's Rational Theology*. Ithaca, NY: Cornell University Press.

Wood, A. (1984). Kant's Compatibilism. In A. Wood (ed.). *Self and Nature in Kant's Philosophy*. Ithaca, NY: Cornell University Press, 73–101.

Wundt, M. (1924). *Kant als Metaphysiker*. Stuttgart: Ferdinand Enke.

Wundt, M. (1939). *Deutsche Schulphilosophie im 17. Jahrhunderts*. Tübingen: J. B. C. Mohr.

Wundt, M. (1945). *Die Deutsche Schulphilosophie im Zeitalter der Aufklärung*. Tübingen: J. B. C. Mohr.

Yong, P. (2014). God, Totality and Possibility in Kant's *Only Possible Argument*. *Kantian Review*, 19(1): 27–51.

Yong, P. (forthcoming). Hegel and the Given: A Phenomenological Interpretation of Hegel's Philosophy. Ph.D dissertation. University of California, San Diego.

Index

Table and figures are indicated by an italic *t* and *f* following the page number.

The manufacturer's authorised representative in the EU for product safety is Oxford University Press España S.A. of el Parque Empresarial San Fernando de Henares, Avenida de Castilla, 2 – 28830 Madrid (www.oup.es/en or product.safety@oup.com). OUP España S.A. also acts as importer into Spain of products made by the manufacturer.

www.ingramcontent.com/pod-product-compliance
Lightning Source LLC
LaVergne TN
LVHW011821200926
848484LV00005B/1271
9780198712626